ROUTLEDGE LIBRARY EDITIONS:
POLLUTION, CLIMATE AND CHANGE

Volume 6

CLIMATOLOGY OF WEST AFRICA

CLIMATOLOGY OF WEST AFRICA

DEREK F. HAYWARD AND JULIUS S. OGUNTOYINBO

Routledge
Taylor & Francis Group

LONDON AND NEW YORK

First published in 1987 by Hutchinson Education

This edition first published in 2020
by Routledge
2 Park Square, Milton Park, Abingdon, Oxon OX14 4RN

and by Routledge
52 Vanderbilt Avenue, New York, NY 10017

Routledge is an imprint of the Taylor & Francis Group, an informa business

British Library Cataloguing in Publication Data
A catalogue record for this book is available from the British Library

ISBN: 978-0-367-34494-8 (Set)
ISBN: 978-0-429-34741-2 (Set) (ebk)
ISBN: 978-0-367-36246-1 (Volume 6) (hbk)
ISBN: 978-0-367-36250-8 (Volume 6) (pbk)
ISBN: 978-0-429-34484-8 (Volume 6) (ebk)

Publisher's Note
The publisher has gone to great lengths to ensure the quality of this reprint but points out that some imperfections in the original copies may be apparent.

Disclaimer
The publisher has made every effort to trace copyright holders and would welcome correspondence from those they have been unable to trace.

The Climatology of West Africa

Derek F. Hayward
Julius S. Oguntoyinbo

HUTCHINSON

London   Melbourne   Sydney   Auckland   Johannesburg

BARNES & NOBLE BOOKS

Totowa, New Jersey

Hutchinson Education

An imprint of Century Hutchinson Ltd

62–65 Chandos Place, London WC2N 4NW

Century Hutchinson Australia Pty Ltd
P O Box 496, 16–22 Church Street, Hawthorn,
Victoria 3122, Australia

Century Hutchinson New Zealand Ltd
P O Box 40–086, Glenfield, Auckland 10
New Zealand

Century Hutchinson South Africa (Pty) Ltd
P O Box 337, Berglvei, 2012 South Africa

First published 1987

First published in the USA 1987 by
Barnes & Noble Books
81 Adams Drive
Totowa, New Jersey, 07512

© Derek F. Hayward and Julius S. Oguntoyinbo 1987

Set in 10/11 pt Bembo

Printed in Great Britain by
Butler & Tanner Ltd, Frome and London

British Library Cataloguing in Publication Data

Hayward, Derek F.
  Climatology of West Africa.
  1. Africa, West—Climate
  I. Title   II. Oguntoyinbo, Julius S.
  551.6966   QC991.W4

Library of Congress Cataloging-in-Publication Data

Hayward, Derek.
  Climatology of West Africa.

  Bibliography: p.
  Includes index.
  1. Africa, West—Climate.   I. Oguntoyinbo, J. S.
II. Title.
QC991.A44H39   1987   551.6966   87–1219

ISBN   (UK) 0 09 1646413
      (US) 0-389-20721-7

Contents

Acknowledgements

It is almost unnecessary to state that a book five years in the making has involved the families of the authors in much sacrifice and inconvenience. Thanks are therefore due to Elizabeth, Margaret and David, not only for their forebearance but also for their help with the tedious tabulation of much of the data analysed in this work, and to Esther, Funke, Ebun, Bola and Ayo for their understanding, endurance and encouragement.

Many hours of research have been spent in the library of the Meteorological Office at Bracknell, UK, and the assistance of the staff there is gratefully acknowledged.

To permit a concentrated period of work in the library, accommodation was provided in the home of Rita and John Bull, near Bracknell. Their kindness and generosity is remembered.

The support of colleagues in the Middlesex Polytechnic, financial assistance towards the cost of transport and typing, and the provision of a sabbatical term, have been much appreciated. Especially are the cartographic skills of Steve Chilton, Chief Technician in the School of Geography and Planning in the Polytechnic, here acknowledged. Nearly all the maps and diagrams have been drawn by him and thus he has made a significant contribution to the book.

Special thanks are conveyed to Mark Cohen of Century Hutchinson, for his patience and guidance over the prolonged period in which the book has been in preparation.

Acknowledgement is made to the following persons or organizations for permission to reproduce, from the works cited, figures and tables:

Figure 1, reproduced in part from Vasic, R. S. (1977), by permission of the Director, Languages, Publications and Conferences Department, WMO, Geneva. **Figure 4a** and **Tables 7a** and **b**, reproduced by permission of Smithsonian Institution Press, from *Smithsonian Meteorological Tables*, prepared by Robert J. List, Smithsonian Miscellaneous Collections, Volume 114, sixth revised edition. Smithsonian Institution, Washington DC 1984 (fifth reprint). **Figure 4b**, reproduced from Hanson, K. J. (1976) in *J. Geophys Res.*, by permission of the author and the American Geophysical Union. **Figure 6**, reproduced from Monteny, B. A. and Gosse, G. (1978), by permission of the authors and Elsevier Scientific Publishing. **Figures 9** and **10**, reproduced from Ojo, O. (1970b), by permission of the author and the *Journal of Tropical Geography*. **Figures 14** and **15**, reproduced from Dorot, G. (1973), by permission of the Director-General, ASECNA, Dakar, Senegal. **Figure 33**, reproduced from Ayoade, J. O. (1980), by permission of the author and the *Singapore Journal of Tropical Geography*. **Figures 34** and **35**, reproduced from Mukherjee, A. K. and Moore, H. G. (1973), by permission of the Director, Meteorological Department, Freetown, Sierra Leone. **Figure 36**, reproduced from Wexler, R. in the *Bulletin of the American Meteorological Society*, **27** (1946), by permission of the society. **Figure 38**, reproduced from Dubief, J. (1979), by permission of SCOPE, Paris. **Figure 41**, reproduced from Ojo, O. (1969), by permission of the author and Springer-Verlag. **Figure 52**, reproduced from Martin, D. W. (1975), by permission of the Director, Languages, Publications and Conferences Department, WMO, Geneva. **Figure 59**, reproduced from Bertrand, J. et al. (1974), by permission of the authors. **Table 16**, translated from Bertrand, J. (1976), by permission of the Société Météorologique de France. **Figure 61**, reproduced from Anyadike, R. N. C. (1979a), by permission of the author and Springer-Verlag. **Figure 83**, compiled from data in Nicholl, B. (1965), by permission of the author. **Table 20**, reproduced from Aspliden, C. I., Tourre, Y. and Sabine, J. B. in *Monthly Weather*

Review, **104** (1976), by permission of the American Meteorological Society. **Figure 85**, reproduced from Obasi, G. O. P. (1974a), by permission of the author. **Figure 86**, reproduced from Eldridge, R. H. (1957), by permission of the Royal Meteorological Society. **Figure 89**, reproduced from Clackson, J. R. (1958), by permission of the Director of Meteorology, Nigeria. **Table 21**, reproduced from Charre, J. (1973), by permission of the *Revue de la Géographie Alpine*. **Figure 93**, reproduced from Palmen, E. (1951), by permission of the Royal Meteorological Society. **Figure 95**, reproduced from Defant, F. and Taba, H. (1957), by permission of *Tellus*. **Figures 96** and **98**, reproduced from Air Ministry (1962), by permission of Her Majesty's Stationery Office. **Figure 99**, reproduced from Erickson, C. O. in *Monthly Weather Review*, **91** (1963), by permission of the American Meteorological Society. **Figures 100** and **106**, modified from Leroux, M. (1976), by permission of the author. **Figure 104**, reproduced from Acheampong, P. K. (1982), by permission of the author and *Geografiska Annaler*. **Figure 105**, reproduced from Bernet, G. (1968), by permission of the Director-General, ASECNA, Dakar. **Figure 107**, reproduced from Musk, L. F. (1983), by permission of the editor, *The Geographical Magazine*. **Figures 109** and **110**, reproduced from Ledger, D. C. (1964), by permission of the Institute of British Geographers. **Figure 112**, reproduced from Sykes, S. (1972), by permission of Methuen & Co. **Table 26**, reproduced from Sidikou, A. H. (1977), by permission of CSER, Ahmadu Bello University, Zaria, Nigeria. **Table 28**, adapted from Lawson, T. L. (1977), by permission of the author. **Table 29**, adapted from Fanger, P. O. (1970) and Koenigsberger et al. (1974), by permission of the authors. **Table 30** and **Figure 122**, reproduced from Gates, D. M. (1972). **Figure 119**, **Tables 32** and **33**, extracted from Adefolalu, D. O. (1984b), by permission of the author and Springer-Verlag.

Table 34, compiled from Lee, D. H. K. (1957). **Table 35**, extracted from UN Demographic Yearbook (1985), copyright United Nations (1985). Reproduced by permission. **Figure 120**, reproduced from Knoch, K. and Schulze, A. (1956). **Table 36**, reproduced from Brazol, D. (1954), by permission of the Director-General, Servicio Meteorologico Nacional, Buenos Aires, Argentina. **Table 37** and **Figure 121** are from Gregorczuk, M. (1967) in Landsberg, H. E. (1972), reproduced by permission of WMO. **Figures 123**, **126**, **127** and **Table 38**, adapted and reproduced from Terjung, W. H. (1967 and 1968), by permission of Swets Publishing Service, Holland. **Figures 124** and **125**, adapted from Ayoade, J. O. (1978), by permission of the author and Springer-Verlag. **Figure 128**, reproduced from Koenigsberger et al. (1974). **Figure 132**, reproduced from Fitch, J. M. and Branch, D. P. (1960), by permission of W. H. Freeman and Co., New York. **Table 39**, reproduced from Oyegun, R. O. (1982b), by permission of the Royal Meteorological Society. **Tables 40** and **41**, adapted from UN Department of Economic and Social Affairs (1971). **Table 42**, adapted from Hudson, J. C. and Stanners, J. F. (1953). **Table 43**, reproduced from Holler, E. and Kerner, G. (1963?), by permission of Afrika-Verein EV, Hamburg. **Figure 133**, adapted from Coursey, P. R. (1944). **Figure 134**, reproduced from Urdahl, T. H. (1952). **Figure 135**, reproduced from Brooks, C. E. P. (1946), by permission of the Royal Meteorological Society. **Table 44**, reproduced from Grundke, G. (1955/6). **Figure 137**, reproduced from Patnaik et al. (1980), by permission of the Director of Meteorological Services, Nairobi, Kenya.

In the entries above where permission is not indicated, it is assumed that it has been granted, no responses to requests to reproduce material having been received from the authors and/or publishers.

Introduction

The relevance of climatology

The impact of weather and climate on the life of man has always been of enormous significance, determining the seasonal cycles of food availability and type, the existence of many diseases, and controlling the range of human habitat. In his ignorance man has feared the elements, quaked beneath the fury of a storm, and seen in it the anger of the spirits. The Masai word for god also means rain; words for sun and rain also have the secondary meaning of god in parts of Cameroon. Even today, despite increasing urbanization and the level of knowledge accessible to the educated, our dependence on favourable weather conditions is barely diminished. Consider the relevance still of climate to agriculture anywhere in the world, to aviation, road transport and shipping, to building design and construction, to the use of fuel, to industry, water resources, sport and many leisure activities, and to physical and mental health.

The development of meteorological observations in West Africa

Before the twentieth century, however, the sciences of climatology and meteorology barely existed. The study of the atmosphere was conducted by a few interested amateurs, mainly in Europe. Climatic records were non-existent for much of the world and descriptions of weather and climate, especially in accounts of journeys in foreign lands, were often inaccurate, exaggerated or romanticized, reflecting the lack of understanding at the time:

> Pedro da Cintra, a Portuguese explorer [sailed to West Africa] in 1462 . . . the sailors . . . named the mountain Sierra Leone, or the Lion Mountain, on account of the continual roaring of thunder on its summit, which is always enveloped in cloud (Ramusio, 1550).*

> On the night of the 19th we had the first tornado of the season. . . . a light-grey shadowy fog seemed creeping stealthily down from Mount Oriel, as if to overwhelm our hill. Some moaning gusts came and went, and then all was obscured in the black cloud of the whirlwind, as with its wrathful voice it swept over us, shaking the poor house unmercifully in its giant teeth (Melville, 1849).

> For the air . . . I do not know what fault it has; 'tis extreme hot, 'tis also subtile and piercing, and I believe enters a man's body easier than that in England. . . . it corrodes iron much more, not by the moisture, for it is not so moist, and besides it does it in the dry weather too (Hillier, 1707).

Hillier resided at Cape Coast, Ghana, in the late seventeenth century and noted qualitatively the weather every day for the year 24 November 1686 to 30 November 1687, but such attention to even this level of detail was rare indeed.

In fact in West Africa the first meteorological observations and regular instrumental readings were, apparently, made on the Île de Gorée, Dakar, in 1855 and at St Louis in 1856, with a meteorological station permanently established there in 1862, although it did not function regularly in the years 1863–8 and 1883–92. Nevertheless in 1901 records and means for a 23 year

---

* Full references quoted in the text are cited in the bibliography beginning on p. 240.

9

period were published and compared with those for Paris (Constantin, 1901). But this work was ahead of its time. Even a network of rainfall stations was not well-established before 1950 (Figure 1) and between 1900 and 1920 only one synoptic station (see glossary) existed in the then French West African Sahelian countries. A rainfall station was set up at Korhogo in Ivory Coast in 1905, which functioned until 1926 then produced no records until 1945. Otherwise there were no other stations in that country before 1919. Then followed a steady increase in rainfall stations to 1950, a marked increase to the 1970s then a levelling off. Synoptic stations on the other hand multiplied sharply in the 1930s, with

remarkably little growth since (Vasic, 1977; Hubert, 1934).

The situation in anglophone West Africa was much the same and is illustrated for the period up to 1959 for Nigeria, which then included the former British Cameroons. Here the first rainfall records commenced in 1891 at Lagos racecourse; there was a steady increase through the 1920s and 1930s, followed by a dramatic rise in the 1950s (British West African Meteorological Services, Technical Notes 16 and 17, 1959, 1960). However, many of these stations were established in schools and institutions that have failed to maintain prolonged and accurate readings. The data graphed refer only to stations for which there were, in 1959, complete records for at least ten years, or, if the stations commenced operations in the 1950s, were still in existence in 1959. Elsewhere in West Africa, south of the Sahel,

Figure 1 Development of rainfall recording and synoptic stations in Chad, Niger, Upper Volta, Mali, Mauritania and Senegal 1890–1975, and in Nigeria (and British Cameroons) 1890–1959

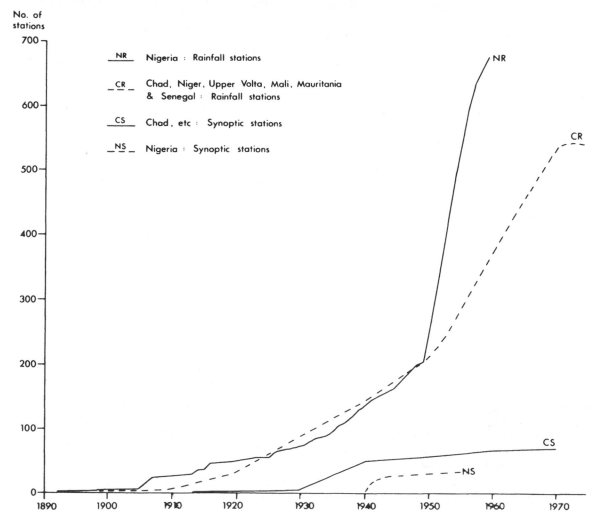

meteorological records commenced at least as early as 1885 at Douala, 1889 at Yaoundé, 1886 at Accra, 1881 at Freetown, 1907 at Banjul, 1905 at Conakry and 1918 at Bolama. Technical Note 17 suggests that there were no synoptic stations in Nigeria before 1941 when 15 were established. This number doubled by 1952 and then remained unchanged, except for the addition of Bamenda in 1953, until Cameroon's independence in 1960 and the reorganization of the national frontiers in 1961. By 1980 the number of synoptic reporting stations in Nigeria had risen to 30 and in Cameroon to 20 (Table 1). In Nigeria the earliest radiation instruments (Gunn-Bellani integrators) were in use in 1959, recording rain gauges in

1948, sunshine recorders in 1939 and tank evaporimeters in 1953.

The World Meteorological Organization's publication No. 9, 1981, lists observing stations in West Africa which report for synoptic purposes. Their distribution is mapped in Figure 2 and Table 1 provides further information. The total number of synoptic reporting stations in 1980 was 192, of which 144 were principal land stations (see glossary), 61 investigated the upper air by pilot balloons, 13 by radiosonde, 19 by radiowind. Over 170 stations maintained sunshine records and 38 submitted radiation data, though half of these stations were in the one country, Ghana.

Table 1 *Number of official observing stations in West Africa (WMO 1981)*

Country (area within W. Africa)	Number of observing stations	No. of extra official rainfall stations	Number reporting data for		
			Sunshine	Radiation	Upper air
Algeria	1	nil	1	1	1
Benin	6	25	6	nil	3
Bourkina Faso	7	20	7	5	2
Cameroon	20	28	10	1	1
Gambia	7	nil	7	nil	2
Ghana	21	nil	19	19	6
Guinea	12	nil	6	nil	1
Guinea-Bissau	3	nil	3	2	nil
Ivory Coast	14	31	14	nil	5
Liberia	2	nil	1	nil	nil
Mali	18	31	18	3	9
Mauritania	11	11	11	nil	6
Niger	12	17	12	nil	6
Nigeria	30	nil	29	3	7
Senegal	13	11	13	nil	4
Sierra Leone	6	nil	6	2	1
Togo	8	30	8	3	2
Western Sahara	1	nil	1	nil	nil
Totals	192	204	172	39	56

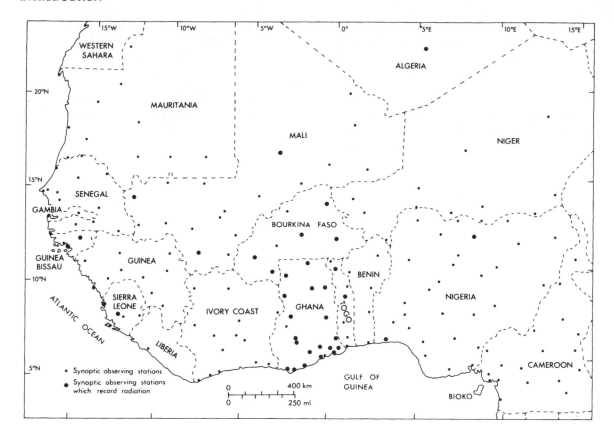

Figure 2 Distribution of synoptic observing stations in West Africa

In addition there are many stations, for example in schools, colleges, universities, agricultural and research stations, that are properly maintained and keep detailed records, operating even as principal climatological stations and supplying data to their national meteorological offices, but whose data are not used internationally for synoptic purposes; thus they are not included in Figure 2 and Table 1.

The basis of an understanding of weather and climate had to be the assemblage of reasonably accurate and prolonged records of climatic data for extensive areas. The main maritime nations sought to establish a programme of weather recording at sea in the mid nineteenth century. The International Meteorological Organization was created in 1878. Two world wars led to a rapid expansion of weather observation and research into atmospheric processes and the World Meteorological Organization was recognized as a special agency of the United Nations in 1951. The Organization essentially seeks to serve the needs of member nations which, at the time of writing, number 150 and exchange meteorological information often using internationally standardized procedures and codes. In West Africa only Western Sahara is not a member of the WMO.

Modern approaches to meteorological research

The availability of radiosonde and rawinsonde balloons, radio-wave investigation, rocket and satellite flights has in recent years transformed a science that was essentially descriptive and empirical into the modern, quantitative discipline of meteorology, concerned with experimental physics and fluid dynamics, atmospheric chemistry and radiation. Numerical modelling and increasingly accurate weather prediction are now made possible by powerful computers.

If these advances are initially taking place in

the world's wealthier nations, they are also providing a basis for a more thorough understanding of climate in the less developed world, that is, essentially, in the tropics. Recent projects sponsored by the WMO include, for example: World Weather Watch (WWW) designed to expand the observing network and improve communications and data collection; the Global Atmospheric Research Program (GARP), aimed to increase understanding of the general circulation of the atmosphere and develop the physical and mathematical bases for methods of extended range weather prediction; the GARP Atmospheric Tropical Experiment (GATE), the Barbados Oceanographic and Meteorological Experiment (BOMEX), the Monsoon Experiment (MONEX) and the West African Monsoon Experiment (WAMEX), all of which have been particularly concerned with subtropical and tropical regions. WAMEX, of course, is particularly relevant in this study and it is interesting to note that although few of the West African nations possess the technology to investigate the atmosphere in depth, for this project national programmes of observation and research were to be supported by Meteosat, Tiros N, Meteor and Nimbus G satellites, by the special systems established for the first GARP Global Experiment, by tropical wind observing ships, ordinary commercial shipping, aircraft dropwindsondes, tropical constant level balloons, southern hemisphere drifting buoys and by Aircraft Integrated Data Systems (AIDS).

The aims of this book

A thorough understanding of climatic processes, to permit accurate and long-term weather prediction, is essential in the tropics. It is here that two-thirds of the world's population live, here that the most rapid population growth is taking place and here that many new nations have emerged with the most fragile of economies, the most rapidly growing cities and the most disturbed social systems as modern technology is implanted into traditional societies.

The situation is exemplified in West Africa, a region set apart by history, culture, traditions and languages and by physical geography, especially climate. Little of the past or present can be understood without reference to the all-pervading climatic environment. Little can be sensibly planned for the future if this environment is ignored. West Africa has a population of some 160 million people, forecast to double in 20 years, where possibly 70 per cent at present still depend directly on agriculture for their livelihood and thus on the vagaries of the weather. With economic and food production problems increasing, a greater awareness of the total environment is needed. The climate should be regarded as a resource, like water and soil, gold, diamonds, oil and cocoa, or the people themselves. It is hoped that this book will provide a useful introduction to an appreciation of West Africa's climates, what they are like and why, and the extent to which man is governed by them and in turn can affect them. The book is essentially aimed at first year undergraduates of climatology and geography, but it is also for those involved in agricultural, planning and development issues in West Africa, all, in fact, who have the well-being of the West African peoples at heart.

West Africa defined

Over many years West Africa has come to mean the present non-Arab, if in some cases strongly Islamic, states of Mauritania, Senegal, Gambia, Guinea-Bissau, Guinea, Sierra Leone, Liberia, Ivory Coast, Ghana, Togo, Benin and Nigeria, all with an Atlantic coastline, and the land-locked republics of Mali, Bourkina Faso and Niger. Climatic elements and patterns pay no heed to political boundaries however, so for the purposes of this book West Africa is defined as that part of the continent that lies south of the almost permanently arid central Sahara Desert and experiences a seasonal change of climate related to the advance northwards, each northern hemisphere summer, of humid air from the Atlantic and Gulf of Guinea. This advance may reach as far north as 24° in Mali and Algeria in July and August, but more commonly approaches 20° across western Mauritania and Niger. Thus the 24th parallel is conveniently used for the northern limits of the region (Figure 2). The eastern and western bounds are arbitrarily fixed at 15°E and 18°W and the southern edge of the map is delimited by the 2°N parallel. Thus nearly the whole of the Cameroon Republic and Bioko (Equatorial

Guinea) are included and the southernmost parts of Western Sahara and Algeria, but the Cape Verde Islands are excluded.

This is not to say, of course, that there need be no discussion of meteorological conditions outside the region if one is to seek to understand the climates within it, but descriptions of climate in Part One of this book are confined to the land area mapped. Part Two seeks to explain the climate, but in terms that a non-specialized readership may appreciate. Part Three aims to show the relevance and importance of climatology to life and development in West Africa.

There is now abundant literature on, and evidence for, Quaternary and recent climatic change in West Africa, but this interesting subject is only briefly considered in this text, partly for economic reasons, partly because it was felt by the authors that this topic was more appropriate to a book aimed at a more specialized readership.

PART ONE

Description *The Climate of West Africa*

Introduction

To most people the meaningful climatic elements
are those felt and experienced, that is the sensible
elements of temperature, sunshine, precipitation,
humidity and wind. Underlying all meteorological activity however is solar radiation, of which
visible sunshine is but a part. It is solar radiation
that drives the great atmospheric engine, that
provides the fundamental energy which has to be
transformed or transferred in various ways to
create climates as we know them. It is the net
radiation (see glossary) which basically determines atmospheric and soil temperatures. In turn
temperature will primarily determine global patterns of air pressure and thus air mass movement
and wind flow. Together these elements will
largely control precipitation, evaporation and
evapotranspiration. Each of these components of
climate is discussed below with additional comments on visibility, on the seasonality of West
African climate, on storms and on the climatic
regions of West Africa.

1 Radiation, sunshine and temperature

Radiation

The sun transmits energy mainly in the form of electromagnetic waves, with a spectrum as in Figure 3. About 45 per cent of the total radiation emitted is in the visible light range, 9 per cent in the ultra-violet and 46 per cent in the infra-red (heat) ranges. Depending on the length of day and the elevation of the sun in the sky, which alter seasonally and with latitude (Table 7, p. 26), the total daily solar radiation that would be received at the earth's surface within tropical and sub-tropical latitudes *if there were no atmosphere* is indicated in Figure 4a.

However, the earth does possess an atmosphere (for composition see Table 2), and radiation reception at the surface is more realistically shown by Figure 4b. In the winter months at the

Table 2 *Mean sea-level dry air composition of the earth's atmosphere by volume*

Gas	Volume (%)
Nitrogen	78.08
Oxygen	20.95
Argon	0.93
Carbon Dioxide	0.03
Neon	0.0018
Helium	0.0005
Krypton	0.0001
Xenon	0.00001

Note: In addition air generally contains water vapour, hydrocarbons, hydrogen peroxide, sulphur compounds and particulate matter in small but very variable quantities.

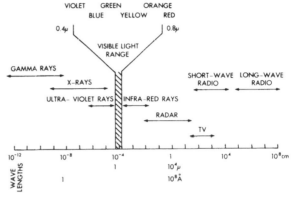

Figure 3 The solar radiation spectrum

higher latitudes less than 50 per cent of the short-wave radiation impinging on the outer atmosphere can be recorded at the surface. Figure 5 and Table 3 explain why. The key words in Figure 5 are 'absorbed', 'scattered' and 'reflected'. Atmospheric absorption accounts for about 19 per cent of the inflowing radiation. The gases CO_2, O_3, O_2 and water vapour are most important. Ozone (O_3) is especially significant for absorbing ultra-violet radiation between 0.22 and 0.29 microns (μ). In fact between them O_3 and O_2 absorb almost all radiation at wavelengths less than 0.29μ, which is fortunate for human health. Scattering is essentially achieved by the particulate matter in the atmosphere, including water droplets. Microscopic dust and water particles deform the wave flow, dispersing or diffusing in all directions the energy propagating in a single direction. Generally the less 'pure' the atmosphere the greater is the scattering and the paler the sky. Thin stratiform cloud, harmattan dust and smoke pollution from major bush fires will produce pale skies, whereas a rain-washed atmosphere after a shower will be clear and blue.

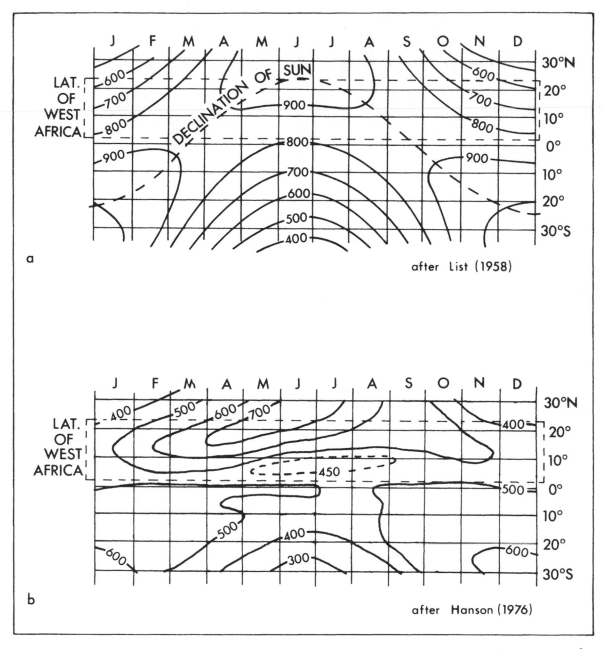

Figure 4 a Total daily solar radiation at the top of the atmosphere in tropical and sub-tropical latitudes (cal cm^{-2}); *b* Daily solar radiation at the earth's surface in tropical and sub-tropical latitudes (cal cm^{-2})

The atmospheric gases more readily scatter the shorter-wave light (blue) than the longer (red), hence the blue colour of the sky. Beyond the atmosphere the sky is black. On average 12 per cent of solar radiation reaching the atmosphere is scattered, 7 per cent to space and 5 per cent to the earth.

In an experiment conducted off the coast of West Africa on a 'typical' hazy day in August 1972, it was estimated that the net depletion of solar energy due to harmattan dust was about 100 langleys (ly) (see glossary and Appendix 1) per day. This amount equates with that lost when the skies are cloud-covered in the same region (Carlson et al, 1973). At Tamanrasset in southern Algeria on 20 January 1942 a pyrheliograph recording of 1.713 langleys was made, the highest value in an experimental period running from March 1939 to March 1942. Even so it was estimated that the radiation received was reduced by the atmospheric components, proportions and amounts shown in Table 4 (Castet, 1942).

Monteney and Gosse (1978) illustrated the level of atmospheric transparency and the transmission of energy through the atmosphere above Adiopodoumé, 20 km west of Abidjan, for 10-day periods between November 1974 and March 1976 (Figure 6). It is interesting to note the falls in radiation reception in January and February when the influence of the harmattan and its dust is most prominent. Even when skies are cloudless atmospheric turbidity may significantly attenuate solar radiation. The lowest values are in the period June to September, despite the zenithal sun, for this is the time of greatest cloud development and precipitation and significant quantities of marine aerosols, mainly NaCl, are brought in from the Gulf of Guinea by the dominant south-westerly winds.

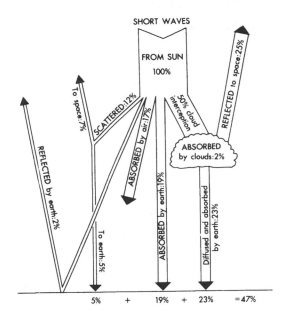

Figure 5 Short-wave radiation transfers

Cerf (1980) noted that in a period of harmattan at Niamey direct solar radiation at 0.5μ was reduced 75 per cent, whereas at Abidjan it was only 40 per cent. The explanation for the difference related to variability in aerosol types and quantities. Maritime aerosols extinguish the long-wave (red) radiation much less than does inland dust.

The direct and diffuse radiation together received on a unit horizontal surface is termed the *global radiation*. If the clouds are poor absorbers of radiation they are the main reflectors, reflecting on average half of what they intercept. Thus, of the total short-wave radiation reaching the atmosphere some 49 per cent reaches the earth directly or after diffusion or scattering, but of this, 2 per cent is reflected back to space (see

Table 3 *Short-wave radiation transfers*

	%	%
Radiation from the sun to earth's outer atmosphere	100	
Radiation intercepted and absorbed by clouds		2
Radiation reflected from clouds		25
Radiation diffused from clouds and absorbed by the earth		23★
Radiation absorbed by atmospheric gases and particulate matter		17
Radiation scattered by atmosphere and returned to space		7
Radiation scattered down to, and absorbed by, earth		5★
Radiation directly absorbed by the earth		19★
Radiation reflected back from earth and so not absorbed		2
Totals	100	100

Note: ★ Total absorbed by earth = 47 per cent

Table 4 *Reductions in radiation at Tamanrasset*

	Dry air %	Water vapour %	Dust %	Total %
Summer 1941	12	11	14	37
Winter 1941	12	9	6	27
Summer 1942	12	11	9	32
Winter 1942	12	9	3	24

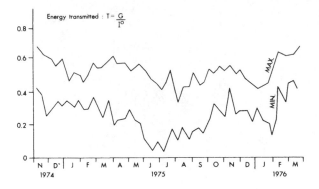

Figure 6 Variation of atmospheric transmission (maximum and minimum values) for 10-day periods, Nov. 1974 to March 1976, at Adipodoumé, Ivory Coast (after Monteny and Gosse 1978)

Notes: T is the atmospheric transmission of energy; G is the global radiation; I° is the fraction of G received at the surface.

Figure 5). Now the reflection of radiation does not normally involve a change of wavelength, but if absorbed radiation is re-radiated from most natural objects it is in the infra-red range, peaking at about 10μ, as absorption involves the creation of heat by molecular conduction.

The re-radiated energy, being longer in wavelength than solar radiation, is more readily absorbed by the atmosphere, the clouds absorbing, at all wavelengths, ozone at 9.4 to 9.8μ, CO_2 at 13.1 to 16.0μ, and water vapour at 5.3 to 7.7μ and beyond 20μ. Only about 9 per cent of re-radiated energy escapes to space, through the narrow waveband in which there is no absorption by water vapour, clearly the most important of the gases. The larger the amount of water vapour in the atmosphere the greater is the absorption, hence rapid night cooling when skies are clear and air dry, as in much of West Africa in the 'dry season'. The absorbed longer-wave radiation is again re-radiated after absorption by the atmosphere, to produce the so-called 'greenhouse effect', by which surface temperatures are maintained at a higher level than they would otherwise be. Cloud layers alone may increase counter-radiation by more than 25 per cent over clear sky values, water vapour levels remaining unchanged. The lower and thicker the clouds the more they contribute to counter-radiation (Table 5).

Absorption of radiation by the earth is influenced by a number of factors including the angle of incidence of the solar radiation, relief, aspect (though this is not so important in the tropics where sunshine comes from all directions during the year), soil moisture and porosity and the temperature and emissive power of the surface. Moist soil can more readily and more deeply absorb short-wave radiation than can dry soil.

Albedo

The emissive power (see glossary) of a surface depends upon its temperature and nature. Dark surfaces will have a greater power as they will absorb more radiation than paler surfaces, which will reflect more. Also, as noted above, the reflected radiation is of short wavelength and this is not readily absorbed on its outward journey and so does not contribute much to atmospheric heating. The ratio of reflected to incoming short-wave radiation (ΣR/ΣT where ΣR is the total reflected short-wave radiation for a stipulated period and ΣT is the total incoming radiation for the same period) is termed the *albedo*. This has a most important influence on the absorption of radiation and so on air and soil temperatures, micro-climate and even regional climate. Albedo is mainly a function of surface colour, texture and the angle of incidence of the incoming solar radiation. For example the taller the vegetation the greater will be the trapping of radiation by reflection between the leaves. Minimum values of albedo, especially in grasslands, will therefore

Table 5 *The importance of cloud in counter-radiation*

Cloud type (Ch. 4)	Height (m)	1–k*
Cirrus	12,000	0.84
Cirro-stratus	8,400	0.60
Alto-cumulus	3,600	0.34
Alto-stratus	2,100	0.20
Strato-cumulus	1,200	0.12
Stratus	460	0.04
Nimbo-stratus	92	0.01
Fog	0	0.00

Note: * 1–k is the ratio of outgoing radiation from cloud-covered skies to that with clear skies. Overcast = 1.

be when the sun is high in the sky, when there is more scattering between leaves and so less total reflectivity upwards. Desert albedos are high so that the actual net energy distributed to the lower atmosphere is less than that imparted over tropical forest. The albedo of bare soil depends on the water and organic matter content, on soil texture and, again, on the angle of incidence of the radiation. Reflectivity decreases with increasing wetness as radiation is trapped by internal reflection in soil pores.

Albedos obviously vary with time and place, which, along with a lack of standardization in methodology and in description of environment, may explain the apparent discrepancies in Table 6. The data are drawn from a small selection of sources and research carried out over many years. The work of Oguntoyinbo is especially interesting. The values quoted for vegetation and crops are the means of a number of measurements taken at different times of the year, for the canopies and colours of the leaves of the various plants alter with age and health and the level of available moisture. The Burroughs data were obtained by Nimbus 6 satellite.

The high albedos of clouds are particularly important. At any one time 50 per cent of the earth is cloud-covered, accounting for 75 per cent of the earth's total albedo. Burroughs discusses the importance of changes in albedo on climate, which are due to increased air pollution, overgrazing, forest clearance and desertification, all significantly on the increase in West Africa. Note from Table 6 how the albedo will alter if tropical forest is reduced to bush, if savanna is reduced to desert. It is probable that over 50 per cent of the forest cover of West Africa has been removed (up to 70 per cent in Liberia, Ivory Coast and Ghana) and that the consequence has been a change in the albedo in the region of 40 per cent in the last 100 years. The significance of this to human occupancy is discussed in Part Three of this book.

If the mean albedo values for Nigerian plant communities as published by Oguntoyinbo (1970a, 1970b, 1974, 1979, 1983) are valid across West Africa, one might attempt to utilize maps of vegetation to deduce patterns of radiation reflectivity across the whole region. Figure 7a has been drawn up by this method from a highly generalized, small-scale vegetation map, an attempt being made to allow for the fact that most remaining forest in West Africa is secondary and more than half of it has been replaced by

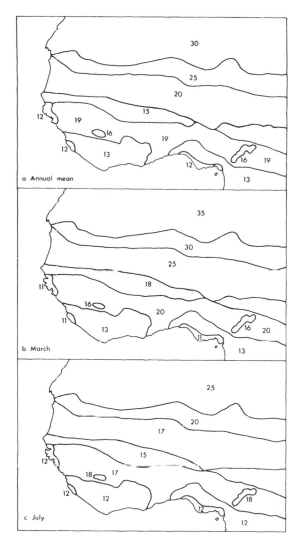

Figure 7 Albedos (per cent) in West Africa

bush or crops, the latter generally having higher albedos than the original vegetation. Figure 7b, using the same vegetation map base, provides a crude picture for March, the month with highest values in Oguntoyinbo's tables, and Figure 7c illustrates the situation in July when mean values are usually at their lowest, when cloud cover is most complete over West Africa.

The crudity of this approach to assessing reflectivity is well-appreciated (the values indicated in the July map for the forest and savanna areas are nearly 60 per cent higher than those noted by Krishnamurti in an even smaller-scale map published in 1977) and the patterns produced are grossly over-simplified. The map of albedos for

21

Table 6 *Mean albedos (per cent) for clouds and various surfaces in West Africa*

Feature	List Smithsonian Met. Tables (1958)	Albedos %				
		Sellers (1965)	Ojo (1970b)	Oguntoyinbo (1970a)	Krishnamurti (1977)	Burroughs (1981)
High cirrus						<20
Cirro–stratus	49–64	44–50				
Alto–stratus	39–59	39–59				50
Cumulus		70–90				70
Strato–cumulus	56–81					
Stratus	31–75	59–84				
	The Smithsonian and Sellers values are for overcast conditions					
Dry concrete		17–27				
Dry sand dunes		35–45				
Desert		25–30	28		30	35
Dry season savanna		25–30	20			
Wet season savanna		15–20	18			
Sudan savanna				15		
Guinea savanna				19		
Derived savanna			25	15		16
'Bush'				10		
Mangrove and swamps				12		
Rain forest			16	13		7
Jos plateau vegetation				16		
Cassava				19		
Cocoa				16		
Cotton				21		
Ground nuts				17		
Maize				18		
Wet rice				11		
Sugar cane				15		
Tobacco				19		
Yam				19		

Note: The Food and Agriculture Organization (FAO 1966) recorded the albedo (per cent) of some West African soils as follows: Dry ferruginous soils on acid rocks 15; Dry red ferrisols on sands 17; Saturated black alluvial soils 9; Dry dark grey silts 12; Damp alluvial soils on marine deposits 12; Dry semi-arid brown and red-brown soils 17; Mean value 14.

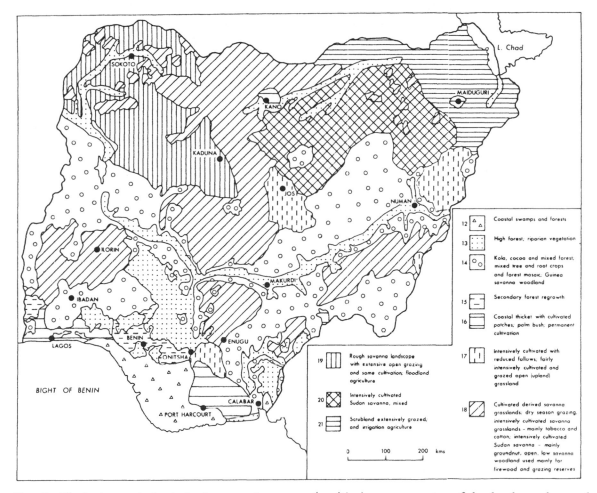

Figure 8 Albedos (per cent) of major land-use types in Nigeria (after Oguntoyinbo 1979)

major land use types in Nigeria compiled by Oguntoyinbo (1979) makes this abundantly clear (Figure 8), and this map in turn is, of course, also a highly generalized picture of a dynamic and, in places, rapidly changing land use scene.

The radiation balance

The albedo or reflection coefficient is clearly of fundamental importance in determining the radiation balance over a surface. The radiation absorbed at the surface (the net radiation) is the sum of the upward (negative) and the downward

(positive) components of both the solar and terrestrial radiation:

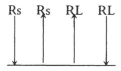

- ↓ Rs is the incoming short-wave radiation
- ↑ Rs is the reflected short-wave radiation
- ↑ RL is the long-wave (infra-red) radiation from the surface
- ↓ RL is the counter infra-red radiation from the atmosphere

At night, of course, there is no ↓ Rs but continuous ↑ RL, which may be reduced by ↓ RL, especially in cloudy conditions. So although the overall net radiation in the tropics is markedly positive, at night it is usually negative and thus to the regular annual variation in radiation noted above, related to seasonal climatic change, there is added a second rhythmic pattern, the diurnal variation. For daytime conditions the net radiation is simply expressed in the following equation:

$$Rn = (Q + q) (1 - \alpha) - I$$

where Rn is the net radiation, Q is the direct solar radiation, q the diffuse radiation, $(1 - \alpha)$ the albedo and I the effective outgoing radiation, which is itself the long-wave radiation emitted by the surface minus the infra-red counter radiation from the atmosphere to the earth depending upon air temperature, water vapour content and cloud cover.

The linear relationship between daytime net radiation and direct solar radiation Q has been expressed by Davies (1967) for West Africa thus:

$$Rn = 0.612Q - 28 \text{ (gcal cm}^{-2} \text{ day}^{-1})$$

over surfaces with albedos from 20 to 30 per cent. On average, he claims, Rn is 55 per cent of solar radiation in the region. Instruments for measuring radiation and albedo are briefly described in Appendix 2.

For the correct assessment of (Q + q) direct observations by pyranometers are really required, but such instruments are sparsely distributed in West Africa (Figure 2 and Table 1). However the relative abundance of sunshine recording stations has encouraged a number of researchers to seek to calculate radiation inputs utilizing sunshine records. Glover and McCulloch (1958) expressed the relationship between sunshine and radiation thus:

$$Q/Qo = 0.29 \cos\alpha + 0.52 \text{ n/N}$$

where Q is the global radiation (Q + q), Qo is the global radiation without atmospheric depletion, cos is the station latitude and n/N is the percentage recorded of maximum possible sunshine. Davies (1965b, 1966b) produced radiation estimates for Nigeria utilizing the equation:

$$Q' = Qo (a + b (S/So))$$

where Q' is the observed global radiation (Q + q), Qo is the maximum possible global radiation (i.e. the amount that would be available if there

were no atmospheric depletion), a and b are monthly constants, S is the number of sunshine hours recorded and So is the maximum number of sunshine hours possible per day. Ojo (1970b) utilized this approach as the basis of his work which produced the most detailed maps yet of global and net radiation for West Africa (Figures 9 and 10).

Ezekwe and Ezeilo (1981) concluded from their work with a pyranometer at Nsukka, Nigeria, that in any modelling or estimation of radiation it is necessary to take account of relative humidity, maximum air temperature, geographical declination, latitude and altitude as well as sunshine hours. They refer to the work of 14 different researchers and tested the equation models of five against their pyranometer records. They came to the conclusion that the model of Swartman and Ogunlade (1967) in West Africa gave the best agreement with measured data, departing by only ± 6 per cent. The parameters utilized here were sunshine hours and relative humidity. This is not to say, of course, that all pyranometers necessarily record with great precision. Stanhill et al. (1971) tested four different types of pyranometer both in the laboratory and the field and found that the comparative error between the instruments in the field reached 25 per cent with the sun's angle of incidence at 60° and 40 per cent with an angle of 40°.

The distribution of radiation in West Africa

Of the maps of radiation published by Ojo the two offering the greatest contrast, those for January and July are, with permission, reproduced here as Figures 9 and 10. Despite differences in radiation levels there are distinctive patterns of distribution common to both months, and indeed to all months in the year. The lowest values are recorded along the coast with an increase inland to about 10°N, then a decrease towards the central Sahara. In the January map of global radiation maximum values in excess of 475 langleys per day outside the Sahara occur in a band between 8 and 12°N across the region and especially at these latitudes in the uplands centred on the Fouta Djalon, the Atacora Mountains and the Jos Plateau. In July the zone of maximum

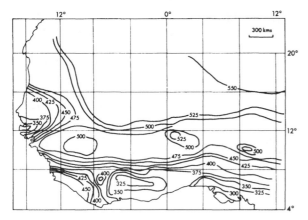

Figure 9a　Global radiation over West Africa in January (langleys per day) (after Ojo 1970b)

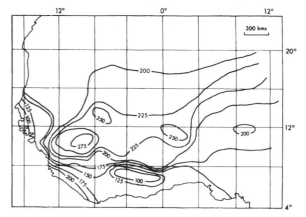

Figure 10a　Net radiation over West Africa in January (langleys per day) (after Ojo 1970b)

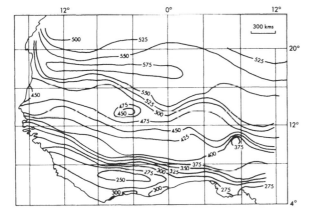

Figure 9b　Global radiation over West Africa in July (langleys per day) (after Ojo 1970b)

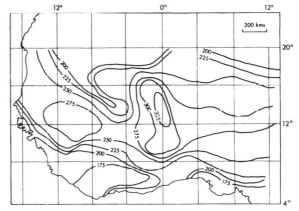

Figure 10b　Net radiation over West Africa in July (langleys per day) (after Ojo 1970b)

values (over 550 ly day⁻¹) has shifted north-wards, near to the latitude of the overhead sun. In both months the fall off in radiation from the bands of maximum levels occurs sharply to the south, and especially clear is the steep gradient at 8 to 9°N. In January there is a similar steep gradient, as it were, encircling Senegal. In both maps a rise in radiation values is marked on the Guinea coast between 0° and 2 to 3°E longitude, an anomalous area that will be referred to in this text as the Togo Gap.

Ojo suggests that the patterns mapped seem to indicate the importance of cloud, the steep radi-ation gradients correlating with a marked change in cloudiness, the most being along the coast, especially in January. Figure 54a (p.75) perhaps supports this view. However, he further suggests that onshore winds bringing cloud account for the low values in Senegal and, conversely, the implication must be that the upland areas named

above, the Togo Gap and the lowlands of Sierra Leone and Liberia, are less cloudy in January. With the exception of the Jos Plateau this situation is not borne out by the map of mean cloudiness for this month. It is possible that the steep gradi-ent reflects not only a marked latitudinal change in cloudiness but also in humidity (see Figure 48a (p.69)). The importance of water vapour in deter-mining net radiation has been mentioned above. Nevertheless, in July mean cloudiness is pro-nounced in West Africa south of about 16°N, except for the area between Kano and Maiduguri in Northern Nigeria, and the pattern of global radiation mirrors this fact. So, despite the possi-bility of more radiation in this summer season, reflected in the higher values of 575+ ly day⁻¹ at 18°N, except in Senegal July global radiation levels are generally lower than those for January, by as much as 125 ly day⁻¹ in the Fouta Djalon, 150 ly day⁻¹ in Liberia and some 70+ ly day⁻¹

elsewhere, except in the Niger Delta. The steep radiation gradient that persists at about 8°N is not matched by any equivalent change in the maps of mean cloudiness or relative humidity however.

Turning now to the maps of net radiation, Figures 10a and 10b, the three-fold zonation of radiation from south to north is immediately apparent again. However values are markedly reduced: in January by 60 per cent in the north, 50 per cent in the middle zone and by as much as 70 per cent in parts of the south; in July by as much as 65 per cent in the north and commonly 25 to 30 per cent elsewhere. The more widespread cloud in July, combined with the 'greenhouse effect' and low albedos of the rain-soaked bush and forest and swamps of the southern areas, produces lower global and net radiation values in July and diminishes the differences between them. Further north, however, with decreasing cloud amounts and increasing aridity, there is an increase in mean albedo and an in-

creased outflow of long-wave radiation in clear sky conditions. Thus net radiation values decrease from the middle zone towards the Sahara. There is also, despite clear skies, a reduction in received radiation due to high levels of atmospheric turbidity. This turbidity affects the 99.9 per cent of radiation that comes within the infra-red to ultra-violet ranges. Light intensity is often less than one might imagine it would be in this tropical environment with an overhead sun, deceiving many an inadequately equipped photographer into producing under-exposed material.

Sunshine

The intensity of direct radiation from the sun is a crucial factor when it comes to recording sunshine as opposed to radiation. Sunshine recorders may be of two types, the first utilizing the heat-

Table 7a *Approximate mean duration of daylight in hours and minutes for each month of the year*

°N	J	F	M	A	M	J	J	A	S	O	N	D
5	11.54	12.00	12.06	12.12	12.18	12.24	12.24	12.18	12.12	12.00	11.54	11.50
10	11.36	11.48	12.05	12.18	12.36	12.42	12.36	12.30	12.12	11.54	11.42	11.30
15	11.18	11.36	12.00	12.30	12.48	13.00	12.54	12.36	12.12	11.48	11.24	11.12
20	11.06	11.30	12.00	12.36	13.06	13.18	13.12	12.48	12.18	11.42	11.12	10.54
25	10.42	11.18	12.00	12.42	13.18	13.36	13.30	13.00	12.18	11.30	10.54	10.36
50	8.37	10.07	11.50	13.45	15.23	16.18	15.55	14.30	12.40	10.50	9.07	8.12

(Paris: 48°50′N; London: 51°30′N)

Table 7b *Approximate mean duration of civil twilight in minutes for each month of the year*

°N	J	F	M	A	M	J	J	A	S	O	N	D
5	22	21	21	21	22	23	22.5	21.5	21	21	22	23
10	22.5	21.5	21	21	22	23	23	22	21	21	22	23
15	23	22	21	22	23	24	23.5	22	21.5	22	22.5	23
20	24	22.5	22	22.5	23.5	25	24	23	22	22	23	24
25	25	23.5	23	23	25	26	25.5	24	23	23	24	25
50	37	33	32	34	39	45	42	37	32	33	36	39

ing power of the radiation, the second the chemical action produced by visible and ultra-violet rays. The former type is by far the most common used in West Africa and one example is described in Appendix 3. The instrument in question cannot record until the intensity of direct solar radiation reaches a certain minimum value, about 0.2 to 0.4 cal cm^{-2} min^{-1}, and so even on clear mornings it may not record until some time after the sun has risen. High levels of atmospheric pollution may prevent the recording of sunshine even when conditions are cloudless and diffuse radiation is particularly prominent. Surrounding obstacles should not, but may, shadow the instrument from sunshine for part of the day. Flaws inherent in the equipment itself may also lead to under-measurement.

Nevertheless, even allowing for these factors, hours of sunshine for West Africa are fewer than residents in higher latitudes usually expect. Tables 7a and 7b indicate that, in the first place, no part of the region can experience more than approximately 14 hours of daylight (see glossary) per day including the well-recognized short periods of civil twilight (see glossary) that occur in the tropics because of the high angle at which the sun sinks below or rises above the horizon. (Astronomical twilight lasts between 50 and 56 minutes longer than civil twilight between 0° and 25°N latitude.)

Although WMO No. 9, 1981 lists over 170 synoptic reporting stations in West Africa that record sunshine, and although there must be many more non-synoptic stations with the appropriate instrumentation, published sunshine records are especially scarce and attempts to produce maps of sunshine duration for the region are therefore very limited. A major problem is that even when the data are published and means calculated, as for example in the splendid Annales des Services Météorologiques de la France d'Outre-Mer, the records for many stations are for such short periods that the figures are hardly meaningful. Sunshine variability can be quite dramatic in most parts of West Africa and at any time of the year. For example in January 1973 at the University of Ibadan, Nigeria, 229 hours of sunshine were recorded, but in the same month in 1974 only 154 hours were registered. In the month of August in 1968, 1969 and 1970 some 99, 71 and only 53 hours of sunshine were recorded respectively, the mean for those three years being therefore 74 hours. But in August 1972, 1973 and 1974 the hours recorded were 91, 131 and 124,

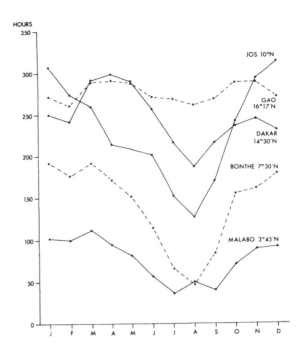

Figure 11 Mean monthly sunshine hours for selected stations in West Africa

giving a three-year mean of 115. In July 1968 at Bolama, Guinea-Bissau, 230 hours of sunshine were recorded, but in July 1972 only 130.

From records made accessible to the writers, means have been compiled for 70 stations and are incorporated in Appendix 4. To show more clearly the seasonal and latitudinal variation that occurs in West Africa, a selection of the data has been utilized to produce Figure 11 and most of the mean daily values have been used in the production of Figure 12 (a–d). Here means based on periods shorter than six years have only been used where the short and longer term values are not too discrepant or where the lack of their use would have left too large an area without plotted data. As it is there are over-large areas without adequate coverage in the north and, sadly, also in Guinea and Liberia.

From the maps a number of clear features emerge, matched, not surprisingly, by the patterns observed in the maps of distribution of net radiation.

1 There is a clear latitudinal variation in sunshine, the Guinea coast being much less sunny than the Sahel and Saharan margins. The greatest contrast is seen in the summer (August map) when the north is up to 80 per

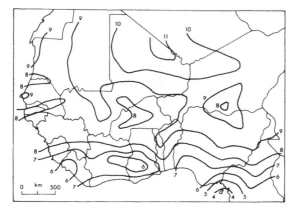

Figure 12a Mean daily sunshine hours in February

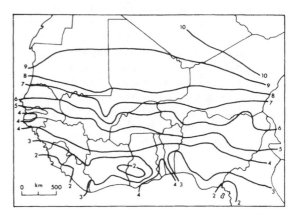

Figure 12c Mean daily sunshine hours in August

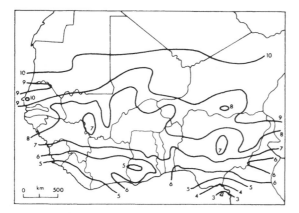

Figure 12b Mean daily sunshine hours in May

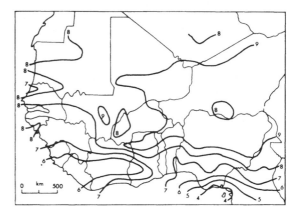

Figure 12d Mean daily sunshine hours in November

cent more sunny than the coast between Nigeria and Liberia. The contrast is least in November and December. However, the gradation from south to north is not regular, markedly less sunny areas in the far south-east and in east-central Ivory Coast and west-central Ghana producing a marked bunching of the isohels just north and south of these areas.

2 With the exception of the August map, north of about 10°N the latitudinal variation in sunshine hours is less clear. This is in part due to a small number of stations having seemingly anomalous values. In all months except August, Zinder in Niger (means based on nine years of records) has significantly less sunshine than its nearest neighbours. Consistently Ségou in Mali (records only for 1953–58) has much more sunshine than 'nearby' Bamako or Mopti (1950–58). The map brings out other

apparent anomalies which perhaps a long run of records might remove or confirm. The local site conditions for each station are unknown, as are the care, maintenance and functioning of the instruments, the completeness and accurate publishing of the records, the expertise and conscientiousness of meteorological assistants. There is no reason, however, to doubt the veracity of the published records from which the maps have been compiled. The causes of the apparent anomalies are not so easy to explain.

3 Of all the regions of West Africa, however, the most complex is that along the western coast. Consistently there is here a marked reduction in sunshine across the Guinea-Bissau and Senegal border, then a marked rise northwards to Dakar, a sharp decline across northern Senegal, followed by an increase again along the coast of Mauritania. Despite the paucity of

data for northern Senegal and Mauritania this distribution seems quite marked and the pattern of isohels is complicated. Generally coastal Senegal and Mauritania are less sunny than the interior, which probably reflects not only increased cloud but also sea mist along the coast, but why Dakar and Thiès should have such markedly higher values than stations immediately to the north and south is less easy to explain.

4 Most well-known of all the exceptional areas in West Africa is that referred to above as the Togo Gap which is clearly indicated in the maps of sunshine duration. Each map shows a zone of increased sunshine between Ghana and Nigeria, destroying the regular pattern of east-west trending isohels across southern Nigeria and Ivory Coast.

5 A further distinct area indicated by this original attempt to map the distribution of sunshine is the southernmost part of the south-west facing coastline. Southern Sierra Leone and Liberia clearly stand out as the third least sunny part of West Africa after the region bordering the Bight of Biafra and central Ivory Coast and Ghana.

The data in Appendix 4 and Figures 11 and 12 all indicate clear seasonality in the West African climate, the northern hemisphere summer period being less sunny than the winter. But even the small selection of data represented in Figure 11 shows significant regional variation. The graph for Gao (Mali) shows two periods of minimum sunshine (February and August) and two maxima (April and October–November). Although the periods of least and most sunshine are not of equal duration, the actual mean levels of sunshine recorded in February and August are the same, as are those in April and November. At Dakar (Senegal) the situation is different. Again there are two minima (December and August) and two maxima (January and April), but the April and August conditions are clearly dominant. Gao and Dakar represent the state of affairs in the northern populated part of West Africa. The graphs for Jos (Nigeria) and Bonthe (Sierra Leone) illustrate conditions in more central latitudes. There is here quite clearly one major sunny season and one markedly less sunny. Coastal Bonthe is less sunny than inland, upland Jos in the same month. Malabo in Bioko has obviously the least sunny climate, and al-

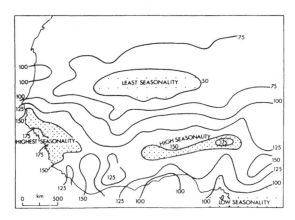

Figure 13 Seasonal variation in sunshine hours

though the general pattern of seasons observed for Bonthe and Jos is repeated, there is an interesting difference in that there is a slight but significant improvement in conditions in August, whereas this month is the least sunny at Dakar, Bonthe and Jos.

The graphs further indicate that seasonal contrasts are greatest at Jos (December 313 hours, August 127 hours) and least at Gao (February 261 hours, April 291). This difference between maximum and minimum sunshine levels has been plotted for 72 stations in West Africa to produce Figure 13. On the basis of sunshine alone it can be seen that the Sahelian region has the least seasonality followed by the most southerly areas, whereas the most prominent seasonal change is experienced along the south-west facing coastline (Senegal to Liberia) and in the 'middle belt' from northern Ghana to central Nigeria. The reasons for this will be discussed in Part Two.

Temperature

Just as there is a relationship between levels and patterns of radiation and sunshine, so there is also a relationship between these two elements of climate and temperature, meaning in this chapter the temperature of the lower atmosphere, of the air near the surface of the ground as recorded by dry bulb thermometers in screened conditions. But there is also a relationship between ground temperature and conditions and air temperature.

Factors determining temperature

Solid ground heats mainly by molecular heat conduction, the degree of conduction depending upon moisture content and porosity. Heat is spread through fluids, including the air, primarily by dynamic convection or turbulence, which are slower processes than conduction. From Table 8, column 1, it can be seen that the thermal conductivity of granite is seven times greater than that of still water, though this, in turn, is 30 times greater than the thermal conductivity of still air. Hence dry sand, highly porous but lacking water as a conducting medium in the pores, has a thermal conductivity only one-tenth of that of wet sand. Columns 2 and 3, however, show that the wetter the material the more heat is required to raise its temperature. These facts are highly relevant in climatology. Air temperature over land depends on the fraction of net radiation that is used in heating the air via the ground, plus any energy brought in by the air motion (advection). The relationship between radiation and air temperature depends, then, on (1) the albedo; (2) the effective outgoing long-wave radiation (i.e. the difference between the upward long-wave radiation from the surface and the downward counter radiation from the atmosphere); (3) the heat flux (flow) to the air; (4) the heat flux into the ground; and (5) advection.

An increase in albedo means a decrease in the net radiation input at the surface of the ground. If albedo was the only factor to consider, the higher albedo areas of northern West Africa (Figure 7)

Table 9 *Soil temperatures (°C) at Tamanrasset (1933–40)*

		Depth	
	Surface	38 cm	59 cm
January absolute minimum	−13.8	—	—
August absolute maximum	+69.5	—	—
Mean minimum (January)	12.8	—	—
Mean maximum (July)	32.8	—	—
Annual mean	22.8	24.7	24.7
Annual range	20.0	14.1	12.0

Source: after Castet 1942.

should experience lower temperatures than the southern, given the net radiation levels indicated in Figure 10.

But the temperature of the air is determined by the proportion of absorbed radiation that is returned to the atmosphere in the form of long-wave radiation and is absorbed by the atmosphere. This proportion depends on a number of factors:

1 The first is radiation efficiency, which is the fraction of global radiation that is retained by the earth (i.e. global radiation minus albedo and effective long-wave radiation). From Table 8 we can appreciate why water bodies, and thus wet soils, take longer to heat up but also to cool down. Wet soils absorb more radiation which can penetrate more deeply. Radiation efficiency decreases from 60 to 70 per cent in humid, coastal West Africa to 30 per cent in the southern Sahara. What radiation is being retained is not being quickly released. The dry soils of the Sahel and Sahara will receive high inputs of radiant energy during cloudless days, but rapidly release the energy during clear nights, i.e. the thermal inertia is low (see Table 9). The result is a marked diurnal range of air temperature. That the absolute maximum screen temperature recorded at Tamanrasset in that same period, 1933–40, was 39°C (July) and the absolute minimum was −7°C (January) reflects the low thermal conductivity of the air. When the soil is dry only contact heating of the air at the earth's surface provides for heat exchange.

Table 8 *Thermal conductivities of surface materials*

	Thermal conductivity cal deg C cm^{-2} s^{-1}	Specific heat cal g^{-1} deg C	Thermal capacity cal deg C cm^{-3}
Bare granite	0.011	0.2	0.52
Wet sand	0.004	0.3	0.48
Wet marsh soil	0.002	0.8	0.70
Still water	0.0015	1.0	1.0
Dry sand	0.0004	0.2	0.3
Wood	0.00035	0.3	0.18
Still air	0.00005	0.24	0.00024

2 A crucial factor, to be discussed as a separate element of climate in Chapter 3, is evaporation and evapotranspiration. A proportion of the radiation absorbed by the ground will not be available as a source of heat for the air if the ground is moist, for it will be used in evaporation and evapotranspiration. That proportion used to raise the air temperature is the sensible heat (H); that which is utilized without affecting temperature is the latent heat (LE). The ratio of the amount of the one to the other (H/LE) is known as the Bowen ratio. In humid, southern West Africa the Bowen ratio is low, as 80 to 90 per cent of the net radiation is used in evapotranspiration. Thus, if only 10 to 20 per cent is available to heat the lower atmosphere, temperatures are moderated. In fact over well-watered surfaces it is rarely possible for shade temperatures to exceed 34°C. (Ocean surface temperatures cannot rise above 29°C.) In arid conditions, however, the average annual Bowen ratio exceeds one, indicating more sensible heat flux than latent heat flux. If the soil is dry, as it is for the greater part of the year in northern West Africa, the net short-wave radiation may significantly exceed that changed to long-wave and used in evapotranspiration. The surplus energy is then transformed into convection and stored in the soil, and the greater the surplus the more rapid is the heating. The heat flux into the soil, as seen, is determined by the thermal conductivity of the medium. In West Africa the daily heat flow to and from bare soil will rarely exceed 5 per cent of the net radiation, but it will be even lower if the ground is vegetated.

3 The role of advection in determining temperatures in West Africa must not be overlooked, but its importance will become apparent in discussions to follow on the impact of the sea on West African climate and on the circulation of the atmosphere over the region.

Marine influences

Coastal West Africa, being usually more humid and cloudy than the north, not only receives a smaller input of net radiation, but also experiences much counter-radiation. The effective long-wave radiation in langleys per day probably ranges from some 196 ly in the north to 188 ly at the coast in January and from 234 to 185 ly in July. Hence diurnal temperature ranges are much less in humid West Africa than in the more arid north. Furthermore, the south is usually under the influence of airflow from the ocean and advective heat fluxes are more important when radiation fluxes are reduced. When the surface water cools by releasing heat to the atmosphere it sinks to be replaced by upflowing warmer waters. So the whole mass of heated water needs to be cooled before the surface layers reach minimum temperatures. Hence the large water masses, into which temperature penetrations may reach 600 metres, maintain a regular temperature for long periods of time. They thus regulate the air temperatures above them, enhancing them when they might otherwise be low and vice versa, and thus the Gulf of Guinea influences the air temperatures of coastal West Africa, reducing extremes and maintaining a limited range.

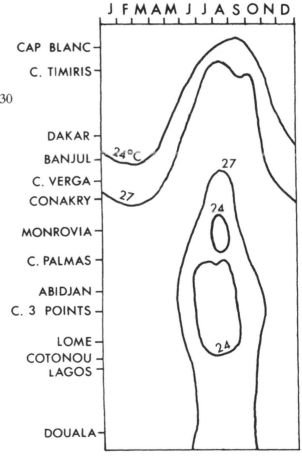

Figure 14 Mean monthly sea surface temperatures along the coast of West Africa (after Dorot 1973)

The annual variation of temperature experienced along the Guinea coast is, however, far from uniform. Maximum temperature departures from the annual means for a selection of West African coastal stations are noted in Table 10. The stations are listed in the order of their distribution along the coast of West Africa, from Dhakla in the north-west to Kribi in southern Cameroon. They fall into four groups, II and IV having smaller departures from the mean and so smaller ranges of temperature than groups I and III. These differences reflect important variations in sea surface temperatures along the coast. Figure 14 implies that temperatures of the Atlantic surface off the coast of Western Sahara are per-

sistently below 24°C, owing primarily to the presence there of the cool Canaries Current. Here temperatures reach their highest level, as one would expect, some two to three months after mid-summer. Dhakla, almost on the Tropic of Cancer, records its highest mean temperature in September, but it is only 22.9°C which further indicates the control of marine conditions over the coastal climate of this territory, the windiest part of West Africa with winds dominantly from the north. Nevertheless air temperature ranges are marked (Table 10).

Between Capes Blanc and Verde sea surface temperatures reach a high 27.5°C in September/October, but are even 10°C lower in January to

Table 10 *Air temperature departures from the annual mean and sea surface mean temperatures along the West African coast (°C)*

		Air temperatures				Mean sea temps for two monthly periods						Annual
		a	b	c	d	J/F	M/A	M/J	J/A	S/O	N/D	range
I	Dhakla	+6.4	−7.0	13.4	5.4	<20	<20	<21.1	<23.3	24.0	<23.3	4.0
	Nouadhibou	+9.0	−8.6	17.6	6.5	20	20	21.1	24.2	24.4	23.3	5.6
	Nouakchott	+8.8	−12.6	21.4	8.3	20	20.6	22.5	27.2	27.1	23.0	7.2
	Dakar	+7.4	−7.1	14.5	7.4	17.5	17.1	21.25	27.2	27.5	21.8	10.4
	Banjul	+9.1	−10.3	19.4	6.6	24.4	24.4	27.2	27.5	27.5	25.5	4.5
	Bolama	+5.7	−7.7	13.4	3.0	24.4	23.3	26.7	26.7	26.7	25.6	3.4
	Conakry	+5.8	−4.2	10.0	2.9	26.8	27.0	27.5	26.8	27.5	27.3	1.1
II	Freetown	+4.1	−3.7	7.8	2.6	25.6	26.1	26.7	26.1	26.1	26.7	1.1
	Monrovia	+4.9	−3.5	8.4	2.6	28.5	26.7	26.7	26.0	21.1	26.7	2.5
	Tabou	+4.4	−3.6	8.0	2.8	26.7	26.7	26.7	24.4	25.6	26.7	2.3
III	Sassandra	+4.4	−4.8	9.2	3.2	26.7	26.7	26.7	23.9	25.0	26.7	2.8
	Abidjan	+5.1	−5.0	10.1	3.6	26.6	27.7	27.4	23.1	23.8	27.8	4.7
	Takoradi	+5.9	−4.7	10.6	3.4	26.7	26.7	26.7	23.9	25.0	26.7	2.8
	Accra	+4.6	−4.8	9.4	3.5	26.7	26.7	26.7	23.9	24.4	26.7	2.8
	Lomé	+5.1	−4.7	9.8	3.5	27.6	28.4	27.9	23.6	25.3	27.5	4.8
	Cotonou	+4.9	−4.4	9.3	4.5	28.0	29.2	28.5	25.3	26.2	28.0	3.9
	Lagos	+5.2	−3.7	8.9	3.3	28.0	26.7	26.7	24.4	24.4	26.7	2.3
IV	Kríbi	+4.4	−3.8	8.2	2.4	27.0	>27.0	>27.0	24.4	24.4	>27.0	4.0

Notes: **a** departure of highest mean monthly maximum temperature from the station annual mean temperature; **b** departure of lowest mean monthly minimum temperature from the annual mean; **c** temperature range, i.e. value in column a minus value in column b; **d** range of mean monthly temperatures.

April. This is due mainly to an upwelling of cool water off this coast at this time (Figure 15), most pronounced when the influence of the Azores anticyclone is most felt. The clockwise circulation of air flow from the anticyclone results in north to north-east winds over this part of West Africa which will therefore blow parallel to the coastline. Sea surface currents generated by winds tend to flow with a force only 1 to 10 per cent of that of the wind and to be diverted by some 30° to the right of wind direction in the northern hemisphere (left in the southern). Thus when the north-east trade winds are most influential there will be a tendency for the warmer surface waters to diverge from the coastline to be replaced by upwelling colder waters from below. So the north-west coast of West Africa not only experiences air flow from outside the tropics in winter, but also air flow that, right to the shores themselves, passes over abnormally cool waters.

South of Dakar the upwelling is less marked, but cool water conditions in winter persist as far south as Conakry, lowering air temperatures

Figure 15 Zones of upwelling along the West African coast (after Dorot 1973)

accordingly by a degree or two and thus increasing coastal temperature ranges. Between Conakry and Cape Palmas (Tabou) however, a zone of essentially onshore south-westerly winds, a period of cold upwelling is not experienced. Sea surface temperatures remain high all year, only falling slightly in the coolest, cloudiest season, July to September, when still imparting some influence on the coastal climate. Here is experienced, along with the other permanently warm water area of West Africa, the Bight of Biafra (Group IV, Table 10), the lowest temperature ranges.

Between Tabou and Lagos (Group III) there is again a marked seasonal change in the temperature of the surface of the sea, but here the lowest values are recorded in July and October (Figure 16) another period of upwelling. Pople and Mensah (1971) suggested that the upwelling is the result of evaporation which increases surface salinity and causes the upper marine layers to sink to be replaced by cooler waters from below in a thermohaline convective cell. However Houghton (1973) observed no significant difference in evaporation from a Class A pan onshore at Tema and a floating container in the harbour in

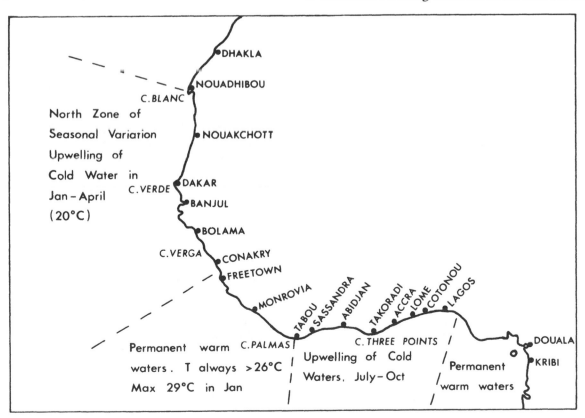

the period April to September, 1972. Then ocean surface temperatures offshore ranged from 29°C to 27°C from April to June, dropped to 22°C in early July and then ranged from 19 to 22°C into September. He noted how coastal the cold upwelling was. When the sea surface temperature was only 20.3°C three kilometres from the coast, it was 22°C, 20 km out to sea and 25°C, 100 km offshore. He believed no satisfactory explanation for the upwelling was available.

Four years later, however, O'Brien and Adamec (1977) suggested that the explanation lay across the Atlantic Ocean, easterly winds off Brazil exciting equatorial trapped Kelvin waves (see glossary) that travel eastward to impinge upon the African coastline. In association with west to south-westerly winds paralleling the coastline between Liberia and Nigeria, most prominently in the summer, they lead to another zone of upwelling. The consequence to temperature ranges is shown in Table 10 even if a comparison of mean temperatures for Monrovia, Abidjan, Lomé and Douala gives little indication that the sea temperatures significantly influence one part of the coast of southern West Africa more than another.

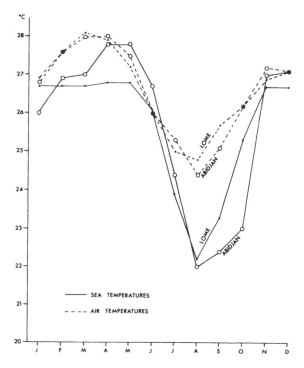

Figure 16 Air and sea surface temperatures off Lomé and Abidjan

Temperature distribution

These facts are clearly brought out in Figures 17 and 18. All the maps show the importance of latitude in the determination of the patterns of temperature distribution, in that the isotherms generally trend east to west. However there are differences in the patterns of detail between the northern and southern regions. Maps of mean temperature for four months have been chosen to illustrate the annual picture.

The January map shows a clear diminution of temperature from south to north, with the influence of the sea seemingly at a minimum, though observable north of Dakar and south of Cape Verde to Cape Palmas. The temperature gradient is more pronounced north of 15 to 18° latitude however and the smooth pattern of parallel isotherms is disrupted by the local influence of altitude. Most marked are the low temperature areas of the Fouta Djalon in Guinea centred on the station at Mali at 1464 m (19.1°C), the Jos Plateau (Jos, 1285 m and 20.8°C) and the Cameroon/Adamawa Highlands (N'Gaoundéré, 1098 m and 21.6°C). Altitude also has a significant effect on the climate of Tamanrasset in the far north of West Africa, which is in the Hoggar Massif at 1400 m. It is this fact together with the sparseness of meteorological stations elsewhere in southern Algeria, northern Mali and northern Niger, which leads to distinctive and apparently anomalous patterns in the far north-east of the region and the steep isothermal gradient there in all the mean temperature maps. On the January map the warmest place is Bohicon in Benin (28.2°C) and the coldest is Tamanrasset (12.8°C). This month is clearly the coolest in the year in the north and, following the discussion above on the determinants of air temperature, the reason should be clear.

However January is not the coolest month of the year in the south. Here the lowest temperatures are experienced in August (Figure 17c), in the middle of the cloudiest season with marine influences at a maximum. The influence of altitude is again clear, the lowest temperature being at Mali in Guinea (18.3°C) while Arouane in Mali has a mean temperature of 33.9°C, a range of 15.6°. The bunching of isotherms at about 15° latitude is unrelated to altitude and so reflects a change in climatic influences to be discussed later. In August the marine influence north of Dakar and especially north of Nouadhibou is most marked.

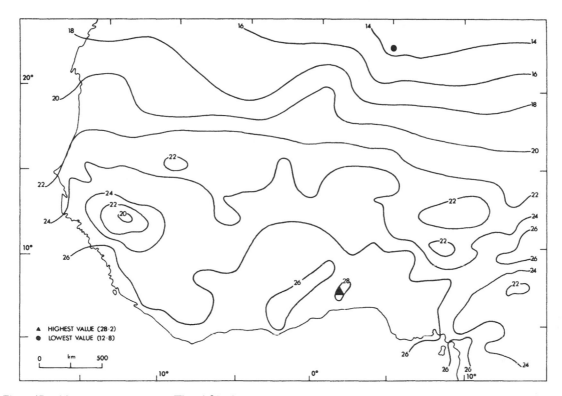

Figure 17a Mean temperatures over West Africa in January

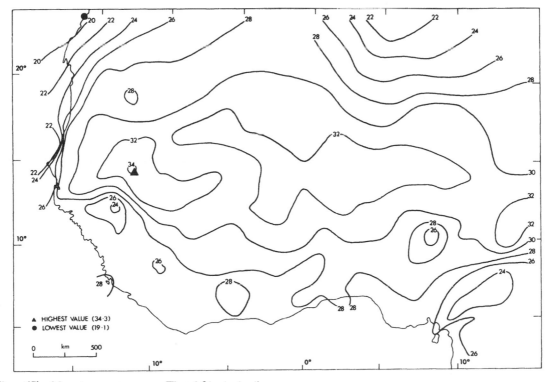

Figure 17b Mean temperatures over West Africa in April

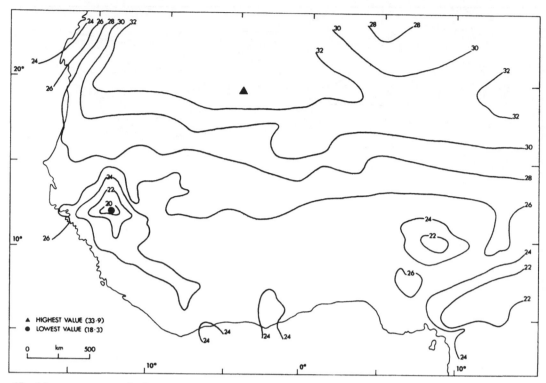

Figure 17c   Mean temperatures in August

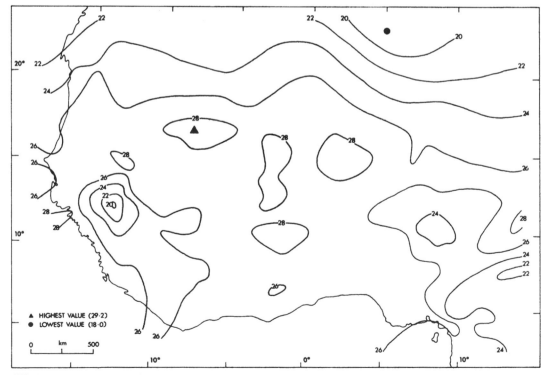

Figure 17d   Mean temperatures in November

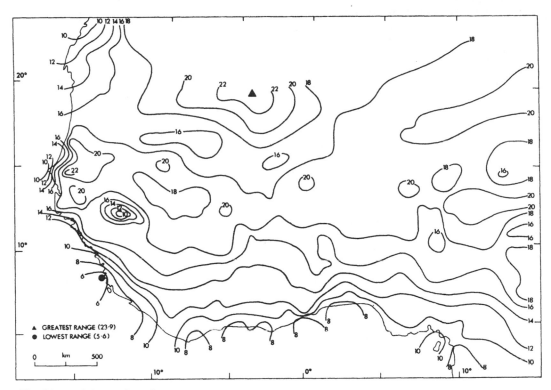

Figure 18a   Mean minimum diurnal temperature ranges

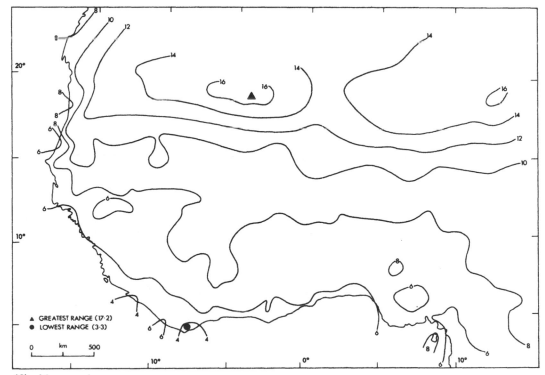

Figure 18b   Mean maximum temperature ranges

In the maps we thus see two 'seasons' of minimum temperatures, one in the north in January and December, the other in the south in August (with July). So clear is this pattern that of the 128 stations whose data were utilized to create the maps of mean temperatures, only two fail to register their lowest mean temperature in one of these four months (Figure 19). Furthermore, when the distribution of the stations having their lowest temperatures in December/January or July/August is plotted (Figure 20), the north–south pattern becomes fully apparent. The 'north' is seen to be north of approximately the 11th parallel. Only one station with lowest temperatures in July/August lies north of the heavy line drawn on the map, and that again is the anomalous Mali in the Fouta Djalon at 1464 m. The validity of the line as a climatic boundary is further suggested by its correlation with clear divisions on other maps in this section of this book.

The authors chose to map the mean temperature patterns for January and August after Figure 19 had been compiled. To complete the annual picture maps were further drawn for April and November, the latter month because it is the most uniform. Both highest and lowest temperature stations are in the north in November (Néma in Mauritania with 29.2°C and Tamanrasset again with 18°C, a range of only 11.2°). A pattern of isotherms paralleling the parallels no longer exists in the south. Local conditions, particularly altitude, dominate. Tamanrasset's low value, because of the effect of altitude, produces an exaggerated pattern of temperature gradient in the north-east. The influence of the Canaries Current and northerly winds are still to be seen in the north-west.

If there are two short 'seasons' of minimum temperature (December–January and July–August), there is only one 'season' of highest temperatures, commencing abruptly and most markedly in March and fading steadily thereafter to July (Figure 19). April, midway between the 'low temperature' months of January and July, was chosen to show the 'high temperature' situation (Figure 17b). Again the four high-altitude areas influence the scene locally and are responsible for steep temperature gradients in the far north, the south-east and south-west. Again the ocean off the north-west coast, now at its coldest, makes its presence felt and the temperature of 19.1°C at Dhakla is the lowest on the map. Most dramatic, however, is the temperature

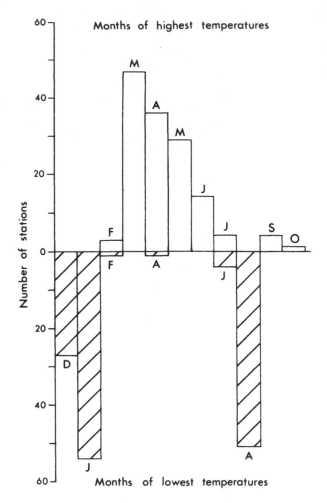

Figure 19 Months of highest and lowest temperatures

gradient along the Senegal coast. For it is inland from here, between about 15 and 17°N, that the highest temperatures in West Africa are found. The value of 34.3°C at Kayes in western Mali is the highest on the map. Here at these latitudes, between the north with its post-winter cold, arid conditions, clear skies and high albedos, and the south with its marine-influenced conditions and growing cloud cover, is West Africa's 'hot belt'. This is well demonstrated in Figure 21. Once again the division between north and south is clear and the boundary drawn in Figure 20 when reproduced on Figure 21 is seen to be significant once more. South of the heavy line, i.e. south of approximately 10 to 11°N, the highest temperatures are experienced mainly in March (see also Figure 19), a month in advance of the zenithal sun, but mean temperatures across the region are only 27.8°C. North of the boundary the 'north' falls into four sub-regions, elongated west to east, except on the west coast where the cool marine influence is seen again, lowering mean temperatures. The highest temperatures are experienced progressively later in the year, reading from south to north and now coincide, except along the coast, with the period of overhead sun; highest temperatures, as noted above, are not found in the Sahara but rather in the Sahel.

The greatest diurnal ranges of temperature are however experienced in the far north, but away from the influence of the Hoggar Mountains and of the sea. Figure 18 a shows the distribution of

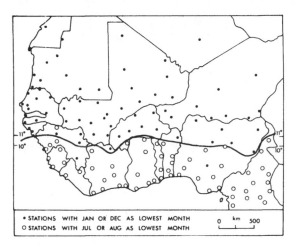

Figure 20 Distribution of stations having either December/January or July/August as the months with lowest mean temperatures

Figure 21 The seasonal pattern of the highest mean temperatures (the mean maximum temperatures for each delineated region are indicated, and the dates of the northwards moving overhead sun at the parallels marked)

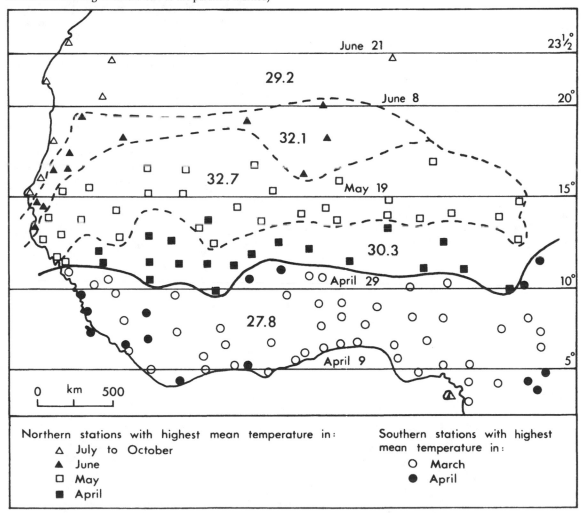

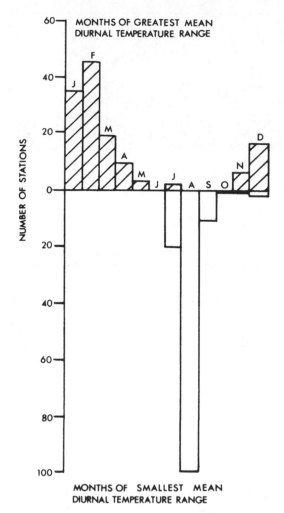

Figure 22 Months of greatest and smallest diurnal temperature ranges

mean minimum diurnal temperatures irrespective of the month of occurrence. The lowest values are recorded along the south coast, especially facing south-west, where sea surface temperatures are high all the year and direct on-shore winds from a warm ocean are important, and highest values are centred on Arouane (285 m above sea level) in northern Mali. Essentially this is the picture for August when marine influences and cloud are most marked throughout the region (Figure 22).

Conversely the least cloudy, most sunny period peaks in February, and Figure 18b shows the distribution of mean maximum temperature ranges. The lowest values are still along the coast being only some 2 to 4°C greater than in August, and in the Fouta Djalon. The highest values are again centred on Arouane furthest from the influences of altitude or the sea at that latitude. The role of the cool waters, now at their lowest temperature between Conakry and Dhakla, is clearly seen, especially in Senegal where, over a distance of but 135 km from Dakar to Djourbel, the mean maximum diurnal temperature range increases from only 9.5°C to 22°C.

It is hoped that the relationships between radiation, sunshine and temperature have been demonstrated clearly, along with the patterns of their distribution in West Africa. The significance of these elements of climate as determinants of atmospheric pressure and thus wind flow in the region, is considered in the following chapter.

2 Atmospheric pressure and wind

Pressure

The global mean atmospheric pressure at sea level fluctuates around the value known as the normal or standard atmosphere, which at 45° latitude and at 0°C is about 1.1 kg cm^{-2} or just over 1 bar or 1013 millibars (mb). Variations in atmospheric pressure are closely related to air temperature, water vapour content and vertical and horizontal air movement, The horizontal motions we call winds can exert considerable, even destructive, pressure, but in this section of the chapter we are concerned with vertically exerted pressure, the result of air density. It is generally appreciated that cool or cold air will subside, increasing its pressure on the air and the earth beneath it, but warm air expands and rises relative to its surroundings, thus decreasing pressure locally. Deviations below the normal atmosphere only abnormally exceed 31 mb, or above it exceed 20 mb. In the tropics sea level pressure tends to be lower than the normal or standard, but except in storm centres does not deviate far from the mean. However, even small differences are highly significant. Altitude, of course, has an important impact on locally recorded pressure, which diminishes on average about one thirtieth of its value each 275 m, that is about 1 mb per 10 m in the lower atmosphere, the rate decreasing at greater heights. It is a common practice in meteorology to describe altitude in terms of pressure rather than in metres or kilometres. To show large regional patterns of pressure the actual pressure recorded by barometer or barograph is often adjusted to the value that would have been recorded at the temperature of the time at sea level. This practice has been adopted in this chapter although the effects of altitude were not excluded when considering temperature.

Variations of pressure horizontally are smaller than they are vertically, in West Africa not often exceeding 1 mb per 150 km, but these form the crucial patterns mapped. They vary further, of course, with time, not only seasonally, reflecting annual oscillations in radiation and temperature, but also even diurnally. Figure 23 shows the diurnal pressure wave at Malabo (Bioko) which is characteristic of the tropics generally with its amplitude of about 2 mb, with maxima near 10.00 and 22.00 hours, minima near 04.00 and 16.00 hours. These regular fluctuations are perhaps analogous to the oceans' tides, but whereas they are caused by the gravitational pull of sun and moon, the pressure wave is possibly a response to the daily rhythm of heating and cooling.

Seasonal patterns of pressure

Although the diurnal wave may have some local influence on the weather, to be hinted at below, we are more concerned with the major seasonal pressure changes, illustrated in Figures 24 and 25. Figure 24 shows the mean monthly pressure adjusted to sea level at four stations in West Africa.

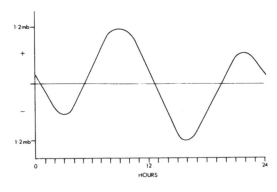

Figure 23 Diurnal pressure wave at Malabo

Two, Tessalit and Cotonou, are chosen to represent the situation that prevails in the far north and far south of the region. Kabala and Sansanding have been chosen as representative of the area between the interior and the coast. Of all the stations whose records have been used to compile the maps (Figure 25), Tessalit in Mali has the greatest pressure range in the north and Cotonou in Benin (with Ilorin, Nigeria) in the south. It is immediately apparent that annual pressure changes are greater in the north than in the south. It is further to be noted that both Tessalit and Cotonou have one maximum and one minimum period each year, but that these periods do not

coincide. Tessalit's highest pressures are experienced in the northern hemisphere winter, when they tend to be above the standard 1013 mb, and lowest pressures in summer. On the other hand Cotonou has its highest values in summer, lowest in March. The reasons for this are not far to seek if the periods of maximum and minimum pressure are compared with those of highest and lowest temperature (Figures 17 and 19).

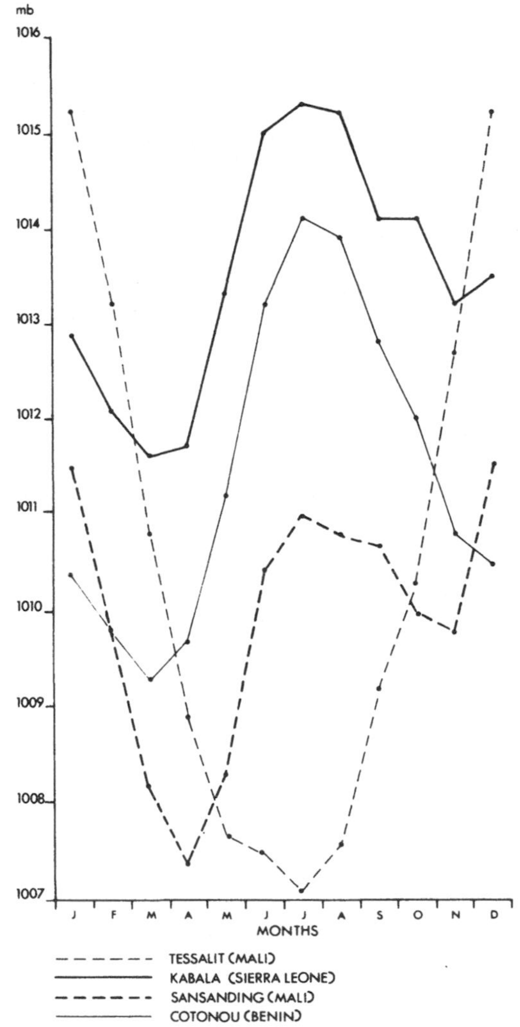

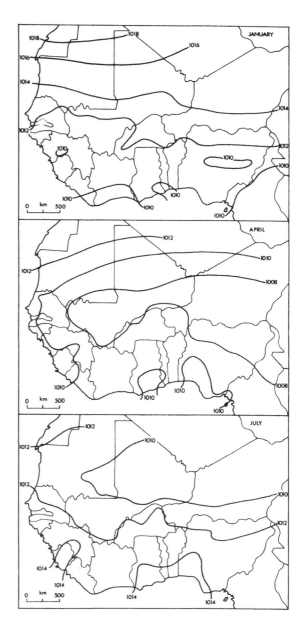

Figure 24 Mean monthly sea level pressure at four selected stations

Figure 25a Mean sea level pressure in January; *b* Mean sea level pressure in April; *c* Mean sea level pressure in July

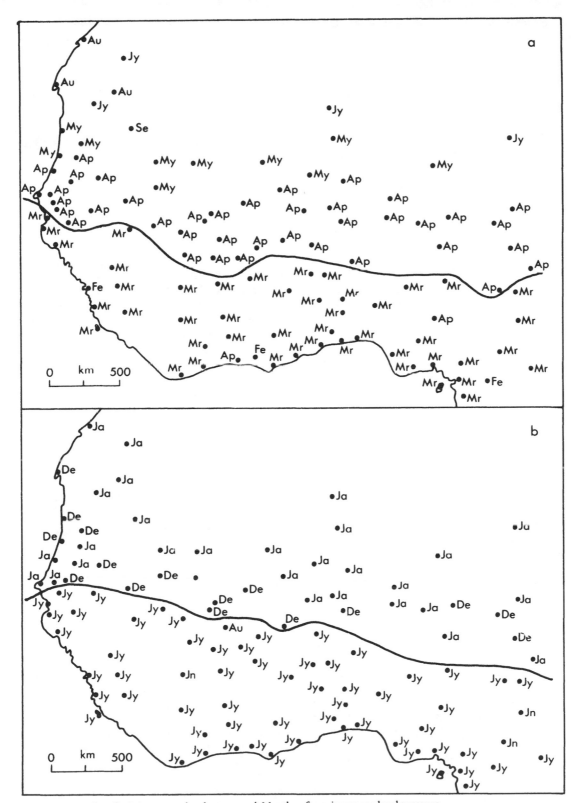

Figure 26a Months of minimum sea level pressure; *b* Months of maximum sea level pressure

Somewhere between north and south there must be a transition zone, which, it might be expected, should be rather broad and vague. This is not so, however, as Figure 26 shows, and the coincidence of the dividing lines drawn on the maps with those indicated on the temperature maps, separating north from south, is remarkable. Minimum pressure throughout the south is in March. For some 500 km north of the dividing line it is in April, thereafter progressively later in the year northwards. This is to be expected, given the temperature conditions, and accounts for the contrasts between the graphs for Cotonou, Kabala and Sansanding and that for Tessalit. Greater contrasts are to be seen, even between Sansanding and Kabala, with regard to month of highest pressure. At Sansanding highest pressure is in December; at Kabala it is clearly in July. However, other distinctive features at these stations are the two troughs and two peaks in each graph (Figure 24). Kabala, though with that obvious maximum in July, also has a subsidiary high pressure period in December–January and so has two troughs in March and November. Sansanding has its highest pressure in December/ January, another high in July and so two lows in April and November, that in April being much more pronounced.

These features are related to temperature and humidity changes that occur annually in West Africa, as humid air from the Gulf of Guinea extends across the region from the south from March, reaching the far north in August before retreating southwards to clear all but the coastal areas by the end of the year. Thus the far north tends to experience little of the monsoonal invasion with its related cloud, rain and cool summer conditions, and the far south experiences little of the Saharan air which spreads south as the humid air retreats. So Tessalit and Cotonou have a single period of high pressure coinciding with their coldest season, and a single period of low pressure coinciding with the warmest season. The central area of West Africa comes significantly under the influence of both air masses, however. As the temperature maps indicate, the Sahelian zone is coldest in December and January but, with the northwards advance of the overhead sun, attains its highest temperatures in April, hence the mean pressure change. By midsummer however the air mass from the ocean has impinged on the central region, bringing down the temperature, accordingly raising the pressure. By November the humid air has

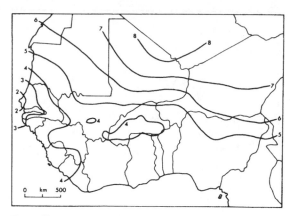

Figure 27 Mean annual range of sea level pressure

retreated again, providing a slight rise in temperature and fall in pressure before the winter cool returns. Kabala has a similar pattern of temperature and pressure change, but being more markedly under the influence of the south-west airflow with its cloud and rain, temperatures do not attain levels as high as those experienced in the Sahel, except in January, and so pressures tend to remain higher all year.

In Figure 25b the coincidence of the lowest pressure area with the highest temperatures (Figure 17b) is obvious, as is the cooling influence of the ocean at this time. Such is the moderating influence of the sea, reducing temperature ranges and especially along the coast of Senegal, that it is here and on the Grain Coast further south that minimum annual pressure ranges are experienced. Conversely greatest ranges are characteristic of the north (Figure 27).

Surface wind

Wind direction

It is the imbalances between the atmospheric pressure of one area and another that generate the wind, the pressure-gradient forces directing the air movement from the regions of greater air density to those of lower density and pressure. But the Coriolis effect and friction cause the air to flow obliquely across the isobars at a narrow angle (for further discussion see Part Two, Chapter 8). Winds in the northern hemisphere appear to be diverted to the right of the most direct track to the centre of lowest pressure, whereas winds in the southern hemisphere appear diverted to the left. The isobars in West Africa, it can be seen, generally trend west to east across the region, the higher pressures being in the south in the summer and in the north in the winter. Thus one should expect south-westerly winds to predominate in the summer and north-easterlies in the winter: that is the advance of a maritime tropical air mass, an extension of the St Helena anticyclone, across West Africa in the middle of the calendar year; and the retreat of this airmass, to be replaced by continental air from the Sahara, from the North African anticyclone, with the onset of winter.

This regional pattern of air flow is clearly shown in Figure 28a–d. Once again the change from one set of conditions in the south to another set in the north is quite abrupt, even allowing for the scale of the map, and the pulsation of the airmasses north and south is obvious. Only the coastal region north of Dakar again fails to conform to the general pattern, this area coming under the influence of northerly winds that are circulating out of the anticyclone centred on the Azores. The situation is crudely summed up in Figure 29.

The greater detail in Figures 28a–d must be used with caution. Wind is a particularly variable climatic element. The wind roses have been drawn from data from a number of sources and mainly from records of twice- or thrice-daily readings of wind direction, as registered by wind vanes at the time of observations. Different stations make their observations at different times. Those that note wind direction between midnight and 06.00 hours will record many more calms than those that observe at 09.00, noon or 15.00 hours, as winds characteristically lessen at night, especially in the south. Thus the roses for Cape Coast, southern Ghana, are based on daily readings at 09.00 and 15.00 hours, those for Tamale, northern Ghana, on readings taken at midnight and noon. In July, for example, over 70 per cent of readings at midnight at Tamale register calms, whereas at midday the value falls to 12 per cent. Hence the sharp contrast between some 'neighbouring' stations concerning the percentages of calms. At Ouagadougou, where readings are taken three times daily, calms were noted 101 times at 06.00 hours in 1960, but only 26 times at noon. At Mopti (Mali) in the same year, calms at 06.00 hours totalled 169, at noon 23 and at 18.00 hours 89.

Variations in the roses also reflect local con-ditions, the sites, exposures and altitudes of the wind vanes or anemographs, the crudity of observation by using wind vane alone, the quality of equipment and the experience of observers. Wind flow, of course, is not steady but occurs in gusts and eddies. For all these reasons one should expect the variations illustrated, realizing also that none of the stations is closer to any other than 200 km (Dakar to Banjul).

The lengths of the lines on the roses have been carefully drawn to indicate the percentage frequency of total wind observations for eight directions according to the scale indicated. That winds from the north and east are dominant in January, and winds from the south and west are dominant in July, is clear. The percentage frequencies for these wind directions and for these months were plotted and isopleths interpolated to produce Figure 30 a–d, which provides an alternative method for illustrating much of the information incorporated in Figure 28. The coincidence of the position of the surface discontinuity marked and labelled ITD (Inter-tropical Discontinuity) in Figure 28 between the Atlantic and Saharan air masses in January and July with that of the bunched isopleths in Figure 30 (a and b) is noticeable and provides further confirmation that marked changes in dominant surface wind direction between north and south occur over a short distance.

Wind velocity

What Figures 28 to 30 do not provide is any indication of *wind velocity*. The relevance of a knowledge of this element of climate is discussed in Part Three and storms and squalls are considered in Chapter 6. The 'run of the wind', for example the distance the mean wind flow traverses in a period of time, from which mean wind speed can be calculated, is measured by cup-counter anemometers (Appendix 5). Their distribution in West Africa is not widespread (for instance in 1980 there were only 20 in the whole of Nigeria) and records of *mean* wind speed for the region are not readily available. Figure 31 has been compiled from the data of only 54 stations across the region. Nevertheless, it is possible to deduce the following points:

1 Through the year as a whole (Figure 31) mean wind speeds tend to be low, the greater part of the region experiencing velocities less than

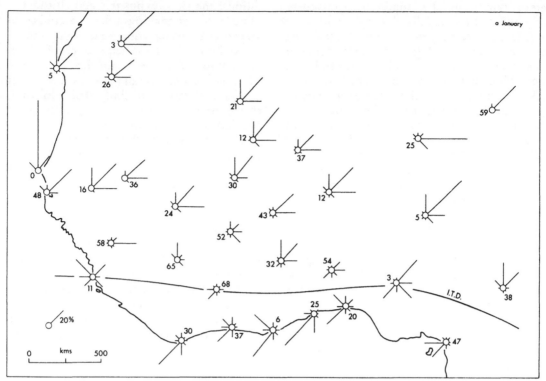

Figure 28a　Mean surface wind direction in January

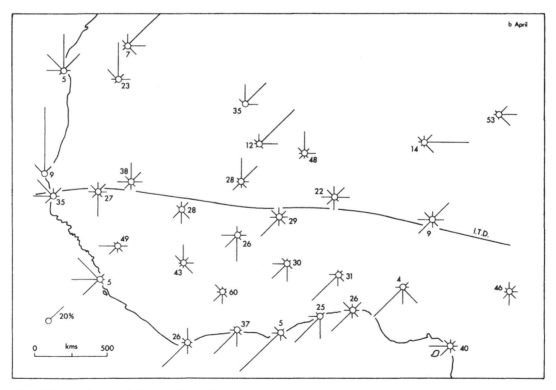

Figure 28b　Mean surface wind direction in April

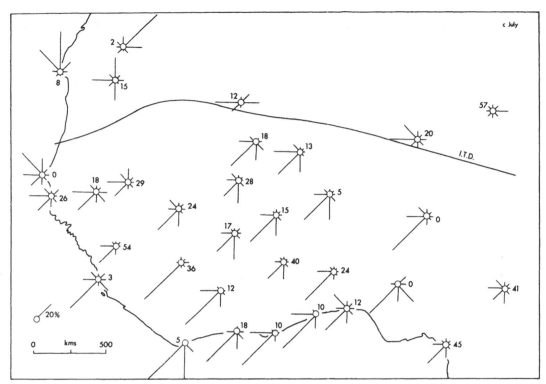

Figure 28c Mean surface wind direction in July

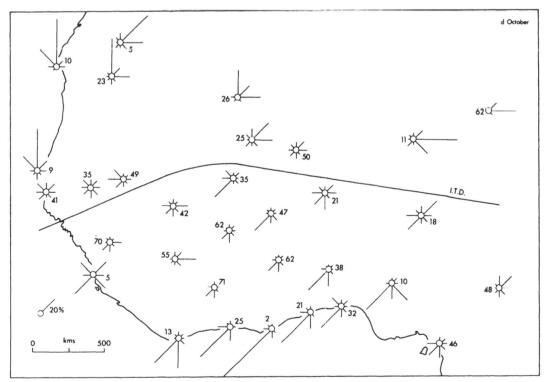

Figure 28d Mean surface wind direction in October

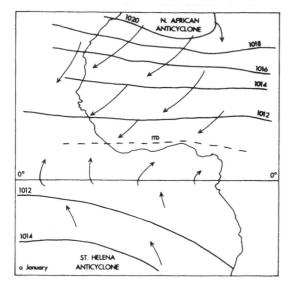

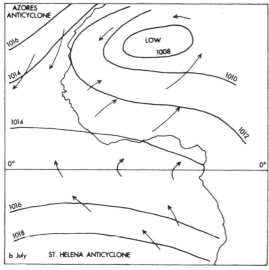

Figure 29 Pressure and wind patterns in *a* January and *b* July

movement, i.e. sea breezes, which are discussed below. It is noticeable how the overwhelmingly dominant north winds of the winter season, when screened air temperatures a few kilometres inland and sea surface temperatures off shore are most similar, diminish in importance and give way to winds with a more direct onshore component in the summer when land and sea temperatures are most diverse (Figures 14, 17 and 28).

2 However, maximum wind velocities along the north-west coast tend to be recorded in the transition period from coolest to hottest seasons, and vice versa.

3 Elsewhere in northern West Africa the season of highest mean wind speeds is associated with that of lowest temperatures (Figure 32a), when the pressure gradient from the cool north to the warmer south encourages the strongest push southwards of the harmattan across the sparsely vegetated region. Southern areas have their lowest mean wind speeds in this season.

4 It is with the advance northwards across West Africa of the maritime airmass in the summer that the Guinea coastlands have their season of highest mean wind speeds (Figure 32c). Seasonal variations in wind speed in Nigeria have been illustrated by Ayoade (1980) (Figure 33).

5 The most uniform conditions across West Africa as a whole are experienced therefore in the seasonal transition periods about April and especially October, even though these same periods are characterized by the development of storms and line squalls (Chapters 6 and 7), whose violence is hidden in the mean values used as the basis of Figure 32a–d. It is during such storms that the greatest wind speeds are normally encountered, speeds comparable perhaps with those commonly observed during the passage of depressions over land in middle latitudes, though not with those experienced in typhoons or hurricanes in other tropical areas. Table 11 indicates maximum gusts for a number of stations in West Africa. On the whole it seems that wind speeds increase both with latitude and altitude (Kortright: 335 m above sea level; Jos: 1285 m).

The frequency with which dangerous gusts occurs is low. The probability of 52 knots being exceeded at any one place in Ghana is about once in 10 years on the coast, once in two years inland. Gusts of over 43 knots may be expected twice a

3 ms^{-1} (about 11 km, or 7 miles, per hour; 6 knots). The Saharan margins and the Guinea coast west of Nigeria, particularly in the Togo Gap, have mean wind speeds in excess of this figure, but clearly the windiest part of West Africa is the coastal area from Banjul to Dhakla. The way in which the bunched isotachs parallel the coastline suggests, however, that the explanation for the windiness relates not only to the persistent flow of air around the Azores anti-cyclone (producing the northerly winds in Figure 28), but also to direct onshore air

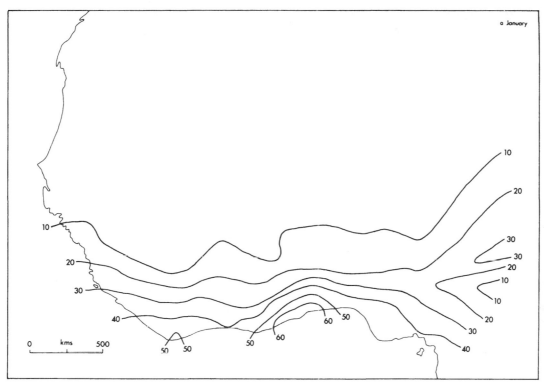

Figure 30a   Percentage frequency of wind from the south-west in January

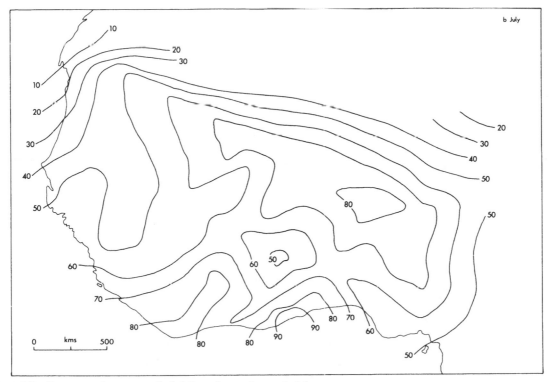

Figure 30b   Percentage frequency of wind from the south-west in July

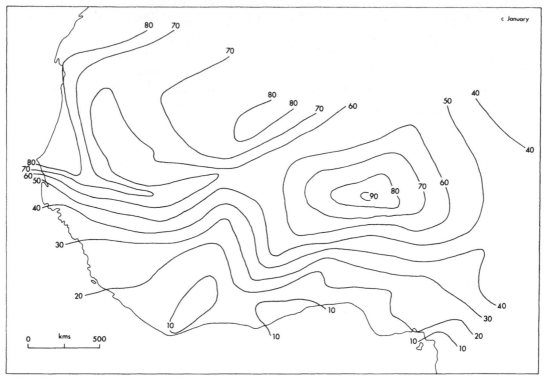

Figure 30c   Percentage frequency of wind from the north-east in January

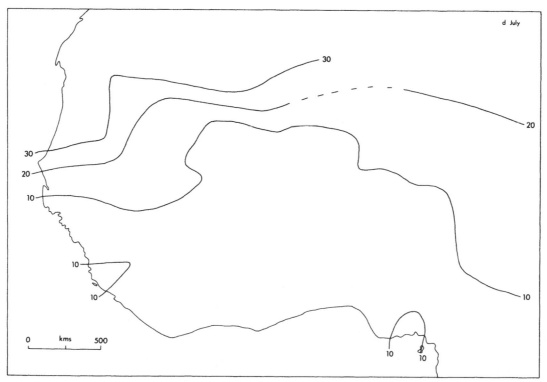

Figure 30d   Percentage frequency of wind from the north-east in July

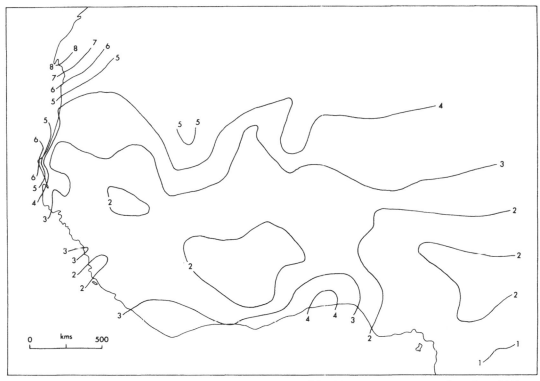

Figure 31 Mean annual wind speed (ms⁻¹)

year on the coast, three times a year north of the forest zone (Walker, 1962). Ayoade has compiled for Nigeria the most extreme wind with a 100-year recurrence to be 112.6 knots at Warri, 55.8 knots at Ilorin. Okulaja (1968) has noted that about 75 per cent of daily maximum gusts are between 12 and 18 knots at Ikeja (Lagos) and that the mean maximum gust is 15.9 knots.

Diurnal wind variation

A consideration of wind speed in terms of mean values also hides the diurnal variations in wind flow that commonly occur in West Africa. These also fluctuate with season and with latitude, depending upon the location of the Inter-tropical Discontinuity (ITD) between the maritime and continental air masses. North of the ITD, that is throughout most of West Africa in the winter, the dry season temperature inversion noted in Chapter 1 will produce lowest wind speeds during the night and early morning. As the inversion breaks down following sunrise so the surface wind speed tends to increase, only to fall off again in the afternoon at the time of maximum

Table 11 *Maximum gusts for selected West African stations*

Station	Max. gust Knots	Period	Reference
Kortright (Freetown)	75	1960–7	Williams (1970)
Tamale	60	?	Walker (1962)
Accra	57	?	"
Jos	80	1953–69	Ayoade (1980)
Kano	c.73	"	"
Maiduguri	c.73	"	"
Minna	c.55	"	"
Makurdi	c.55	"	"
Ibadan	c.56	"	"
Potiskum	c.55	"	"
Lagos	c.55	"	"
Enugu	c.55	"	"
Sokoto	c.55	"	"
Warri	c.63	"	"
Benin	c.45	"	"
Port Harcourt	c.45	"	"

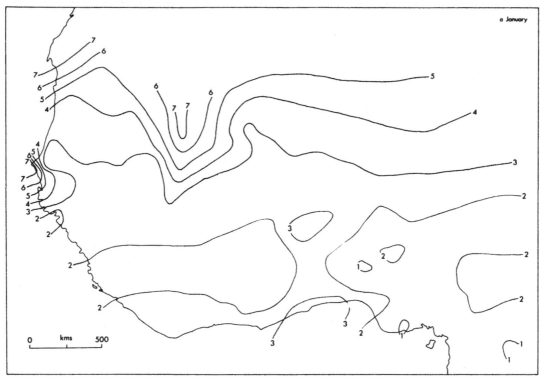

Figure 32a   Mean wind speed (ms⁻¹) in January

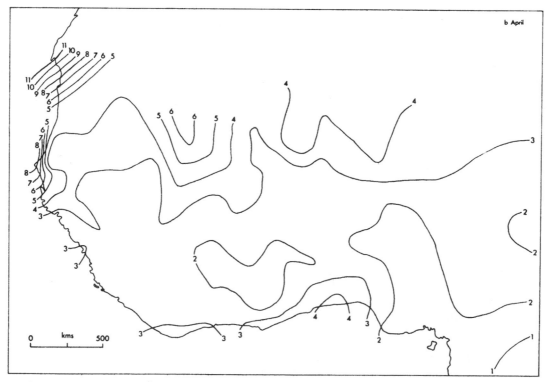

Figure 32b   Mean wind speed (ms⁻¹) in April

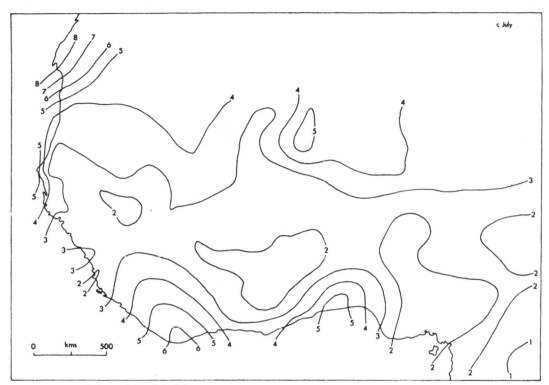

Figure 32c   Mean wind speed (ms⁻¹) in July

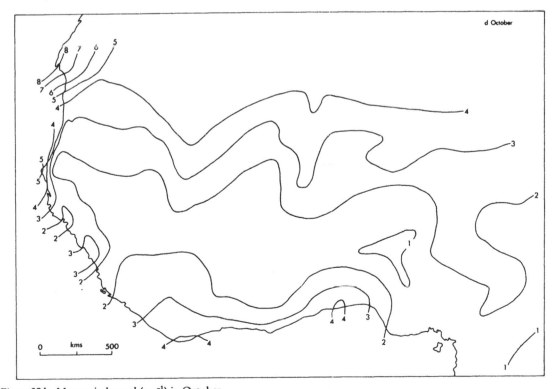

Figure 32d   Mean wind speed (ms⁻¹) in October

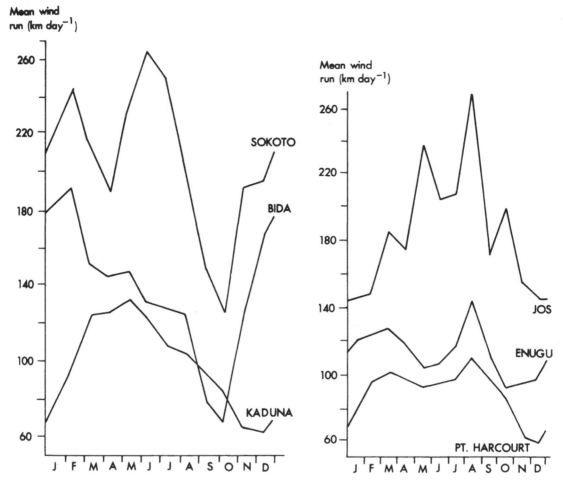

Figure 33 Seasonal wind speed variations in Nigeria (after Ayoade 1980)

convectional mixing. South of the ITD, that is throughout most of West Africa in the summer, with temperature conditions remaining steadier under the influence of marine south-westerly winds and an extensive cloud cover, winds tend to be strongest in the morning but also persistent through the day and night.

Land and sea breezes

In many coastal regions the main causes of diurnal wind variation are land and sea breezes, and these also tend to be most prominent outside the main rainy period, for then temperatures both onshore and over the sea are lower and more uniform throughout the day. Figure 34 shows

the relationship between mean monthly dry bulb temperatures and the duration of breezes from the sea. Figure 35 describes the mean conditions of speed, steadiness and direction for land and sea breezes in January and July in Sierra Leone. In January a breeze from the north-east (60°), i.e. from the land, is prominent after sunrise, though wind speed is only about 4 knots and the breeze is rather erratic. The change from land to sea breeze is sharp; by noon a steady breeze from 240° has replaced the land breeze, with speeds attaining 9 knots at about 14.00 hours, thereafter declining to 7 knots before the flow reverts to a land breeze again in the early hours of the morning. This pattern does not necessarily occur daily throughout the month. Sometimes the land breeze may not be felt, but may simply weaken the south-westerly flow. In July the land breeze is not so marked, the main flow after sunrise being rather from the south-east (120°). Throughout the afternoon and evening, however, the breeze is strongly from the south-west.

Sea breezes form most readily in quiet atmospheric conditions, because wind flow above, say, 15 knots will create turbulence which will reduce the chance of thermal pressure differences between land and sea developing. When the air over the land heats up, rising currents of even 50 metres per minute may be generated, and will initially lead to a flow of air aloft from land to sea. But with the following reduction of air pressure over the land there will commence a surface flow of air from sea to land. Thus a circulatory flow is established. The descending air offshore warms up on descent, so is less dense than the air further out to sea which then moves in under the adiabatically warmed air and so gets caught up in the circulation. Thus, little by little, the flow expands seawards; hence its commencement after sunrise and its fullest development in the afternoon.

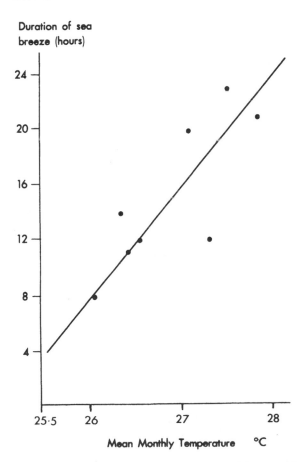

Figure 34 Relationship between temperature and duration of sea breezes (after Mukherjee and Moore 1973)

At night the sea surface may cool 0.3 to 0.6°C in the tropics, whereas the coastal land areas may cool, especially in the dry seasons, by 8 to 11°C. So steeper temperature gradients may be achieved at night. Yet a marked *land* breeze is usually restricted to coastal margins and about 100 metres above the surface where frictional effects are diminished and is strongest where the coast is backed by hills and valleys increasing gravitational effects. Sea breezes, however, may extend up to 1200 metres in altitude and even reach 3400 m. In West Africa they are likely to be least marked along coasts backed by swamps, lagoons or creeks, especially if well-vegetated, and so most prominent along the coast of Senegal and Mauritania and in the Togo Gap. They may penetrate many kilometres inland and so influence, as well as being a result of, local weather conditions. A sea breeze remarkably constant in direction may extend across the Accra Plains to rise sharply at the Akwapim Hills, 32 kilometres from the coast, over which a magnificent line of cumulus cloud may develop, whilst the plains remain cloud free or may have only a scatter or some streets of fair-weather cumulus. Along the coast of Senegal the onset of the sea breeze in mid-afternoon may markedly affect both temperature and humidity (Figure 36).

Dust devils

If coastal West Africa has its distinctive sea and land breezes, the interior, the Sahel in particular, has its own local phenomenon in the dust devil. This is a whirling column of air, some 6 metres in diameter, about a vertical or steeply inclined axis. It is analogous to a tornado in size, but the rotation rate is much less and it is shorter-lived. It markedly differs from a tornado in that it forms in cloudless conditions and so cannot derive its energy from the release of latent heat. It is a feature of local turbulence, small bubbles of heated air rising and being carried forward by the light, but not strong, breeze that is required. On reaching an obstruction, such as an isolated hill or clump of trees, the air takes on a rotary motion and the dust devil is created. It develops a warm, low-pressure core about which circulate strong winds which whip up the dust and can even flatten huts. Up to 20 or 30 per day may be recorded locally in the dry season (Smith, 1937). If the winds are stronger, attaining 16 to 20

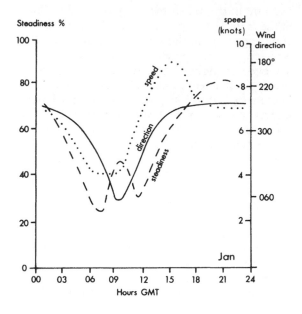

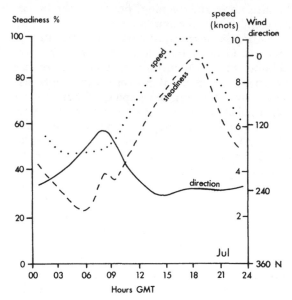

Figure 35 Wind speed, direction and steadiness in coastal Sierra Leone (after Mukherjee and Moore 1973)

knots, and aided by strong convection, dust may be lifted not only locally but over a wide area to create sand or dust storms, but these are discussed in greater detail in Chapter 4.

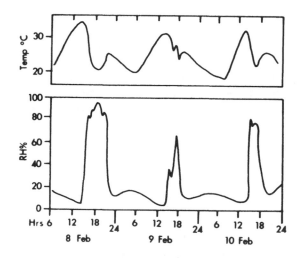

Figure 36 Temperature and humidity on the Senegal coast with the onset of sea breezes (from Wexler 1946)

Upper tropospheric winds

Most of the wind velocities noted above were recorded at, or estimated for, a height above open, clear ground of about 9 metres. At only 2 metres above the actual surface the wind speed would be reduced by about one-fifth, due to increased friction. Average wind speeds increase approximately logarithmically with height above the ground. However, they do not necessarily increase regularly, for reasons to be noted in Part Two. They may differ at the same altitude both spatially and temporally.

It is difficult to present a picture of *mean* wind flow above West Africa for various periods and altitudes as measurement of upper air movement has not been undertaken with the same regularity and in the same detail as has the air flow at the surface. Although there is now an abundant literature on upper air conditions in the region, it almost invariably includes illustration of wind directions and speeds for the specific times or short periods when detailed investigations were made.

Because a selection of such illustration will be utilized in Part Two of this book, the pattern of the mean upper air flow is indicated here in the simple but highly legible form shown in Figure 37; the idea but not the information borrowed

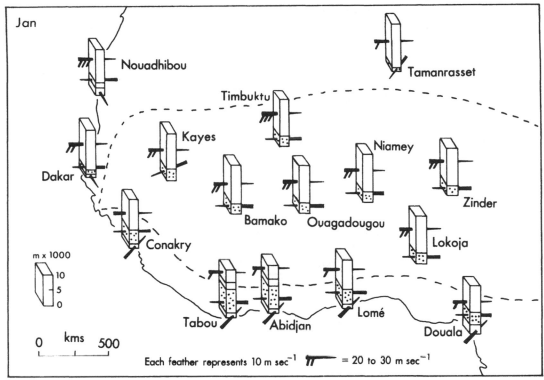

Figure 37a Mean wind flow with altitude in January

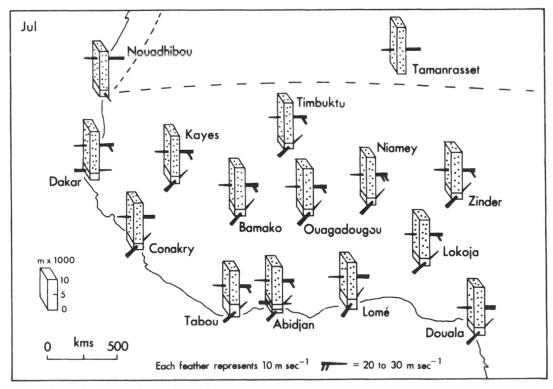

Figure 37b Mean wind flow with altitude in July

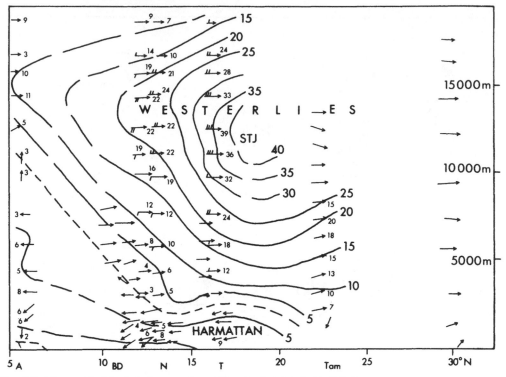

Figure 38a Wind flow with altitude in January along 0° longitude (after Dubief 1979)

Notes: A – Abidjan; BD – Bobo Dioulasso; N – Niamey; T – Timbuktu; Tam – Tamanrasset

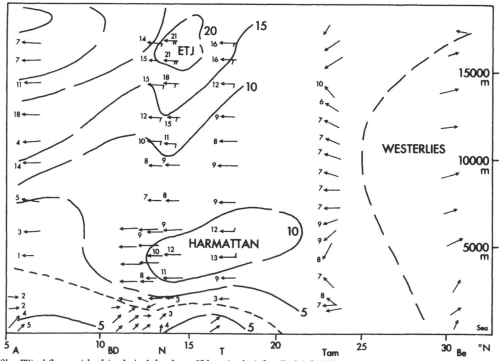

Figure 38b Wind flow with altitude in July along 0° longitude (after Dubief 1979)

Notes: A – Abidjan; Be – Béchar; BD – Bobo Dioulasso; N – Niamey; T – Timbuktu; Tam – Tamanrasset

from Hubert (1938). Each column represents a block of air 15,000 metres high, each block subdivided into layers of differing wind direction and speed. It has already been shown in Figure 28 that in January along the Guinea coast the surface wind, with a mean speed of 2–3 ms⁻¹, is predominantly from the south-west, whereas along the Atlantic coast in the north-west it is mainly from the north. Elsewhere at the surface north-easterlies are dominant. However (Figure 37a), the south-westerly airstream impinging on the Guinea coast is only some 1000 metres deep. Over it ride easterly winds in a layer about 3000 metres thick, which in turn are overridden by westerlies attaining velocities in excess of 20 ms⁻¹ Along the coast of Senegal and Mauritania the surface northerlies are overlain also by weak easterlies in a layer about 1500 m thick, surmounted by fast-flowing westerlies. Elsewhere in West Africa there are essentially only two air masses, surface easterlies (the harmattan) in the lowest 2000 metres overridden by strong westerlies.

The pattern in July is very different. The surface south-westerlies have spread far across the region, everywhere overlain now by easterlies with velocities about three to four times in excess of the surface winds, but have only half the speed of the winter upper westerlies which have now disappeared from the scene. In more detail the wind flow from the surface to about 21,000 m is indicated in Figure 38a and b. In January it will be observed that the winds 1000 m above the surface have higher speeds than the westerlies at 5000 m, but at about 20°N and 12,000 m these westerlies achieve the greatest velocities in a narrow zone to be called the subtropical jetstream (STJ), with speeds diminishing again with further altitude. In July the easterlies at about 8000 to 10,000 m are less strong than immediately above and below and velocities at 15 to 20,000 m and 10 to 15°N again achieve jetstream levels, forming the Easterly Tropical Jet (ETJ). The reasons for these seasonal changes of upper wind direction and velocity will be discussed in Part Two, where it will be observed that it is disturbances of the mean pattern described above that have fundamentally important consequences for the climate and weather of West Africa.

3 Evaporation, evapotranspiration and humidity

Of all the constituents of the lower atmosphere water vapour is the most variable, and its total mass is very small at about 1.3 per cent of the air at the surface, diminishing vertically to about 1 per cent at 1000 metres, 0.5 per cent at 3000 m, to 0.1 per cent at 7000 m. Nevertheless it is vitally important meteorologically. With regard to the radiation balance its importance has already been noted in Chapter 1.

Evaporation

Matter is made up of minute particles called molecules, which are assumed to be in constant motion. In a gas the molecules are much more widely dispersed than in a liquid and even more so than in a solid; collisions are therefore less frequent and mobility is greater. Their impact on any containing surface produces pressure on that surface. Most molecules in a mass do not move far from their mean positions in the mass because of the attraction of the surrounding molecules. However, the momentum of some near the limit or margin of the mass may cause them to break away and escape into space, reducing the volume of the mass. This process is evaporation. It is most applicable to liquids but it can occur from some solids, as in the case of solid air fresheners, or especially in the case of ice which may evaporate directly into the gaseous state without first passing into the liquid. In breaking from the mass the molecules do work which requires the utilization of the latent heat energy in the mass; thus evaporation leads to cooling of the mass. Raising the temperature of the mass will increase the rate of evaporation.

The evaporation of water from earth to atmosphere is primarily from the oceans, which oc-cupy 71 per cent of the surface of the globe and account for 65 per cent of the evaporation into the atmosphere. The remaining 35 per cent comes from the soil, lakes and rivers and wet vegetation. But to the quantity of water vapour transferred to the atmosphere by evaporation must be added that transpired by plants through their foliage, essentially in daylight hours. The two processes together are termed evapotranspiration.

The extent of transfer clearly depends upon a number of factors:

1 water availability;
2 radiation inputs affecting soil, water and air temperature;
3 the wind which should be sufficiently turbulent to bring in 'fresh' air which can absorb the evaporating water;
4 the evaporative power of the atmosphere relating to its temperature and the quantity of water vapour present within it at the surface of the evaporating body.

The measurement of evaporation

Of the instruments most commonly used for measuring evaporation none is particularly reliable, and comparing evaporation measurements from stations or areas utilizing different methods is likely to lead to misunderstanding and confusion. The Piché evaporimeter (see Appendix 6), normally retained under shade, probably under-records evaporation in high-humidity areas, for example in much of West Africa in the wet season, but over-records in the dry season or in wind. Pan or tank evaporimeters are manufactured in various sizes and shapes, some to be used sunk into the ground, others to be raised off the ground. Apart from the problems of water displacement by dirt and objects in the tanks, it is possible that tank evaporimeters overestimate evaporation as they can also lose water by wind splash, as a result of soil-pan temperature gradients, or by consumption of water by birds and animals. Measurements may be equally suspect on days of heavy rainfall (Olaniran 1983a). Ayoade (1976a) noted mean annual evaporation for various stations in Nigeria as recorded by the three types of tank known as class A, raised and sunken. At Maiduguri, for example, the evaporation from a class A pan was 4047 mm, from a raised tank 3312 mm and from a sunken tank 2862 mm. The discrepancies are less pronounced farther south, in Port Harcourt, but still marked: class A, 1630 mm; raised, 1520 mm; sunken, 1352 mm.

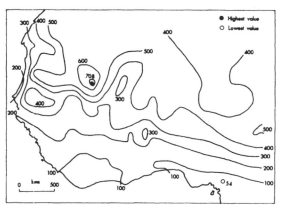

Figure 39a Mean evaporation in March (mm)

Distribution of evaporation

It is perhaps partly because of their unreliability that evaporation data are hard to come by in widely published literature. For 87 stations, mainly in francophone West Africa and mainly utilizing Piché evaporimeters, data were acquired to create Figures 39 a and b and 40 (March is overwhelmingly the month of highest evaporation levels and August of lowest evaporation in the region.) It is not possible to know how accurate the data are. The complicated pattern of isopleths and wide ranges of evaporation over short distances may reflect genuine variation in data or irregularity in measurement. A comparison of Figures 17, 31 and 32 with Figure 40 shows a correlation between areas of greatest mean wind speed, highest temperatures and evaporation. This is perhaps to be expected, especially as the Piché evaporimeter is particularly sensitive to windspeed. The patterns for August and for the year are much less complex than those for March however. There is clearly an increase in the evaporative power of the lower atmosphere north from the Guinea coast and inland from the Atlantic in Senegal and Mauritania. The southern areas are cooler and less sunny than the interior and the cooling influence of the sea is marked in the north-west. It should be appreciated, of course, that the values noted, particularly for the north, are potential water losses, for in the more arid regions the water is not actually available from normal rainfall to be evaporated.

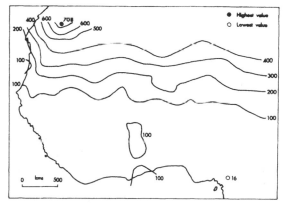

Figure 39b Mean evaporation in August (mm)

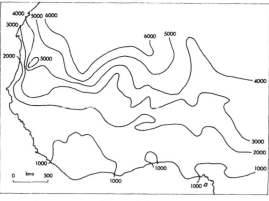

Figure 40 Mean annual potential evaporation (mm)

Evapotranspiration

However, the main reason for an apparently limited interest in assessing evaporation probably relates to the much greater significance of evapotranspiration or potential evapotranspiration because of the relevance of a knowledge of these to agriculture (see Chapter 12). Also potential evapotranspiration (PE) reflects a number of climatic variables: wind, radiation, temperature, as noted above, and thus the mapping and plotting of PE data can provide a valuable insight into regional climatic characteristics. Much of the surface of West Africa is vegetated and considerable quantities of water are passed into the atmosphere through the leaves of plants in addition to that evaporated from the soil and from open water.

The measurement of evapotranspiration

A review of methods for measuring evapotranspiration has been made by WMO, 1966 and up to 1970 by Ward (1971), but the most common instrument used is the lysimeter. Very few of these exist in West Africa, but a description of the type utilized by the authors in Ibadan and Freetown is given in Appendix 6. The accuracy of results obtained by such instruments must be suspect, principally because the grassed area within them is too small to be truly representative of the surrounding vegetation; turbulence over the equipment will vary from that over the surrounds and the contents of the evapotranspirometer represent an artificial environment for drainage. These design weaknesses make the instrument problematical in arid, unvegetated areas, but the writers are also particularly doubtful of the results obtained in the main rainy period in West Africa.

Potential evapotranspiration (PE)

Because of the difficulties associated with the direct measurement or estimation of evapotranspiration, attempts have been made to devise techniques by which *potential* evapotranspiration may be estimated. Consideration of these in detail is beyond the scope of this book, but interested readers can commence investigation of this subject with a general text, e.g. Ward, 1967. Of the methods applied in West Africa, especially in Nigeria, those of Thornthwaite (1954) and Penman (1948) have been most used and tested; see, for example: Garnier (1956 a and b), Chapas and Rees (1964), Davies (1965a), Ojo (1969), Obasi (1972) and Ayoade (1976a). It is generally accepted that Thornthwaite's approach is less suited to the tropics than is Penman's. Thornthwaite's is particularly empirical, utilizing data for mean temperature and hours of daylight only (interestingly two variables that are independent of evaporation rates), and yet the method is complex. Firstly a monthly heat index (i) is calculated from the equation:

$$i = (t/5)^{1.514}$$

where t is the mean monthly temperature in °C. Twelve monthly values are summed to produce the annual heat index I. Monthly potential evapotranspiration is then calculated:

$$e = 1.6b(10t/I)^a$$

where e is the monthly potential evapotranspiration in centimetres, b is a factor to correct for unequal day length (a minimal consideration in the tropics and one of the reasons why the method tends to lead to overestimation of potential evapotranspiration in southern West Africa), t is the mean monthly temperature in °C and a is a constant, a cubic function of I. Penman's method basically utilizes data for net radiation, air temperature, vapour pressure and wind speed in a complicated formula, thus limiting its ready application to areas that can provide such a data base and, if this is large, also computer facilities. By Penman's methods Obasi (1972) produced detailed maps of actual evapotranspiration for Nigeria, noting that in times of water surplus (precipitation exceeding evapotranspiration) actual evapotranspiration equals potential evapotranspiration, but in times of water deficit the latter exceeds the former, as one would expect. However his calculations for the actual mean annual evapotranspiration for southern Nigeria produce values ranging from 1000 mm in the vicinity of Lagos and in the Niger delta to a maximum of 1200 mm around Nsukka and Enugu, and these values equate with those for potential evapotranspiration estimated by Chapas and Rees (1964), who claim actual annual evapotranspiration in southern Nigeria to be about 700 mm.

The distribution of PE in West Africa

In 1974 the State Hydrological Institute of the USSR issued an Atlas of World Water Balance, later printed in English with an explanatory text (State Hydrological Institute, 1974). Maps of actual and potential evapotranspiration incorporated in this atlas were produced from formulae devised by M. I. Budyko, taking into account radiation balance, air temperature and air moisture content. Potential evapotranspiration was computed by a formula expressing 'the proportionality between the evaporation from the moist surface and air humidity deficit that is determined by the temperature of the evaporating surface', this in turn computed by 'using the equation of heat balance of the land under the conditions of sufficient moisture content'. It is claimed that the margin of error is only 4 to 5 per cent for the annual picture. The maps of potential evapotranspiration for Africa were based, however, on data from only 215 stations for the whole continent, and within West Africa the locations of only 26 towns are shown. If the data utilized were from only these 26 stations then the data base was really very small and it is not surprising that the Russian maps show a neat pattern of east–west trending isopleths, indicating a steady decrease northwards of annual actual evapotranspiration and a steady increase northwards of potential evapotranspiration.

The values for southern Nigeria for actual annual evapotranspiration range from 800 to 900 mm, approaching much more closely the values of Chapas and Rees than those of Obasi. On the other hand the range of annual potential evapotranspiration, from 1250 mm in southern Nigeria, southern Cameroon and between Abidjan and Monrovia to 2500 mm in eastern Mauritania and western Mali between Tidjikja and Arouane, equates well with the regional pattern noted by Ojo (1969). He utilized a modified version of the Penman method to produce maps of potential evapotranspiration for the whole of West Africa, which are reproduced here with permission (Figure 41). The seasonal change in potential evapotranspiration is quite obvious, as is the correlation between areas of highest evapotranspiration and highest net radiation and temperature. The contrast between coastal Senegal and Mauritania and the interior of these lands is again marked, and the bunching of the isopleths, especially in the July map, is closely related to the location of the ITD.

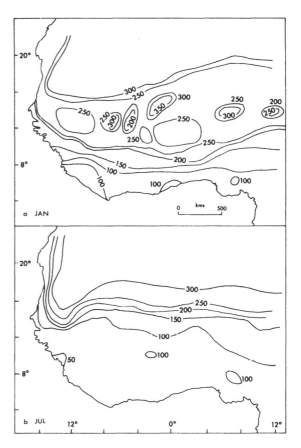

Figure 41 Potential evapotranspiration in *a* January; *b* July (after Ojo 1969)

If discrepancies between the various attempts to assess potential evapotranspiration for West Africa or for parts of the region have been noted, they do not compare with those observed between potential evaporation and potential evapotranspiration. Whereas the values equate for southern areas and in the wet season, which is reasonable, the values for annual and dry season potential evaporation are considerably higher than for potential evapotranspiration in the north. Herein are epitomized all the difficulties inherent in utilizing evaporimeters or in calculating potential evapotranspiration for arid areas where there is a markedly limited and only short-term data base. Walker (1962) claims that open-water evaporation and potential evapotranspiration should be approximately equal over lengthy periods, for example a year, and that both should be about 2000 mm per annum at the Sahara margins, 1676 mm in the savanna zones and about 1370 mm in the forest zones. Rodier (1964) and Davies (1966a) also note that sunken-pan evaporation effectively equates with evapotranspiration in the tropics. Clearly, if they are correct, the maps of evaporation (Figures 39 and 40) must be used with great caution and greater reliance can be placed upon the maps of potential evapotranspiration.

Humidity

There is a limit to the amount of water vapour that can be contained in the air at any given temperature. If saturation has been achieved and there is a fall in temperature, condensation will normally follow. The temperature at which this process occurs is called the dew-point temperature. Water precipitated onto surfaces chilling the air in contact with them forms dew. If the process occurs in the atmosphere it forms mist or fog at ground level, cloud aloft. Just as energy was utilized for the evaporative process, so it must also be released for condensation, a most important source of energy in the atmosphere, of rising currents of heated air in clouds for instance.

Absolute humidity

The *absolute humidity* of the atmosphere refers to the actual amount of water vapour present, which is most simply expressed in terms of the weight of the water vapour in a given volume of air. Table 12 indicates the maximum amounts that can be contained by the air at given temperatures. However, as any given body of air may expand or contract, on rising or sinking for example, the absolute humidity is not a constant for that body. A more usual way, therefore, in meteorology is to express the water vapour content in terms of the contribution it makes to the atmospheric pressure. This vapour pressure cannot be measured directly but can be calculated on the basis of physical theory from wet- and dry-bulb temperatures and atmospheric pressure. Again it will be appreciated that the lower the air temperature the lower will be the vapour pressure (or, conversely, the higher the air temperature the higher the water vapour pressure can be) and as there is a limit at any given temperature to the possible water vapour density, so there is a saturation vapour pressure at any temperature (Table 13). (If the atmospheric pressure

Table 12 *Weight of water vapour (v. density) per cubic metre of air at given temperatures*

°C	grams
15	12.7
20	17.1
25	22.8
30	30.0
35	39.2

Table 13 *Saturation vapour pressure at given temperatures*

Temperature (°C)	Saturation vapour pressure (mb)
10	12.28
15	17.04
20	23.37
25	31.67
30	42.43
40	73.80
50	123.40
100	1013.30

is reduced to any of the values in the second column the boiling point of water will be at the temperature noted in the first column.)

Relative humidity

The ratio of actual vapour pressure or vapour density to the saturated vapour pressure or saturated vapour density is the *relative humidity*, probably the most frequently used but most misunderstood measure of atmospheric humidity. Relative humidity is expressed as a percentage:

$$\frac{\text{Observed water vapour content}}{\text{Maximum w.v. content at the given temperature}} \times 100$$

Thus a relative humidity of 85 per cent with a temperature of 30°C implies a high total atmospheric water vapour content, whereas 85 per cent at a temperature of only 3°C implies a low vapour pressure. The warmer the air the more water vapour it can contain.

The measurement of relative humidity

An instrument used for measuring humidity is described in Appendix 7, but probably the most common method for assessing relative humidity involves the use of dry- and wet-bulb thermometers, the latter having the normal thermometer element covered with a clean, crease-free, single thickness of muslin, fed with purified (distilled) water by a cotton wick. We have noted how cooling will take place during evaporation, as some of the latent heat of vaporization is drawn from the air. If all the heat of vaporization comes from the air then the lowest level to which the air can be cooled by evaporation into it is the wet-bulb temperature, the temperature that a small wet body will take up. The wet-bulb muslin will lose water by evaporation unless the adjoining air is saturated, and thus is cooled, so that the thermometer will register a lower temperature than the dry-bulb thermometer which is recording the temperature of the ambient (surrounding) air. The difference between the dry- and wet-bulb temperatures is known as the 'depression' of the wet-bulb. Hygrometric or psychrometric tables or charts utilizing the dry-bulb temperatures and the wet-bulb depression, permit a speedy calculation of the relative humidity (Appendix 8).

The relationship between relative humidity, temperature and vapour pressure

The relationship between relative humidity, temperature and vapour pressure is demonstrated in Figure 42. Two of the eight stations, Nouadhibou and Tessalit, are in the far north of West Africa; two others, Kayes and Niamey, are in the Sahel; two more, Kabala and Kaduna, are south of this zone but still inland; and the last two stations, Sassandra and Douala, are on the Guinea coast.

Nouadhibou, although approximately at the same latitude as Tessalit, is on the cool, windy Atlantic coast. Only the mean temperature for September exceeds 25°C. The marine influence delays the attainment of maximum temperatures until two to three months after the period of the zenithal sun. With the increase in temperature comes a related increase in vapour pressure and an associated rise in relative humidity.

At Tessalit, however, far from the ocean, very high mean monthly temperatures are attained with the arrival of the overhead sun. Two months later the advancing humid air from the Gulf of Guinea reaches the latitude of the settlement, producing the highest levels of vapour pressure and, again, associated relative humidity.

At Kayes and Niamey highest temperatures are again achieved some two months before the arrival of the humid air. The cloud and rain associated with the Atlantic airmass appreciably lowers the temperatures but increases the vapour pressure and the relative humidity therefore soars.

At Kabala, 230 km from the ocean, ranges of temperature, vapour pressure and relative humidity are reduced, but again, with the summer season dominated by the oceanic air-mass, temperatures are lowered, but vapour pressure and relative humidity are raised. The pattern is similar in central Nigeria, although the variation in vapour pressure and relative humidity from dry season to wet is more marked. Kaduna is 580 km from the Atlantic Ocean.

On the Guinea coast the picture is rather different, for here with the summer fall in temperature comes a related fall in vapour pressure. Practically at all times of the year the marine influence is dominant, and vapour pressure as the graphs show is always high, steady at 30 mb even in the dry season (winter months). Therefore, as the temperature falls the level of vapour pressure also diminishes, but not now the relative humidity. The vapour pressure is so great, even in the cooler, cloudier season, that the relative humidity rises.

The distribution of vapour pressure

This pattern is well brought out in Figure 43. The month of lowest vapour pressure for a narrow zone along the Guinea coast is August, but throughout the rest of West Africa it is January or February (December in inland south Cameroon). The month of highest vapour pressure is April along the Guinea coast, May within a zone 500 km wide inland from the coast, but

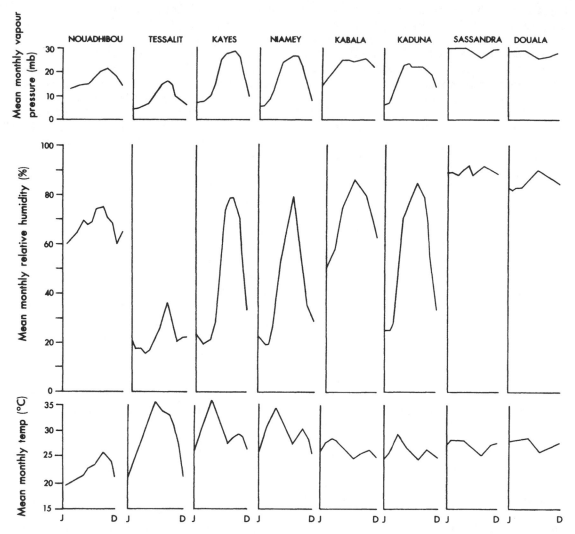

Figure 42 Mean monthly temperature, relative humidity and vapour pressure for selected stations

elsewhere August or September, when, despite the cloud associated with the now present maritime air, temperatures still remain above 25°C.

The distribution of mean vapour pressure for January (Figure 44a), when the harmattan casts its desiccating influence widely across West Africa, shows most areas north of about 10°N with vapour pressure below 10 mb, but with a steep pressure gradient southwards to the Guinea coast which, always humid, attains vapour pressure values of 30 mb in places. In August (Figure 44b) marine influences have spread right across West Africa, slightly lowering vapour pressure at the coast, but significantly raising it inland, to produce a much more uniform scene. If the mean

annual ranges of vapour pressure are plotted (Figure 45), the area of greatest seasonality is clearly seen to be in a zone from eastern Senegal to north-eastern Nigeria.

The difference in the response to temperature change of vapour pressure and relative humidity is also brought out by Figures 46 and 47. Whereas only 36.2 per cent of the 94 stations whose data were used to produce the vapour pressure maps recorded their highest vapour pressure in August, 76.2 per cent of the 126 stations used to map the distribution of relative humidity recorded their highest relative humidity in that month. Although nearly 60 per cent of stations recorded their lowest vapour pressure in January, 15 per cent of stations also experienced their lowest values in August. No stations in West Africa experienced their lowest relative humidity

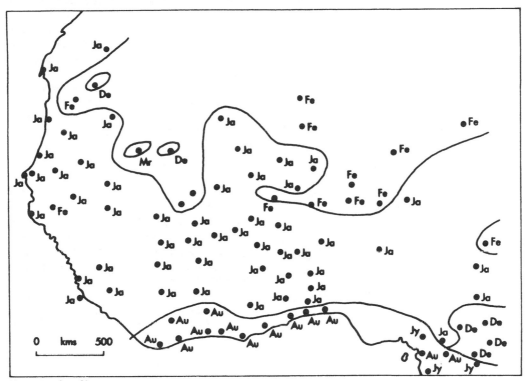

Figure 43a   Months of lowest vapour pressure

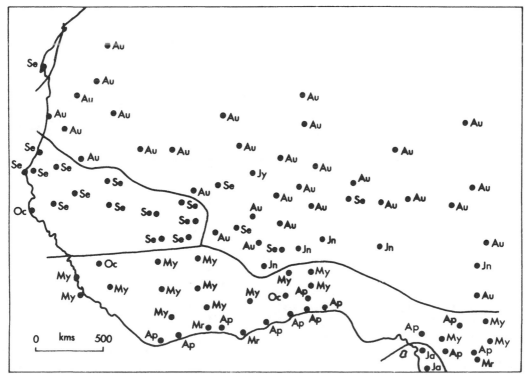

Figure 43b   Months of highest vapour pressure

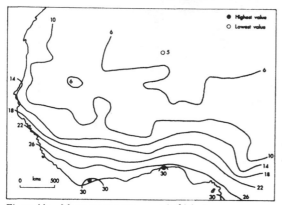

Figure 44a Mean vapour pressure (mb) in January

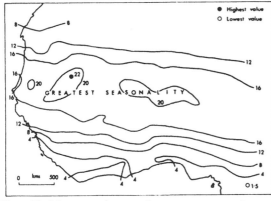

Figure 44b Mean vapour pressure (mb) in August

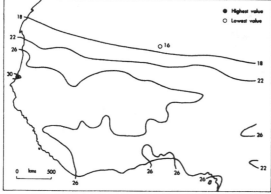

Figure 45 Mean annual range of vapour pressure (mb)

in August, however, but 82 per cent of stations recorded lowest values in January, February and March. Essentially the further the station from the Guinea coast, the later in the year the month of lowest relative humidity, except along the north-east coast where minimum values are noted mainly in January.

The distribution of relative humidity

As is to be expected, when the distribution of relative humidity is considered (Figure 48), the coastal areas show high values throughout the year; values below 40 per cent are widespread in the dry season (February), values over 60 per cent widespread in August. Transitions from high to low values occur over distances not greater than 500 km, the bunched isopleths paralleling the coastlines and essentially being associated with the location of the ITD in the months concerned. As in the case of Figure 45 greatest annual ranges of relative humidity are seen to be across the central parts of West Africa and lowest seasonality is along the Guinea coast (Figure 49). It follows that if seasonal changes in temperature produce variations in relative humidity, the diurnal oscillations in temperature will also induce fluctuations in relative humidity. This is well brought out in Table 14 for a selection of stations from the Guinea coast to southern Algeria. It will be noted that the early morning values of relative

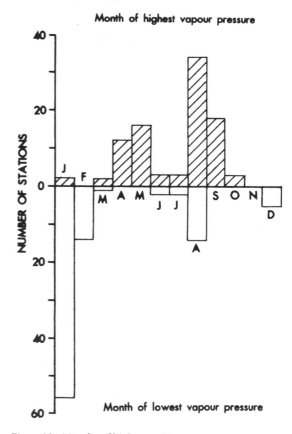

Figure 46 Months of highest and lowest vapour pressure

humidity are in every month higher than those for the afternoon, especially in the Sahel and in summer.

Humidity zones in West Africa

The term humidity has been used to describe a particular water vapour property of the air, but to most people a stated value relating to vapour pressure or even relative humidity has little

meaning. And yet there is no person who cannot *feel* whether conditions are humid or 'dry'. The humidity of the air is a property to which human beings are very sensitive. A further discussion of the relationship between humidity and human comfort and health is left to Chapter 13, but here

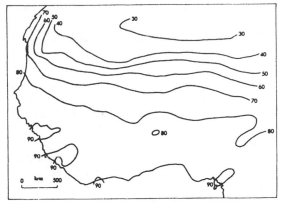

Figure 48a Mean relative humidity (per cent) in February

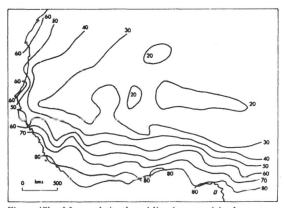

Figure 48b Mean relative humidity (per cent) in August

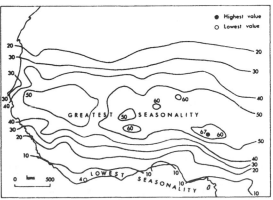

Figure 49 Mean annual range of relative humidity (per cent)

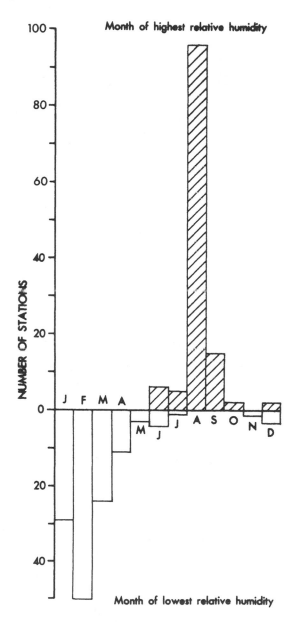

Figure 47 Months of highest and lowest relative humidity

we can note that nearly everyone finds conditions oppressive when wet-bulb temperatures exceed 30°C. Dry-bulb temperatures below 16 to 18°C are invariably felt to be 'cool' regardless of relative humidity. The most comfortable dry-bulb temperature is in the region of 18 to 21°C and the most comfortable relative humidity at those temperatures in the range of 30 to 70 per cent.

On the basis of the analysis made of Figure 42, it might be thought reasonable to divide West Africa into four humidity regions. However, this chapter is brought to an end with an attempt to delimit humidity regions on the basis of what is physically experienced by people, rather than on the basis of variations in vapour pressure or relative humidity. Few parts of West Africa experience mean monthly dry-bulb temperatures below 21°C, but most for varying periods experience relative humidity levels above 70 per cent. Figure 50 has been created on the basis of the mean number of months per year with relative humidity values higher than 70 per cent. The influence of the sea both in the north-west

and along the Guinea coast is clear again, but the influence of altitude has been overlooked in the case of the Fouta Djalon. Mali in Guinea at 1464 m altitude, for example, despite the fact that only the months of February to May have mean dry-bulb temperatures over 21°C, has no month in the year with mean relative humidity values over 70 per cent.

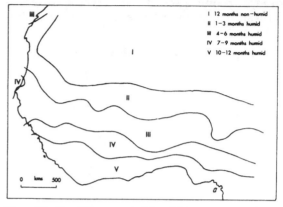

Figure 50 Humidity zones in West Africa

Table 14 *Mean monthly relative humidity (per cent) at different hours at three stations (values rounded off)*

Cape Coast, Ghana (1965–76) (5°06′N., 1°17′W. 16 m above mean sea level)

	J	F	M	A	M	J	J	A	S	O	N	D	YR
0900	84	82	81	82	83	86	86	89	87	82	81	84	84
1500	76	77	77	79	81	83	82	85	86	80	78	82	80.5

Ménaka, Mali (1949–55) (15°52′N., 2°19′E. 295 m)

	J	F	M	A	M	J	J	A	S	O	N	D	YR
0600	35	31	25	25	43	55	70	85	80	61	41	36	49
1200	17	16	12	12	22	31	43	55	45	27	17	18	26

Tamanrasset, Algeria (1935–50) (22°42′N., 5°31′E. 1400 m)

	J	F	M	A	M	J	J	A	S	O	N	D	YR
0730	37	34	31	35	34	27	25	29	29	37	40	40	33
1330	21	27	20	18	23	20	17	20	23	21	24	21	21

4 Cloud and visibility

Cloud

Cloud composition and types

The relationship between solar radiation, temperature, evapotranspiration and humidity has been indicated in the preceding chapters. In turn their relationship with cloud development and thus precipitation will also be obvious. A discussion of cloud and precipitation physics is out of place in this text, but normally, it may be said, cloud formation is preceded by the upward movement of humid air, which leads to its expansion against the diminishing resistance of the lowering pressure of the surrounding atmosphere and cooling. This cooling will lead to immediate condensation in saturated air, or to eventual condensation if the cooling is sufficiently prolonged. The transformation of water vapour into ice or water droplets releases energy (heat) which will encourage further ascent, typically 2.5 or 3 ms^{-1} in the base of tropical cumulus cloud. The level of the base of the cloud corresponds to the altitude of saturation temperature.

For the greater part clouds are comprised of water droplets, even, for reasons not thoroughly understood, at temperatures below 0°C. Only at −40°C (which is also −40°F) do all droplets freeze (although there are claims for water existing as liquid even below this temperature), and thus only the tops of the tallest cumulus (see below) or the high-level cirrus, attaining altitudes where temperatures reach these low levels, are glaciated (frozen). In a typical stratus cloud the droplets have a mean diameter of 10 to 12 microns (μ), and there are some 250 to 650 droplets per cm^3. In fair weather cumulus the diameter is 16 to 20μ, the number 250 to 400 per m^3. In a cumulonimbus the mean diameter is about 40μ and the number 70 to 200 per m^3. Despite the minute size

Table 15 *The falling velocities of water droplets*

Diameter (microns)	Velocity of fall
200	120 cm s^{-1}
100	27 cm s^{-1}
60	15 cm s^{-1}
40	5 cm s^{-1}
20	1.3 cm s^{-1}
10	3 mm s^{-1}
5	0.7 mm s^{-1}
1	30 microns s^{-1}

of the droplets a mature cumulo–nimbus cloud may contain 50,000 to 300,000 tonnes of water.

It is because the water droplets in clouds are so small that the clouds appear to float in the sky. Small spheres fall with uniform velocities, and, depending on the size of the water droplets, their falling velocities in still air will be as indicated in Table 15. A droplet of 10 μ will thus fall 180 mm per minute, less than 5.5 metres in half an hour. As descent may also be countered by rising air, necessary for the birth of the cloud, it is not surprising that the clouds do not 'fall'.

The characteristic clouds of West Africa are as illustrated in Figure 51:

1 Convective, that is cumulus with vertical development exceeding horizontal. Cumulus congestus may grow to heights of 6 km, developing into cumulo–nimbus (Cb) with glaciated summits and associated lightning, thunder and heavy rain.

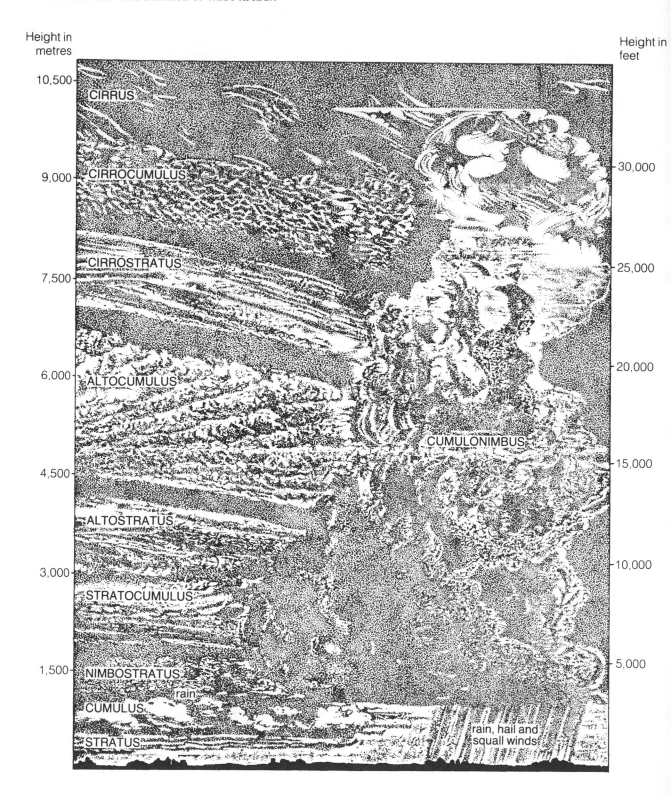

Figure 51 Characteristic cloud types

2 Low-level (0 to 2000 m), for example Trade Wind cumulus with small vertical development (50 to 200 m) and horizontal equivalent diameters of 50 to 500 m. Often aligned into 'streets' and covering about 3 oktas (eighths of the sky), they are most commonly observed in the drier coastal areas of the north-west and of the Togo Gap. They may merge into strato-cumulus (Sc), with bases between 500 to 3000 m, and 100 to 800 m thick, with cumuliform summits. Variations in the temperature and altitude of, and rates of evaporation from, the land surface will produce variable cloud base heights and cloud type.

3 Middle-level cloud (2000 to 8000 m); alto-cumulus (Ac) and alto-stratus (As) in more than one layer are commonly derived from the decay of Cb and the spread of their anvil-shaped tops. Developing Cb may penetrate through these middle layers, although the whole cloud pattern from below may give the impression of stratus.

4 At not less than 8000 m altitude are the high-level cirro-stratus, cirro-cumulus and cirrus clouds, again commonly derived from the spread of anvils from decaying Cb.

Cloud clusters

By satellite imagery and photography the cloud patterns of the world are now continuously monitored, and it has been observed (e.g. Heinricy and Young, 1974; Martin, 1975) how the cloud over West Africa commonly forms in clusters, categorized by Martin into three sizes: A/B covering an area larger than 50 degrees squared, (about 602,400 km^2 at 12° latitude) B covering 10 to 50 deg^2 and C occupying less than 10 deg^2 (about 120,480 km^2 at 12° latitude). He noted in his brief, three month study that the clusters lay generally in ENE to WSW bands with a marked single axis in the pre- and post-rains periods, splitting into a double axis in

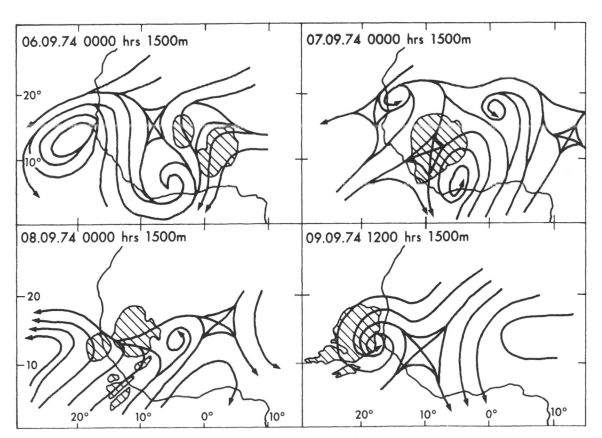

Figure 52 Cloud cluster movement across West Africa (after Martin 1975)

Note: Cloud clusters shaded and streamlines also indicated at 1500 m.

August. The bands moved north and south with the sun and the trajectories of the cloud clusters paralleled the bands. Figure 52 illustrates the typical development, modification and movement of a major cluster. It will be seen that the movement is from east to west. Why this is possible in a region where the surface wind flow is dominantly from the south-west is partly explained by the upper wind pattern in Figures 37 and 38, and by the relationship between the development of the cloud and the passage of westerly moving waves in the upper air south of the ITD (see Part Two).

Heinricy and Young observed in their four-year study that the easterly waves had a dominant periodicity of four to five days. Martin registered the mean movement of the cloud clusters to be over 10 degrees of longitude per day between 10 and 15°N and east of 5°W, diminishing to 6 degrees of longitude per day at the edge of the Sahara and off the coast of Senegal, and to only 5 degrees, or even 4, along the Guinea coast. The vast majority of the clusters lasted less than 24 hours. Small clusters were more numerous than the large, but the latter persisted for longer periods. The time of the onset of cloud cluster development was overwhelmingly in the mid-afternoon in inland West Africa, but along the Guinea coast where markedly fewer clusters were formed, as many developed between midnight and 03.00 hours as in the afternoon.

Regional patterns of cloud distribution

The comments above relate to cloud clusters and short research periods only. The regional patterns of cloud cover, of seasonal and diurnal development, are summarized in Figures 53 to 55. Not unexpectedly (Figure 53) the month of greatest cloud cover in West Africa is August; the months of least cloud December and January. The distribution of the cloud cover is shown in Figure 54. The maps have been compiled from mean data calculated from records maintained for at least eight years for the francophone states, but for only five years for Guinea and Sierra Leone, six years for Ghana and four years for Nigeria, and for a total of 120 stations. Hence some of the rather anomalous isopleth patterns, especially in the January map. It is also asserted (Ananaba, 1983) that ground based observers tend to overestimate cloud amounts by about 7 per cent between latitudes 4 and 6°, by 19 per

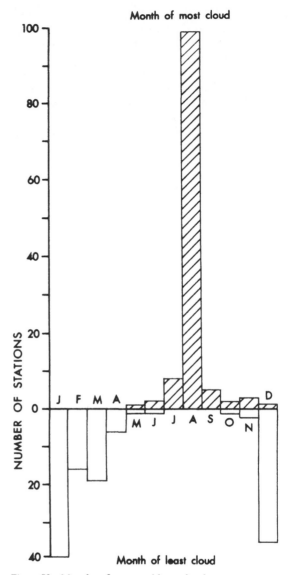

Figure 53 Months of most and least cloud

cent between 6 and 8° where dust in the atmosphere is often confused with cloud. Nevertheless certain regional patterns emerge. The cloudiest part of West Africa is clearly the south-east, followed by the Guinea coast, although here, once again, the Togo Gap stands apart. It is interesting to note the correlation between cloudiness and high land in Guinea and Sierra Leone in August, and in the Cameroon Republic the year round. However, the opposite situation prevails in Nigeria. The least cloudy part of West Africa, the maps suggest, is not the Sahara but northern Nigeria, especially between Kano and Maiduguri. But this area of low cloudiness includes the Jos Plateau,

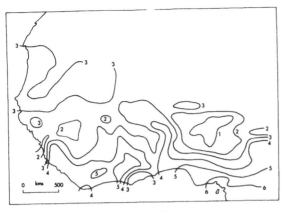

Figure 54a   Mean cloud cover (oktas) in January

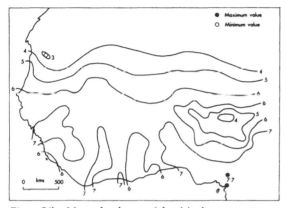

Figure 54b   Mean cloud cover (oktas) in August

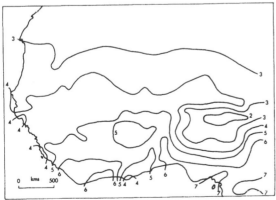

Figure 55   Mean annual cloud cover (oktas)

which does not seem reasonable and is the consequence, perhaps, of utilizing short-term data.

It has been calculated that there is a 20 per cent probability of a clear sky at any time in January

in southern Cameroon, rising to 50 per cent at Timbuktu, but only a 10 per cent probability of a clear sky in July anywhere in West Africa south of 15°N. The percentage probability of an overcast sky in January is 40 to 50 along the Guinea coast, but only 10 at Bamako. In July the values rise to 70 to 80 per cent on the coast, 50 per cent at Bamako and 10 per cent at Tamanrasset (Lebedev, 1978).

Figure 56 bears close scrutiny. It brings out not only regional differences relating to seasonal changes in cloud amount, but also diurnal cloud variability. The lengths of the lines indicate the diurnal variability and the symbol at the end of each line indicates the time of day, according to the key, when the mean maximum and minimum daily values tend to occur each month. For example, at Bilma in June maximum daily cloud cover tends to be nearly 5 oktas and to occur at about 18.00 hours; minimum cover is only about 1 okta and occurs around noon. The graphs are also arranged in sequence, so that to read them from left to right is to see the picture from west to east, and from top (F'Derik, Tessalit, Bilma) to bottom (Douala, Abong-Mbang) is to see the distributions from north to south within West Africa.

The graphs clearly show how less cloudy is the north than the south, yet how Bilma, the station furthest inland, is cloudier than F'Derik or Nouakchott at much the same latitude. But the diurnal change in cloud cover is more dramatic at Bilma than anywhere else in West Africa, Abong Mbang perhaps excepted. Another interesting feature is the marked decline in cloudiness along the north-west coast before the onset of the summer main cloud season. The most northerly stations consistently record their greatest cloud cover in the late afternoon, lowest at noon, except F'Derik outside the midsummer period. In contrast the southerly stations have maximum cloud cover in the early morning, least cover either in the late afternoon throughout the year (Abong-Mbang) or during the wet season (Tabou, Adiaké, Lomé, Meiganga). In the less humid interior cloud develops with increasing convection, but in the persistently humid and cloudier south, maximum development occurs with the period of lowest temperature and highest relative humidity, in areas where convection is not the prime cause of uplift, expansion and condensation.

A further detail, observable from Figure 54 but even more clearly demonstrated in the graphs,

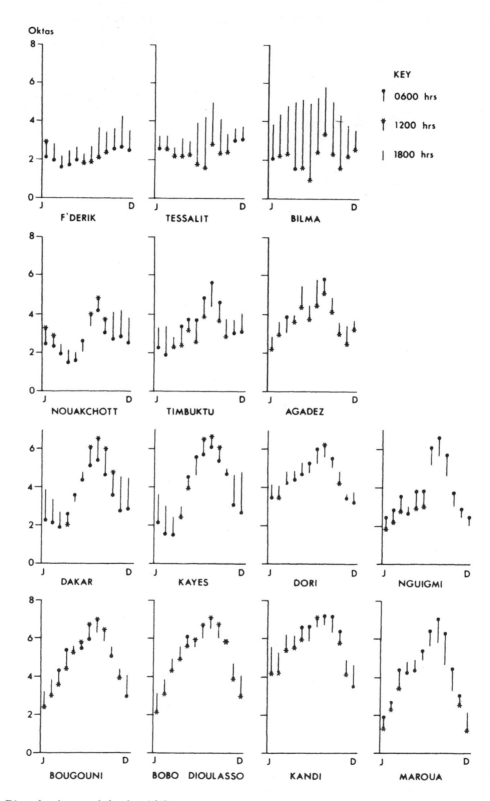

Figure 56 Diurnal and seasonal cloud variability

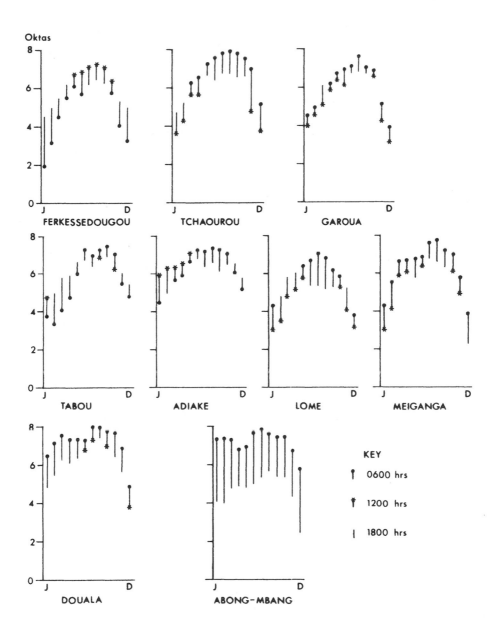

Oktas

FERKESSEDOUGOU

TCHAOUROU

GAROUA

TABOU

ADIAKE

LOME

MEIGANGA

DOUALA

ABONG-MBANG

KEY

0600 hrs

1200 hrs

1800 hrs

excepting that of Dori, is that the stations with the greatest seasonal contrast in cloud cover lie in the middle zone between 10 and 15°N outside Nigeria. The seasonal change in diurnal cloudiness is marked in many stations. For example, the time of least cloud changes from early morning to evening during the rains at Tabou and Adiaké, and from noon to evening at Lomé, whereas the time of most cloud switches from evening, in October to March, to early morning in April to September at Timbuktu.

The data utilized were for francophone states only and for a period no longer than nine years, which helps to explain discrepancies that might be observed between values plotted on the graphs and on the maps.

Visibility

When light passes through the atmosphere its intensity is reduced, due to by scattering by the air molecules, water droplets and aerosols, to diffuse reflection by the larger particles and to absorption by some solids. Thus the maximum distance at which an object can be seen depends mainly on the state of the atmosphere, but also on the sun's elevation, the reflecting power of both object and its background, the general illumination, the size of the object and the quality of the eyesight of the observer. Thus the precise visual assessment of visibility is difficult to achieve even if the objects selected for the assessment are, as they should be, dark in colour and upstanding above the horizon. The use of instruments for measuring visibility is very limited.

Visibility records

Judging from the lack of availability of visibility records it would seem that regular measurements are made at few stations in West Africa and there is a terrible lack of consistency in published data. For example, in the climatic tables for Africa published in Leningrad (Lebedev, 1967), the mean number of days with fog are indicated for just 32 stations in West Africa, the means based on records for only six years, 1951–6; la Direction de l'Exploitation Météorologique at Dakar (ASECNA 1980), for only 17 stations and for the

ten year period 1968–77, published tables indicating the total number of hourly observations per month when visibility fell below seven specified levels; Bertrand et al (1979) published diagrams indicating the mean monthly number of hours with visibility reduced by dust haze below 5 km for 23 stations based on records for just five years! Data for individual stations can sometimes be extracted from the literature, but cannot be assembled to provide a comprehensive record for West Africa as a whole for a particular period. From this confusion the researcher and analyst will make what he can, but regrettably the writers have not managed to acquire a set of data on visibility which permits a regional map of mean monthly or annual visibility to be produced.

Visibility in West Africa

In 1935, however, an interesting set of maps of monthly visibility for the one year 1934 for central and West Africa was published (Barberon, 1935), which is the basis for Figures 57a and b. February was the month of most widespread mist or haze (visibility up to 2000 m) and May the month when conditions were most clear. The outstanding feature of the maps is the poor visibility in southern Cameroon. This area, as can be seen, is the most persistently humid and cloudy in West Africa. The condensation of water vapour into droplets of water, probably about 18 to 30 μ in diameter, which will significantly impair visibility, will readily take place on hygroscopic particles, especially crystals of sea salt derived from evaporation. Such particles can cause condensation long before saturation of the air is achieved, when relative humidity levels, in fact, attain only 75 per cent. To produce fog (visibility less than 1000 m) full saturation may be required, but for mist the ideal conditions pertain in the Bight of Biafra through much of the year. Morning mist is especially prominent with visibility even under 50 m in July and August and remaining below 2 km for a number of days. Such conditions may occur along much of the Guinea coast in the most humid months, June to September, as is well illustrated by Figure 58.

On the north-west coast of West Africa mist due to chilling of the lower air in contact with the cool waters is a common occurrence, a water mist which can extend, however, even 200 km inland.

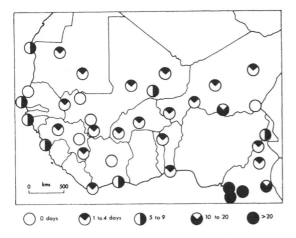

Figure 57a Number of days in February 1934 with visibility less than 2 km (after Barberon 1935)

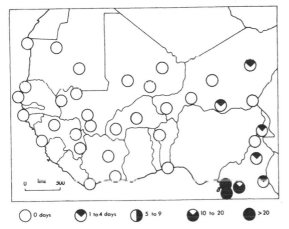

Figure 57b Number of days in May 1934 with visibility less than 2 km

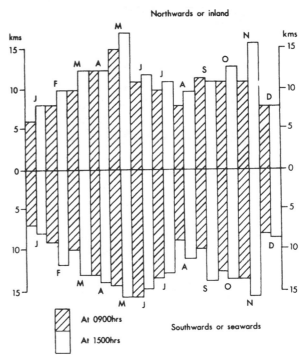

Figure 58 Visibility at Cape Coast, Ghana in 1976 (after Hayward and Addae 1978)

Dust haze

Over the greater part of West Africa, however, especially north of the so-called forest zone, the prevalence of mist, most noticeably in the dry season, cannot be explained with reference to atmospheric moisture. In fact it is more accurate to refer to haze rather than mist, the former term being used to describe visibility impaired by the presence of solid aerosols, not liquid. In West Africa, particularly in the pre-rains period, such aerosols may be contributed by bush fires (Crozat et al, 1978), but by far the greater proportion of atmospheric pollutants is dust from the arid north, associated with the harmattan. The influence of this dust-laden north-easterly air

flow, which may reach even to the Guinea coast in the period December to February, is seen in Figure 58 again. In a ten-year study period of visibility at Lungi Airport, Freetown, it was noted that the months November to February had 12.8, 28.9, 30.5 and 10.3 per cent respectively of the total number of days in the period with visibility less than 1100 yards (1000 m) (Mukherjee and Moore, 1972). To the dust lifted and transported widely by the harmattan must be added that raised by the intrusion of depressions deep into the Sahara from the Atlantic and Mediterranean, which again are more likely to occur in winter. Also that due to possible katabatic winds flowing off the Tibesti and Ahaggar massifs, and to 'dust devils'.

It is possible that the wedge-like intrusion of humid air from the south into the dust-laden continental airmass from the north may actually increase haze levels as water vapour is absorbed by the large ions suspended in the air (Dubief, 1979). Bertrand and Baudet (1976) observed a tendency to increased ice nucleation in the air by the mixing of the oceanic and Saharan airmasses. Kalu (1979), however, conversely believes that the dust acting as condensation nuclei may absorb water, become heavy and so fall from the

79

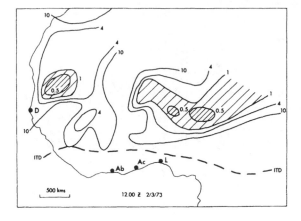

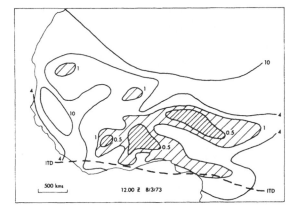

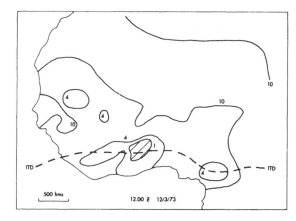

Figure 59 Visibility in West Africa at noon on *a* 2 March; *b* 8 March; *c* 12 March 1973 (after Bertrand et al. 1974)

atmosphere, thus improving the visibility. The writers have recorded intense levels of haze in very humid air at Freetown, and assumed that the dust was in fact descending from the dry north-easterlies riding over the lower level mari-

time airmass. Bertrand et al (1974) illustrate the progress of an area of dry haze from north-east to south-west across West Africa in March 1973 (the basis of Figure 59), explaining this with a map showing winds not at the surface but at 600 m over the region where a 30 to 40 knot ENE air flow dominated. Dust will generally be brought down through subsidence and turbulent mixing, probably explaining why haze is often worse in the morning calm between 06.00 and 09.00 hours. Skies will also be washed clear of all types of aerosols by rain showers. Thus (Figure 58 again), the best visibility is recorded in the two 'shower seasons' (April to May, October to November) that lie between the main wet and dry periods (see Chapter 7).

There are probably two main source areas for the dust, one in the plains between Bilma and Faya Largeau in Niger and Chad, where fine particles are fed to the area by seasonal streams from the Tibesti uplands replenishing the dust that is stripped from the surface by the winds, and the second west of the Ahaggar massif, Tanezrouft in Algeria. From the former source is derived the haze that may extend across Nigeria and the Guinea coast as far as Sierra Leone and Guinea. The dust haze of northern Mali, Mauritania and Senegal probably originates in the area west of Tamanrasset, and it is this dust which may be recorded in Europe and even across the Atlantic.

A relationship between aridity and dust is to be expected. In the Sahelian drought of 1970–74, the mean number of hours with visibility below 5 km increased by 4.5 times at Bobo-Dioulasso, 2.5 times at Bamako over the period 1957–69 (Bertrand et al. 1974). The relationship is summed up in Table 16 (after Bertrand, 1976), where the predominant particle size is under 0.5μ and in the formula:

$$v = K \, \bar{r}/m$$

where v is the visibility, K a constant, $\bar{r}$ the effective radius of the dust particle and $m = 4/3\pi\bar{r}^3\sigma z$, where σ is the density of the particles and z the number density of the particles (Ette and Adewolu, 1980).

Dust sampled at Samaru, Nigeria, 3.5 m above the ground (Samways, 1975) was mainly in the 0.05 to 0.5μ range, 80 per cent less than 0.3μ, and was composed mainly of quartz, but with some kaolinite and montmorillonite clay, an indication that the source region was not local but lay to the

Table 16 *Relationship between visibility and dust concentration*

| | No. particles per litre of diameter above | | | | | | | | Visibility (km) |
| | 0.3 | 0.5 | 1 | 2 | 3 | 5 | 10 | 20 | |
μ									
Abidjan 28.11.73	102,600	44,200	25,300	2760	750	70	0	0	3.1
Niamey 7.01.74	175,000	87,500	50,900	5310	1370	100	0	0	1.6

Source: Bertrand (1976).

north-east. In a large fall-out of dust reported in 1956 (McKeown, 1958), 150,000 particles per cm³ were registered. Dust collected off the coast of north-west Africa, and presumably from the Tanezrouft source, contained a greater percentage of material in the 2 to 8μ class, and the density was less than 10 micrograms per cubic metre of air (Chester and Johnson, 1971).

Hubert (1943) also noted the clay component in the dust and suggested that a wind velocity of only 5 ms⁻¹ (10 knots, 18 km hr⁻¹) can lift the fine particles. As Weisse (1936) claimed wind speeds of only 4 ms⁻¹ could have produced the sand storms at Arouane in 1934 (when storms occurred on one day in four in the period August to February 1935, one day in two in July and one in eight in March to June), it is probable that he did not differentiate between dust and sand. Hubert, however, distinguished between sand storms comprising coarser particles lifted probably not much beyond 1000 m and perhaps pushed ahead of squalls, and dust storms where concentrations may be great even at 3000 m, indicating widespread, strong winds (Carlson and Prospero, 1972). These are produced by the tightening of the isobaric gradient with the incursion from the north of low pressure troughs. The development of such a gradient, from 1026 mb at Sebha, central Libya, to 1010 mb at Kano, was followed 24 hours later by haze in northern Nigeria (Burns, 1961). Surface winds of 30 knots will keep dust airborne. Winds of 45 knots were recorded at Bilma in advance of a major dust

storm across the northern parts of Nigeria, Benin and Ghana in 1974 (Samways, 1976). If 45 knots at the surface, they will be even faster aloft, spreading the dust over a period of days far across West Africa (Figure 59).

The absorption of radiation by the dust and the associated cool, dry air from the Sahara in winter, together produce a significant fall in actual and sensible temperature with the arrival of the harmattan, which is often welcomed in southern, normally humid, areas. The dust may also help to locally improve the structure and the micronutrient content of marginal soils. However, conversely, severe falls of dust might smother young plants. The harmattan also brings an increase in some viral diseases such as pneumonia, tuberculosis and meningitis (see Chapter 13); dust particles smaller than 2μ are trapped in the lungs and respiratory ailments are exacerbated. The dry air leads to cracked skin and lips and nose bleeds. Livestock is similarly discomforted by the harmattan. It increases the risk of damaging fires in field and village. There is often a deterioration of radio signals. And, of course, poor visibility seriously affects transportation, especially by air. Because of dust, visibility was only 300 m at Kano Airport on the morning of 22 January 1973, when 176 people lost their lives in the crash of a Jordanian airliner.

5 Precipitation and the water balance

Precipitation

In this descriptive section of the book it is reasonable to progress from a discussion of cloud to a consideration of precipitation, which in the tropics fundamentally means rainfall, although all the forms of precipitation, from fog drip and dew, to frost, hail and even snow, have been recorded in West Africa.

Fog and dew

Fog is a frequent phenomenon on the north-west coast and dew will be deposited whenever the temperature of the air in contact with the ground reaches dew point, a not uncommon occurrence in the dry season under clear nocturnal skies. However, dew amounts will vary metre by metre over the surface, slight variations in soil or vegetation significantly affecting deposition. Its importance as a source of soil or plant moisture in West Africa is not known, and regular measurements are rarely maintained. They can be made by means of a dew gauge, a specially painted block of wood that is placed outdoors at sunset and must be checked soon after sunrise before evaporation takes place. Not only is it difficult to maintain a regular observation of the gauge at such unsocial hours, but the relationship between the amount of dew registered on the wood and the amount deposited on surrounding natural surfaces is unknown.

Snow and hail

Snow, except on the summits of the Ahaggar Mountains, is barely experienced in West Africa,

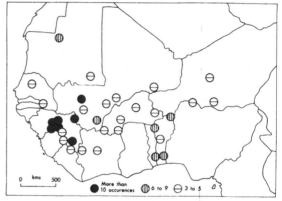

Figure 60   Hail in former French West Africa 1921–34

Note: Boundaries of modern states shown.

but hail is reasonably common, especially in some upland areas. It is almost always associated with the passing of a storm or line squall (see Chapter 6). Picq (1936b) records heavy hail breaking branches from trees on 1 July 1934, near Kédougou, Senegal. Bougnol (1936) reports hailstones up to 40 mm in diameter falling at Mopti, and notes that in the period 1921 to 1934, in what was then French West Africa, 170 stations recorded 321 falls of hail, their distribution as shown in Figure 60. Of these the majority occurred in the Fouta Djalon where 30 per cent of storms produced hail in the Kindia–Siguiri–Mali triangle. Presumably storms are triggered off or rejuvenated by the uplands, where temperatures at altitude remain low enough to prevent melting of the hail before it reaches the ground. Sansom (1966) claims that west central Ivory Coast experiences about one hailstorm per year, as does southern Cameroon and central Nigeria.

Rainfall

Of all the climatic variables in the tropics, however, rainfall is the one most frequently discussed, because, in conditions of reasonably uniform and high temperature, it is rainfall that, by its presence or absence, scarcity or abundance, reliability or variability, determines the seasons, the production or failure of crops or pasture, even life or death. The literature relating to rainfall in West Africa is now abundant, and the Sahelian drought of the early 1970s led to a phase of particular interest in the reasons for rainfall variability. Some of these will be incorporated in Part Two of this book, and the consequences of this variability to life in West Africa are discussed in Part Three. In this chapter spatial and temporal patterns of rainfall are described, with but a limited preamble on causation.

Rainfall mechanisms

If the origins of rainfall are inquired after, it will be commonly suggested that uplift of the air is required, thus leading to condensation of water vapour and the creation of cloud, from which the water droplets fall. The causes of uplift will usually be described as convection due to the heating of the ground, or the forced ascent of moist air over high ground or at fronts, which are the boundaries of opposing airmasses where the warmer air will be lifted over the cooler. In West Africa these explanations are only partially adequate. It is undoubtedly true that much rainfall has a convectional origin, and a consideration of rainfall amounts on Mt Cameroon, over the Freetown Peninsula and the Fouta Djalon will also indicate the relevance of topography. But rainfall produced by these lifting mechanisms will only influence very localized areas. Yet many an air traveller along the Guinea coast in, say, July, will have observed, as have the authors, his aircraft perpetually under, in or above an unbroken layer of cloud on the journey accomplished in a few hours. At each descent of the 'plane, at Lagos, Accra, Abidjan, Robertsfield, Freetown, even Banjul, for example, he will have noted that it was pouring with rain. Such widespread precipitation cannot be explained in terms of convection or topography, and frontal rain such as is commonly associated with higher latitude depressions, although it may affect the far north and north-west of West Africa, does not influence directly the rainfall further south. Some other mechanism is required therefore to explain the widespread rains of the summer monsoon, and this is discussed in Part Two.

Meanwhile, there are further problems for the meteorologist and climatologist in any part of the world. The mechanism that produces cloud droplets may seem obvious, but simple cooling of the air is insufficient; also required are quantities of condensation nuclei, salt, dust and smoke, in the atmosphere on which condensation will take place at low levels of supersaturation. Why is it, too, that the minute, floating, cloud droplets of the order of 1 to 100μ become the drops in excess of 1 mm that are then able to fall from the cloud against the upward moving currents that led to the initial condensation? It is sounder to refer to 0°C as the melting temperature of ice rather than as the freezing temperature of water, for it is obvious that clouds are commonly at altitudes where the air temperature is below 0°C, yet freezing of the droplets has not occurred. They are supercooled, but not frozen. In fact supercooled droplets may exist, as noted in the previous chapter, until temperatures reach −40°C. Thus the tops of the tallest cumulus or the high-level clouds are usually composed of ice crystals, not water droplets. (The mean lapse rate of temperature with ascent in still air is 5.5°C per 1000 m. If the air temperature at the ground surface is 26°C, at 12,000 m it will therefore be −40°C.)

One theory concerning the origin of rain relates to the presence of ice crystals in the cloud. Because the saturation vapour pressure around the ice is lower than around the water droplets, the water vapour will tend to move in the direction of the ice crystals and condense upon them turning directly into ice, a process known as sublimation. The crystals then grow to form snowflakes, become heavy enough to fall as snow and, as they pass through the cloud, gather further water droplets to themselves. As the snowflakes pass into warmer air towards the ground they melt and rain is produced. In higher latitudes, and from tall cumulus with glaciated tops in the tropics, rain may have its origin in snow, but much rain in West Africa falls from clouds in which ice crystal formation has clearly not taken place.

Hence the second main theory suggesting that, especially on salt particles, water droplets of above average size may develop, which, as they move up and down in air currents in the cloud, collide with the minute droplets that normally stay apart. Thus by coalescence the large droplets grow even larger until they are heavy enough to fall as rain.

Rain drops in the tropics are characteristically large. In one research project in West Africa drops were measured ranging from 6.6 to 59.8 mm^3, averaging 20.85 mm^3, with diameters of 2.34 to 4.86 mm (Kowal and Kassam, 1976). In showers of 50 mm per hour intensity or above, rain drops are generally at least 2 mm diameter (Ajayi and Olsen, 1983), although they can be as small as 0.5 mm if the intensity is less than 1 mm per hour. A typical raindrop in temperate latitudes will probably have a diameter of 1 mm.

Yet a further problem remains. More water may fall from the sky than is available immediately above the area of fall in the form of water vapour. Precipitable water in the form of the water vapour content of the atmosphere may be assessed from the equation:

$$PW = \frac{[\bar{r}(p_1 - p_2)]^{10^{-1}}}{g} \text{ mm}$$

(Anyadike, 1979a)

where $\bar{r}$ is the average value of humidity mixing ratios for the atmospheric layer bounded below and above by the pressure values p_1 and p_2 (measured in millibars), and g is the acceleration (ms^{-2}) due to gravity. Utilizing data for only five stations Anyadike (1979a) produced the maps which comprise Figure 61. If the January map is compared with Figure 62a it is clear that during this dry season period, with the possible exception only of Cape Palmas, precipitable water values

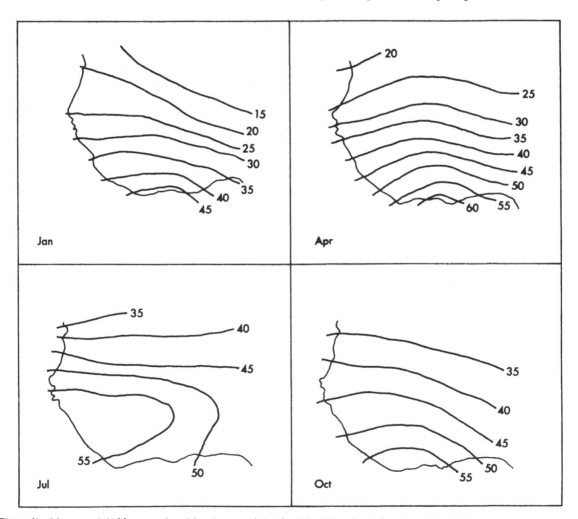

Figure 61 Mean precipitable water (mm) in *a* January; *b* April; *c* July; *d* October (after Anyadike 1979a)

Figure 62a   Mean precipitation (mm) in January

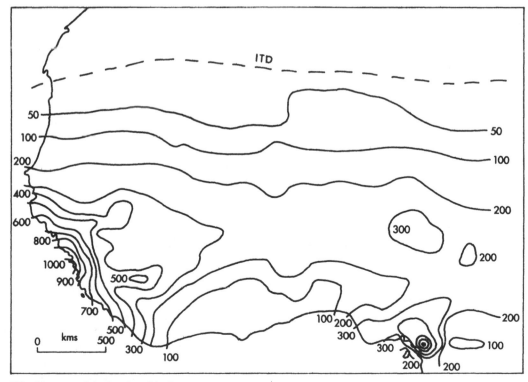

Figure 62b   Mean precipitation (mm) in August

far exceed rainfall, but the mechanisms for rainfall are absent. Conversely, comparing the July pattern for precipitable water with the August rainfall scene (Figure 62b), it is clear that at this time rainfall far exceeds precipitable water. This situation can only exist if there is a strong advection of water vapour into the region, and this, of course, comes with the south- westerly air flow off the Gulf of Guinea. From an analysis of the upper air over Ikeja, Lagos, from 1958 to 1960, Obasi (1965a) observed that whereas there was correlation between months of minimum precipitable water and minimum rainfall (although in 1958 the month of minimum values was July), there was no correlation between months of maximum precipitable water and maximum rainfall. There must, therefore, still be some further factors contributing to precipitation in the region, and these have been discussed by Anyadike (1981).

In addition to the influence of altitude and continentality, of precipitable water vapour (PW), of zonal water vapour flux (ZWVF) and meridional water vapour flux (MWVF), the depth of both the maritime air flow (DMA) and of the convectively unstable layers (DCUL) must also be taken into account. By means of multiple regression analysis Anyadike assessed the relative contributions of PW, ZWVF, MWVF, DMA and DCUL to precipitation at the same five stations referred to in his earlier paper (1979a). Whereas at Abidjan the five variables explained surprisingly only 46.6 per cent of variations in rainfall, at Bamako they accounted for 83.9 per cent (DMA contributing 55.4 per cent and PWC 24.4 per cent), and at Dakar they explained 86 per cent (MWVF 52.7 per cent and PWC 22.3 per cent). At Niamey only 46 per cent of the variance could be attributed to the five factors (DMW 21.4 per cent), so where is the main source of rain here? The answer is in line squalls, discussed in the next chapter. At Nouadhibou only 49.4 per cent of the variance was explained by the five variables, the remaining 51 per cent coming from localized showers in this coastal location. However total annual rainfall levels are very small. It is assumed that over the greater part of West Africa the most important rain influencing factors are: first, the depth of the maritime air; second, the precipitable water vapour content of the air; third, the meridional water vapour flux. Along the Guinea coast PW is more important than DMA. Moisture transportation is vitally important.

Rainfall distribution in West Africa

So vital is precipitation to human well-being, so much is read from or into statistics, maps and graphs relating to the spatial and temporal distribution and quantities of rainfall, that it is essential for analysts to assess the accuracy and reliability of the data they are using. The authors have gone to great pains to utilize the largest data base they could find for the maps and diagrams in this chapter, that is one for 144 stations and, except in eight cases, for periods not less than 11 years. The means for 64 per cent of the stations are based on records of not less than 30 years.

Nevertheless the picture shown is very generalized. Such is the scale at which the maps could be drawn that it is obvious that many local influences upon precipitation have been ignored, especially topography. The isohyets in Figure 62b in the vicinity of Douala have been drawn on the basis of data for four stations, Calabar, Douala, Kribi and Malabo, excluding the influence of Mt Cameroon, (4070 m) and the uplands of Bioko (3007 m). Douala, near sea level and 40 km east of Mt Cameroon has a 47 year mean annual rainfall of 3902 mm. Debundscha, at only 9 m above sea level, on the seaward side of the mountain, has a 70 year mean annual rainfall of 10,617 mm (Chaggar, 1985) with a maximum annual fall of 14,680 mm (Lefevre, 1972). Rainfall increases up to 1800 m, then decreases towards the summit which records only some 2030 mm per year. The windward side of the mountain receives on average 270 days of rain per year, but the lee side only 170. The rain shadow effect is also well demonstrated on Bioko, where Malabo on the northern coast records a mean of 1898 mm per year but Ureka on the south coast 10,451 mm with over 2000 mm in July alone! The mean annual rainfall for Freetown (Falconbridge), where records have been maintained for 90 years, is 3207 mm. However in the mountains south of the city the Guma I raingauge at 366 m altitude has a mean annual rainfall of 6248 mm. On the lee side of the Freetown Peninsula mountains and only 15 km from Guma the mean annual rainfall at Waterloo is only 2557 mm.

It should also be appreciated that the siting of the raingauges can play a crucial part in the determination of the levels of rainfall recorded. It is almost certain that not all the 144 gauges whose records have been used are sited so as to comply with the standard recommendations: any gauge

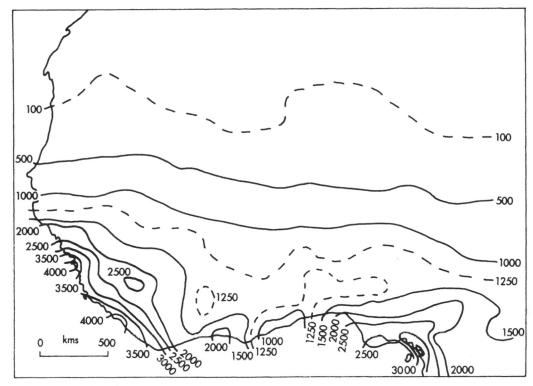

Figure 63   Mean annual rainfall (mm)

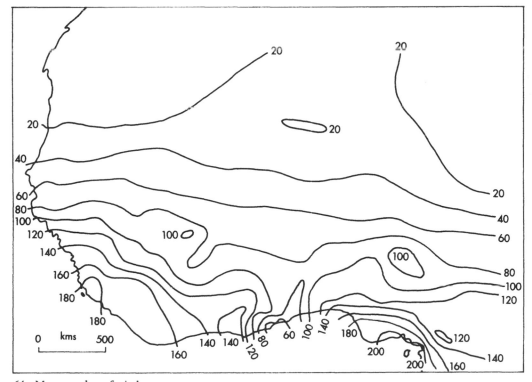

Figure 64   Mean number of raindays per annum

should be on level ground; the land should not fall away steeply on the side from which the dominant wind blows; the distance of the gauge from any nearby object should be not less than twice the height of the object above the rim of the gauge and preferably four times. Not all the 144 gauges will be of the same type, nor all equally coping with the problems of exposure, such as the eddying of the rain-bearing wind over the funnel of the gauge. Not all the records are of equal length. Short periods of records may coincide with a series of unusually dry or abnormally wet years. How reliable have been the readings of possibly many inexperienced meteorological assistants over the years? How regularly and accurately have records been maintained? And finally, if the aperture of each of the 144 gauges is of the standard diameter of 127 mm, the total area from which the rainfall has been collected and measured to represent the conditions prevailing in the whole of West Africa is only a four million millionth part of the area of the region!

However, despite its limitations the map of mean annual rainfall, Figure 63 shows distinctive and important distributions. The greater part of West Africa has a mean annual rainfall of less than 1500 mm, and north from the isohyet of this

value the decline in rainfall is gentle but steady until true desert conditions are reached at about 20°N, with the north-western coastal area not anomalous in this regard. South of the 1500 mm isohyet the increase in annual precipitation is sharp and high values in excess of 3000 mm are reached along the south-west facing coasts of Guinea, Sierra Leone and Liberia in the west and Cameroon in the east. In these areas precipitation rates are some 300 per cent greater than for other parts of West Africa, making them among the wettest places in the world. Eastwards from Liberia and westwards from Cameroon the annual rainfall decreases, at first gently but towards Ghana with dramatic suddenness. From the Ivory Coast/Ghana border to Accra the decrease is of the order of 4.5 mm per km, making this area one of considerable meteorological interest.

The pattern in Figure 63 is effectively repeated in Figure 64. A rainday is defined as a day having at least 2.5 mm of rain. The south-west and south-east areas are again set apart, as is the Togo Gap, with Axim in south-west Ghana having 100 more raindays per year than Accra. The influence of relief is seen in the case of the Jos Plateau, if not elsewhere. On the whole the picture is of diminishing rainfall from south to north, the pattern disrupted only in the south-west and in

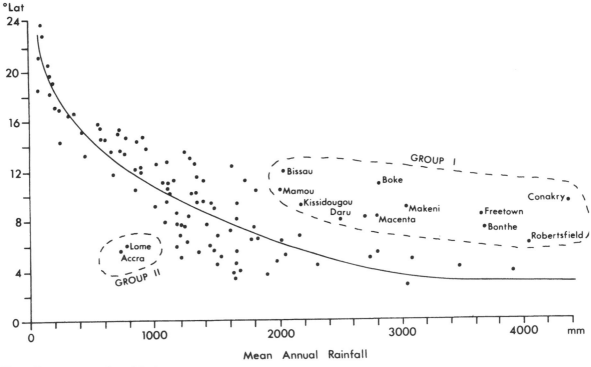

Figure 65 Mean annual rainfall plotted against latitude

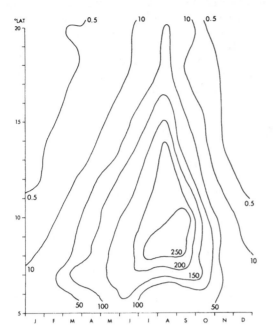

Figure 66 Mean monthly rainfall (mm) along 1°E longitude

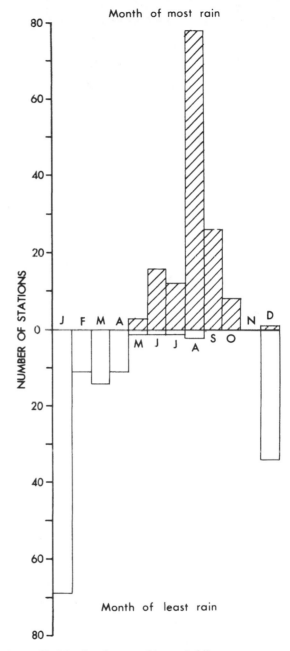

Figure 67 Months of most and least rainfall

the Togo Gap. This is further seen in Figure 65 in which all the stations in anomalous Group I are in Guinea, Sierra Leone and Liberia, and the two stations comprising Group II are Lomé and Accra.

Figure 66 also illustrates this feature of rainfall distribution in West Africa, especially in the months May to October. Once again the Togo Gap is indicated by the decline in rainfall south of 8°N. The latitude of heaviest rainfall along 1°E longitude is between 8 and 11°N. This figure also brings out a marked seasonal contrast, minimum rainfall being recorded in January, maximum in August, except again south of 7° latitude. Across West Africa as a whole this situation is marked (Figures 67 and 68) and the contrast is further demonstrated referring back to Figure 62.

Greatest seasonality, as was also observed in Figures 46 and 50, is in a zone running west to east across the centre of the region, where again the Jos Plateau and the Fouta Djalon are areas of higher precipitation. In January the rain 'belt' has practically cleared the Guinea coast, the most southerly areas alone having a monthly mean above 50 mm. In the winter occasional showers are experienced along the Senegal coast owing to the penetration there of cold fronts passing round the Azores anticyclone from temperate latitudes, but total falls remain very small. In the August

map the Togo Gap is again indicated. In this month the whole of the coastal region between Abidjan and Lagos becomes significantly less wet, whereas elsewhere August rainfall is preeminent (Figures 69 and 70).

It is fairly obvious that the spread of rain across West Africa from the south in the northern

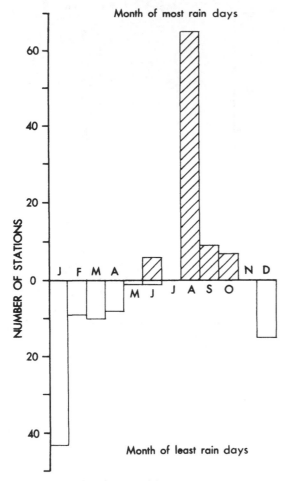

Figure 68 Months of most and fewest raindays

early September, is often referred to as the 'little dry season'. Its significance is illustrated in the individual graphs in Figure 71 where its influence along the coast is seen to end in the west at Sulima (Sierra Leone) and in the east at Warri (Nigeria). Possible explanations for this phenomenon are left to Chapter 10.

Rainfall variability

Seasonality implies temporal variability of course, but as the maps and diagrams are based on mean values it would be unwise to assume that the level of variability is always as shown. Fluctuations of precipitation about means for any period of time can be considerable anywhere, but they are a marked feature of the climate of West Africa. The greater the variability, the less the reliability. One anomalous dry spell in the wet season may be disastrous for farmers; a run of years with below average rainfall may prove nationally calamitous.

At Debundscha, on the seaward slopes of Mt Cameroon, monthly ranges of rainfall have been recorded as follows: January: nil to 636 mm; July: 914 to 2923; August: 568 to 2791; September: 777 to 2372; and December: 42 to 1349 mm. At Dakar for the same months, and for the period 1903 to 1933, the absolute minima and maxima were: January: nil and 10 mm; July: 3 and 290; August: 54 and 476; September: 56 and 318 and December: nil and 98 (Weisse, 1937). The variability that can occur, in terms of absolute maxima and minima, is illustrated in Table 17 which is drawn up for just five stations in one relatively small country, but bears close scrutiny. At Accra there is a 20 per cent probability that once in five years the annual rainfall will be as much as 19 per cent below average or 16 per cent above the mean. At Kumasi the values are 10 per cent below and 9 per cent above (Tandoh, 1973).

Assessments of rainfall variability require reasonably long periods of records. Access to such for all countries in West Africa has not proved easy to achieve. Only for the period 1951–60 have the authors been able to acquire monthly and annual rainfall values for a sufficiently large number of stations (119) spread across the region, to be used as the basis of a rainfall variability map (Figure 72). The coefficient of variability (v) = 100(s/m), where s is the standard deviation and m is the mean rainfall. It is a measure of relative variability. The smaller

hemisphere summer is related to the advance of maritime tropical air across the region. The mean position at the ground in January and August of the boundary between this airmass and the Saharan air to the north is shown in Figure 62. In this book the boundary is called the Inter-tropical Discontinuity or ITD, for reasons that are explained in Chapter 7.

It is to be observed that the main rain zone, however, does not coincide with the boundary on the ground, but lies some distance further south. But the rainfall does not indefinitely increase southwards. When the ITD is furthest north, that is in August, a period of less wet weather, as noted, is experienced between Sierra Leone and Nigeria along the coast and inland to about 7 or 8°N. This less wet period, which usually begins in the last week of July, intensifies to the third week in August and disappears in

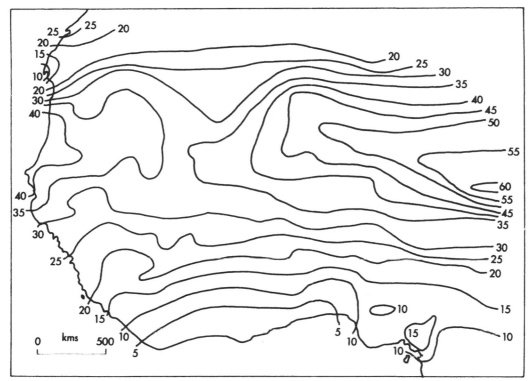

Figure 69   Mean August rainfall as a percentage of mean annual rainfall

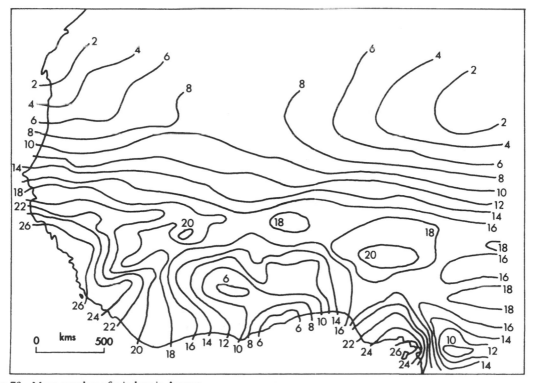

Figure 70   Mean number of raindays in August

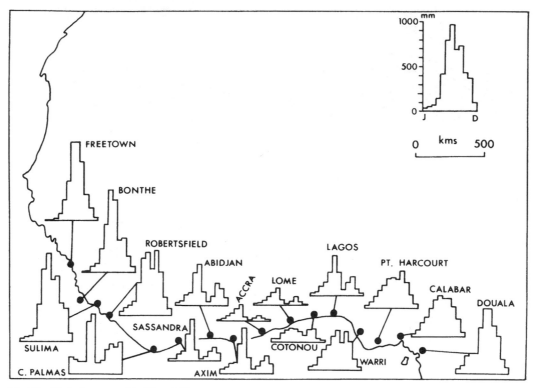

Figure 71 The 'little dry season' in West Africa

Table 17 *Mean, absolute maximum and absolute minimum rainfall (mm) for five stations in Ghana*

	January	February	March	April	May	June
Accra (80)	16 89 0	32 159 0	59 224 0	86 172 8	136 308 7	182 609 1
Axim (41)	49 230 2	73 281 0	112 311 30	122 401 20	336 892 145	510 1190 119
Kumasi (60)	17 124 0	59 150 1	149 268 55	143 259 26	180 296 68	234 396 62
Navrongo (40)	1 12 0	6 79 0	15 105 0	48 147 1.5	111 210 32	142 247 60
Tamale (59)	2 30 0	7 59 0	51 203 0	82 201 3	118 220 27	144 258 36

	July	August	September	October	November	December	Annual
	47 226 0	15 86 0	38 227 0	63 201 0.5	36 139 0	22 134 0	732 1414 333
	177 729 10	52 362 1	84 428 9	147 537 3	201 489 2	110 326 13	1973 3312 1210
	126 446 16	74 400 3	176 356 50	202 394 94	98 220 28	31 127 0	1491 2343 1086
	201 328 49	263 460 56	229 415 120	67 154 0	6 44 0	1 27 0	1095 1397 648
	144 343 41	196 449 51	223 422 86	95 201 19	14 102 0	4 45 0	1082 1567 671

Note: Figures in parentheses indicate the number of years of records.

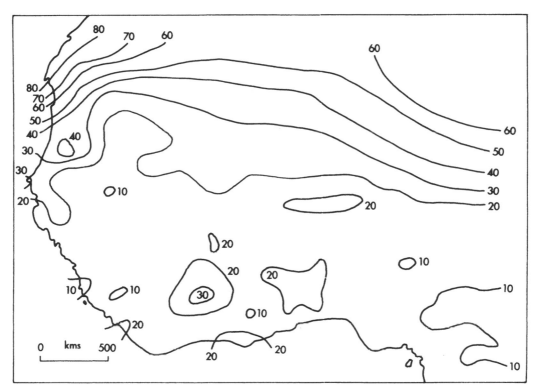

Figure 72 Map of rainfall variability, 1951–60

the value the less the variability. The period 1951–60 was one in which rainfall in the northern half of West Africa was above average generally and in the southern half below average. Nevertheless Figure 72 shows least variability in the Cameroon Republic and in Sierra Leone, considerable variability in central Ivory Coast and otherwise reasonably uniform conditions south of latitude 14°N. Northwards of Nigeria and Bourkina Fasso, however, variability rapidly increases, especially towards the north-western coast.

Much is summed up in Figure 73, which shows how the annual rainfall may significantly vary from year to year. The greater the mean rainfall (the means plotted are for the longest period for which records are available, not for the ten years 1951–60) the greater are the oscillations, but in percentage terms they may be no greater than for the more arid areas. The graphs illustrate the point made above, that in the decade 1951–60 annual rainfall amounts tended to be above average in the northern parts of West Africa, but a general decline is seen through the decade (see the graphs for St Louis, Mopti, Bobo

Dioulasso, Kandi, Batouri and Robertsfield), a decline that was maintained over the following decade and a half.

The exceptional rains in West Africa in 1968 south of 11°N have been described by Hookey (1970). They were far above average in June of that year throughout Ghana and as far east as the Niger delta and Kano, and excessive in July from Conakry to Cameroon. In July Cotonou experienced 665 mm of rain, five times the mean fall, and Accra had 371 mm, seven times above the mean. Severe flooding was recorded in southern Ghana that month. It seems that the diminution of rainfall along the Guinea coast that normally occurs in August was not experienced in 1968, in which year the region as a whole south of 11°N recorded nearly double the mean rainfall. Yet north of 11°N there was below average rainfall!

Long-term variability of precipitation was given considerable attention during the 1970s when lower than average annual precipitation repeatedly occurred in the Sahel, with appalling consequences. The isohyets shifted south by some 1 or 2° of latitude and drought occurred over 5 million km² of West Africa. Other droughts occurred around 1913 and 1942, indicating a possible 30-year cycle and suggesting

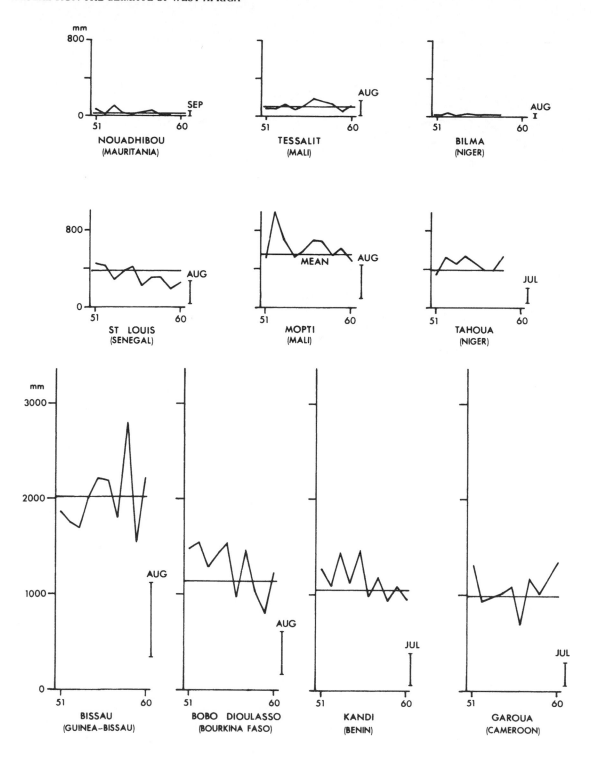

(*Figure* 73 continues on the next page)

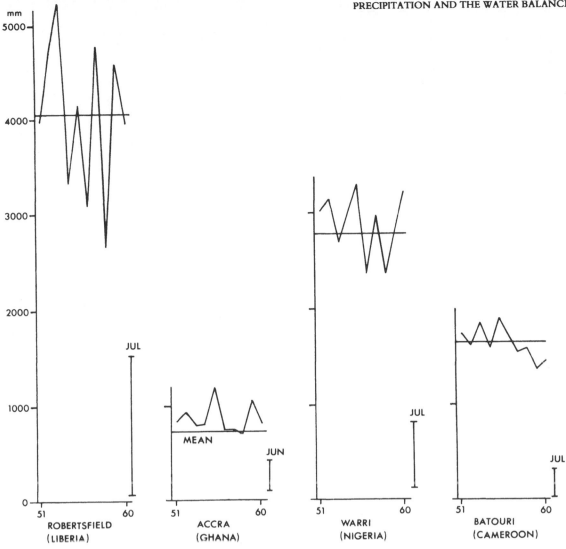

Figure 73 Graph of rainfall variability, 1951–60

Note: The graphs show the annual rainfall in each year 1951–60, and the mean annual rainfall by a horizontal line. The month with the greatest variability is named and the range, absolute minimum to absolute maximum, for that month indicated by the vertical line.

another seriously dry period at the end of this century when the population of the Sahel is likely to be of the order of 54 million, doubling in 20 years. Clearly it would be beneficial to be able to forecast, and so prepare for, periods of marked deviation from the mean, but there is much yet to understand about the mechanisms that produce the variations.

Diurnal variability

Variability of rainfall is also a diurnal as well as a seasonal occurrence, depending upon location,

time of the year and type of precipitation. For example, away from the coast in Ghana there is a tendency for maximum diurnal rainfall to occur in the late afternoon and early evening, especially in seasons of strong convectional activity. However, during the main rains at Tamale there is a concentration of rainfall at night and around dawn, but at Navrongo in mid-morning and early evening. At Kumasi there is least likelihood of rain between 0700 and 1500 hours (Gbeckor-Kove and Dankwa, 1966).

Such variability over short distances makes it difficult to generalize for wide areas of West Africa. At Abidjan 54 per cent of rain falls at night, 46 per cent therefore by day. But of the daytime precipitation 61 per cent falls between 0600 and noon, 39 per cent in the afternoon (Drochon, 1976). At Freetown most rain falls at night and in the early morning, especially in the periods immediately preceding the main rains when the bulk of the precipitation is derived from line squalls which, originating to the east of Sierra Leone (even as far east as Nigeria), most frequently reach the Freetown area at night. In the months of heaviest rainfall and in the 'little dry season' a diurnal pattern is not so obvious. In Gambia, and especially upriver where convectional influences are more marked, rainfall peaks commonly in the early hours of the morning and 12 hours later, in late afternoon. In most of Nigeria the maximum rainfall occurs in the evening and late afternoon.

Rainfall intensity

A further element of variability in rainfall relates to the nature of the precipitation and therefore to its intensity. It is commonly assumed by temperate latitude residents that all precipitation in the tropics comes in the form of heavy downpours of short duration. It is true that much rain in West Africa is derived from storms or as a result of local convection in showers lasting three to six hours, when possibly 100 drops in excess of 2 mm diameter may fall per cm^2, readily assuming the intensity of 25 mm hr^{-1} which will produce runoff and soil loss in open farmland. Throughout Nigeria 50 to 60 per cent of total rainfall comes in showers in excess of 25 mm on 20 to 25 per cent of total raindays (Swami, 1970).

However Kowal and Kassam (1976) estimated that 58 per cent only of annual rainfall was erosive. The Guinea coastal areas particularly are not unaccustomed to low intensity rain lasting in excess

Figure 74   Monthly intensity of rainfall and types of rain

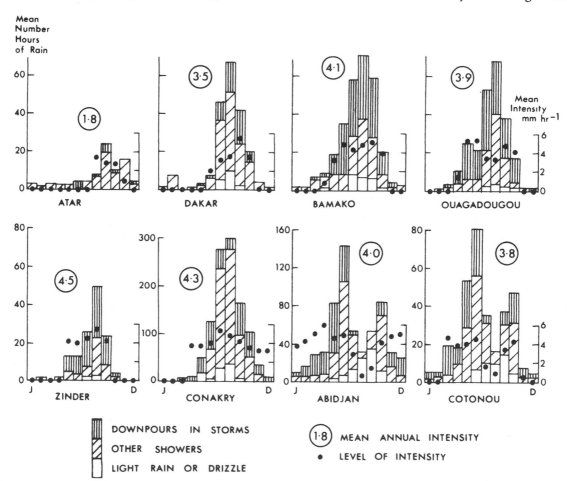

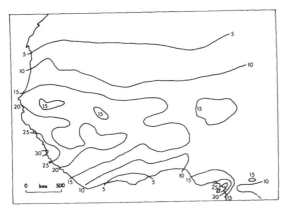

Figure 75 Mean rainfall intensity (mm per rainday) in August

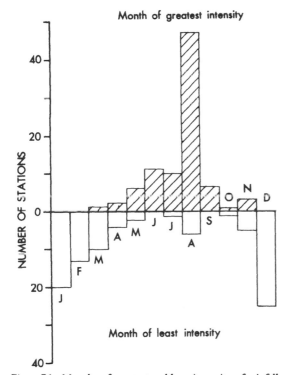

Figure 76 Months of greatest and least intensity of rainfall

for eight stations dispersed across the region.

The graphs do immediately indicate the relative lack of light rainfall, especially in the most arid areas (Atar). However, it is by no means insignificant in the wettest months at Dakar and Bamako, and more particularly along the Guinea coast. The 'little dry season' is again indicated in the graphs for Abidjan and Cotonou. Throughout West Africa the rains of summer take on a markedly different character from those of the drier seasons. In the far north-west the approach and passage of the ITD heralds a relative increase in precipitation from storms, but elsewhere, and again especially in the south and in the wettest areas (Conakry), the rainy season sees a decrease in storms and a marked increase in steady rain or long showers unrelated to line squalls or local thunderstorms.

Nevertheless, in terms of millimetres per hour the rainfall in the less stormy but wetter months is no less intensive. This is further borne out by Figures 75 and 76 where intensity is assessed in terms of millimetres of rain per rainday. The August hiatus on the Guinea coast is seen to reflect not only a decrease in rainfall but also a diminution in rainfall hours and days and in intensity. West and east of the Togo Gap intensities are most marked in the wettest season because of the duration of steady rain. In the Sahel the daily, and sometimes the monthly, total rainfall may come from one single storm.

The relevance of rainfall variability and intensity to agriculture has been hinted at above. A knowledge of maximum rainfall intensity-duration frequencies is obviously of considerable value to those responsible for designing dams, water channels, sewers, or involved in river navigation or flood control. Table 18 indicates maximum 24 hour rainfalls recorded for each month of the year for a selection of stations, and these data are of interest and value. But they are for very short time periods (in the case of Banjul, for instance, only nine years). It is clearly useful to know what is the maximum rainfall one can expect over long periods. There are a number of ways to make this assessment, the most commonly applied utilizing the Gumbel equation:

$$x_T = \bar{x} + k\,(Tn)\,Sx$$

where x_T is the extreme rainfall estimated for the return period T, $\bar{x}$ is the mean of a series of maximum values, k is a frequency factor read

of six hours. In a study of data for the five year period 1951–5 for nine West African stations, Delormé (1963) assessed mean monthly rainfall intensities and the percentage of three different types of precipitation: downpours in storms, showers without storms, light rain or drizzle, that made up the rainfall. The classification is, of course, highly subjective, but his findings are of some interest and are transcribed into Figure 74 from tables, Sx is the standard deviation of the extreme rainfall, T is the return period and n the number of values in a series of maximum values. Applying the formula to Ghana, Dankwa (1974) was able to confirm that the higher intensity-duration frequencies occur in the wetter, not the drier, parts of the country. Tandoh (1973) has shown that Accra (mean annual rainfall 732 mm) may expect to receive an annual fall of 1012 mm in a return period of 10 years, 1147 mm in 50 years, whereas Axim (mean annual rainfall 1973 mm) may expect to record up to 2568 mm in a ten-year period, and 3094 mm in 50 years.

Returning to the 24 hour falls, by a different method (the Hershfield) it has been calculated that Banjul's maximum probable daily rainfall is 813 mm, more than twice the highest (August) value noted in Table 18 (Dept of Hydro-Meteorological Services, Gambia, 1977). Estimates of daily probable maximum precipitation for some Nigerian stations have been made by Ayoade (1976b). The highest estimate is 629.7 mm for Warri (whose mean maximum daily fall is 131.8 mm), the lowest estimate for Bauchi, 243.1 mm (mean: 71.6 mm). We see from Table 18 that Lagos in June has recorded 226.1 mm. It is estimated that in a return period of 50 years 263 mm of rain can be expected in that month, and over 100 years 299 mm.

Table 18   *Maximum 24 hour rainfall (mm)*

	Jan.	Feb.	Mar.	Apr.	May	June	July	Aug.	Sept.	Oct.	Nov.	Dec.	No. of years of records
Sierra Leone													
Freetown (Kortright)	40.0	6.4	44.2	66.5	106.7	233.5	208.3	269.0	182.7	81.5	55.9	53.5	13
Bourkina-Faso													
Ouagadougou	5.1	15.2	38.1	30.5	71.1	55.9	99.1	78.7	68.6	38.1	2.5	0	10
Ivory Coast													
Ferkessédougou	15.2	71.1	83.8	61.0	104.1	58.4	66.0	121.9	86.4	76.2	48.3	27.9	10
Bouaké	38.1	66.0	101.6	83.8	124.5	53.3	94.0	73.7	137.2	58.4	50.8	35.6	10
Abidjan	48.3	150.0	63.5	68.6	142.2	221.0	274.3	94.0	45.7	119.4	104.1	61.0	10
Tabou	66.0	58.4	68.6	142.2	111.8	152.4	104.1	114.3	83.8	180.3	144.8	61.0	10
Ghana													
Tamale	22.9	12.7	71.1	78.7	81.3	66.0	78.7	94.0	81.3	68.6	17.8	20.3	13
Accra	88.9	106.7	109.2	137.2	149.9	302.3	104.1	94.0	114.3	139.7	94.0	76.2	64
Axim	81.3	71.1	53.3	139.7	294.6	429.3	134.6	63.5	68.6	91.4	104.1	53.3	64
Nigeria													
Sokoto	2.5	2.0	7.6	88.9	40.6	55.9	68.6	147.3	61.0	35.6	2.0	0	?
Maiduguri	12.7	7.6	10.2	25.4	53.3	94.0	73.7	116.8	132.1	68.6	7.6	0	34
Kano	2.5	7.6	35.6	30.5	162.6	68.6	91.4	111.8	94.0	45.7	5.1	0	34
Jos	25.4	68.6	68.6	63.5	71.1	96.5	139.7	78.7	83.8	38.1	45.7	33.0	18
Minna	5.1	45.7	68.6	78.7	61.0	83.8	106.7	106.7	134.6	63.5	66.0	25.4	18
Ilorin	33.0	76.2	58.4	99.1	78.7	99.1	73.7	152.4	157.5	88.9	43.2	50.8	18
Lokoja	53.3	63.5	121.9	152.4	96.5	99.1	114.3	137.2	114.3	83.8	53.3	55.9	?
Lagos	116.8	124.5	104.1	116.8	167.6	226.1	190.5	114.3	134.6	190.5	106.7	94.0	34
Port Harcourt	83.8	96.5	91.4	106.7	109.2	165.1	180.3	193.0	139.7	152.4	195.6	38.1	25

(table continues on the next page)

Cameroon													
Loko	55.9	25.4	55.9	53.3	73.7	53.3	94.0	71.1	99.1	76.2	68.6	25.4	10
Douala	147.3	132.1	132.1	127.0	170.2	330.2	279.4	205.7	193.0	223.5	81.3	71.1	28
Lomié	33.0	104.1	63.5	111.8	66.0	43.2	68.6	91.4	124.5	73.7	50.8	78.7	10
Bioko (Fernando Po)													
Malabo	34.7	78.5	52.6	75.6	80.3	88.0	85.0	61.7	83.0	103.2	74.0	72.0	15
Ureka	117.3	129.2	142.7	178.6	359.1	523.8	602.8	425.7	248.6	248.7	204.6	123.5	7
Western Sahara													
Dhakla	5.1	2.0	2.0	2.0	10.2	0	2.5	10.2	88.9	10.2	10.2	43.2	14
Mauritania													
Nouadhibou	17.8	5.1	5.1	0	1.5	1.5	2.0	2.5	25.4	83.8	10.2	66.0	10
Néma	10.2	2.5	2.0	2.5	38.1	81.3	30.5	124.5	33.0	25.4	2.5	17.8	10
Mali													
Arouane	2.5	10.2	0	0	0	30.5	15.2	20.3	61.0	38.1	7.6	2.0	10
Ménaka	2	2	5.1	2.5	33.0	22.9	40.6	53.3	43.2	2.5	0	0	10
Hombori	0	0	2.5	5.1	17.8	33.0	43.2	55.9	45.7	20.3	5.1	0	10
Kayes	17.8	0	0	2.0	45.7	48.3	66.0	121.9	68.6	45.7	2.5	2.0	10
Bougouni	2.0	10.2	15.2	33.0	94.0	76.2	139.7	200.7	170.2	45.7	27.9	2.5	10
Niger													
Bilma	0	0	0	2.0	1.5	0	2.5	30.5	33.0	0	0	0	10
Agadez	0	0	2.0	2.5	33.0	12.7	61.0	43.2	35.6	0	0	0	10
Niamey	2	2.5	53.3	35.6	66.0	51.0	50.8	76.2	53.3	58.4	1.5	0	10
Senegal													
Dakar	2.0	2.5	5.1	2.0	2.0	55.9	150.0	213.4	104.1	96.5	25.4	7.6	10
Gambia													
Banjul	17.8	17.8	2.0	2.0	45.7	50.8	91.4	304.8	119.4	83.8	86.4	7.6	9
Guinea-Bissau													
Bolama	7.6	2.5	2.0	7.6	30.5	111.8	279.4	162.6	221.0	83.8	109.2	20.3	31
Guinea													
Conakry	12.7	12.7	55.9	40.6	78.7	162.6	223.5	358.1	299.7	116.8	96.5	33.0	10

The water balance

In the chapter so far consideration has been given to precipitation: how much falls, when and where. The variability of rainfall has been discussed and its relevance to agriculture mentioned. But how effective is the rain that falls depends upon the amount that reaches the soil and is stored there for use by plants. Some precipitation is evaporated from vegetation and the ground surface, some runs off the surface or flows through and out of the soil. From the soil–vegetation system moisture is further lost by transpiration.

For the world as a whole the total input of water (precipitation in all forms, P) is matched by the outflow (actual evapotranspiration, AE and runoff/throughflow, Q), to create a water balance (P – AE – Q = 0). Locally, however, P – (Q + AE) will produce either a positive result, that is a water surplus, or a negative result implying a water deficit. (The input may be artificially enhanced by irrigation, of course.)

Calculating AE (and potential evapotranspiration, PE, too) is not simple, as was shown in

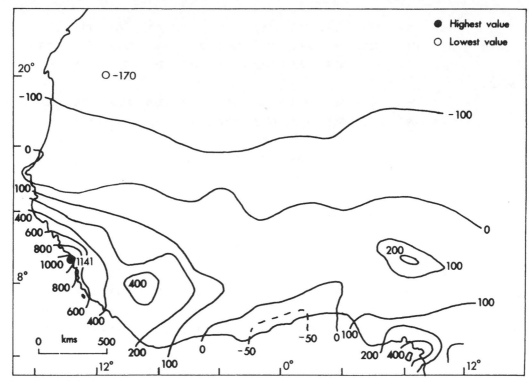

Figure 77a   Water balance (PE–AE) in August

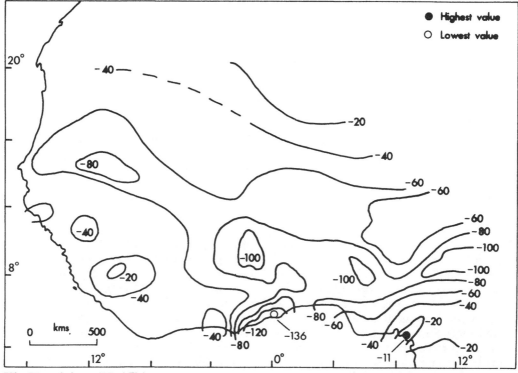

Figure 77b   Water balance (PE–AE) in January

Table 19 *Water balance data (mm) for Bissau, Guinea-Bissau*

	J	F	M	A	M	J	J	A	S	O	N	D	Yr
PE	102	117	141	143	152	149	145	138	135	145	136	107	1610
P	0	1	0	0	13	185	444	682	462	157	22	1	1967
ST	102	69	43	26	17	53	300	300	300	300	204	143	
AE	41	34	26	17	22	149	145	138	135	145	118	62	1032
D	61	83	115	126	130	0	0	0	0	0	18	45	578
S	0	0	0	0	0	0	52	544	327	12	0	0	935

Notes: PE – mean monthly potential evapotranspiration; *P* – mean monthly precipitation; *ST* – mean monthly water storage; *AE* – mean monthly actual evapotranspiration; *D* – water deficit; *S* – water surplus.

Source: Thornthwaite and Mather 1962.

Chapter 3. Calculating Q also presents major difficulties. The Laboratory of Climatology, Centerton, USA, under the direction of C. W. Thornthwaite, worked on the problem of computing climatic water budgets, finally settling on a method involving the use of data for precipitation, potential evapotranspiration, actual evapotranspiration and soil moisture storage (Thornthwaite and Mather, 1957).

The water that is stored in the soil is not being lost from the soil by drainage. The maximum amount of water that a soil can hold after freely draining is termed the moisture-holding capacity or field capacity of that soil. Much of this water can be lost by evapotranspiration however, so that available soil moisture storage varies with soil type, climate and vegetation (with soil use, treatment and crops in agricultural areas). It is calculated that for most soils field capacity is about 300 mm of equivalent precipitation, but the actual amount stored in any period can be calculated given information on precipitation and actual evapotranspiration.

When the soil is at field capacity actual evapotranspiration can equal potential evapotranspiration. But when storage falls below field capacity AE is proportionally less than PE. When the stored water is, say, 60 per cent of field capacity, AE will be 60 per cent of the PE. Now a water surplus will exist in an area when P exceeds PE with the soil moisture storage equalling the water holding capacity. A deficit occurs when PE exceeds AE. Thornthwaite and Mather (1962) published a set of water balance data for 104 West African stations (with some results that signifi-

cantly differ from others produced in the region by, for example, Obasi (1972) and Ojo (1977)), which has been used as the basis of Figure 77.

An example of one set of data for Bissau is reproduced, with knowledge of the compiler, as Table 19. It will be seen that in seven months PE exceeds AE so that in those months a water deficit exists. In June PE = AE. In June also P exceeds PE but as ST has not attained its maximum level (300) there is no water surplus but soil moisture is recharged. In July P far exceeds PE, to the amount of 299 mm, but as in the preceding month, June to July, water storage has been accumulating (rising from 53 to 300) so the 247 mm rise is deducted from P – PE to leave a surplus of only some 52 mm. In August, September and October, as ST is at field capacity so P – PE provides the water surplus.

The water balance in West Africa

Figure 77b illustrates the water balance in terms of PE – AE for January. The whole of West Africa in the dry season is an area of water deficit. Not surprisingly the area with the smallest deficit is around Douala, with its cloud, high temperatures but also high humidities and precipitation in excess of 40 mm in the month. But the north approaching Tamanrasset has a similarly low deficit. This may seem anomalous, but it must be kept in mind what the map is indicating. The difference between PE and AE is not great in the far north for there temperatures are at their lowest level in the year (about 24°C at midday), so

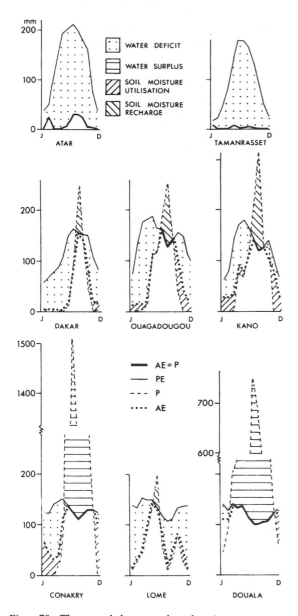

Figure 78 The water balance at selected stations

Gap, so distinctive climatically as earlier chapters have shown, once again stands out (see Lomé). But the reason is not far to seek. PE is going to reflect latitude and compares with that for Douala (or Conakry, the other 'southern' station in Figure 78). But the absence of precipitation for much of the year results in a lack of stored soil moisture and a poor, dry, savanna vegetation. AE effectively equals P therefore and both fall far short of PE.

In the middle of the year the pattern changes drastically, though once again the Togo Gap has a marked deficit as the total rainfall does not normally exceed 200 mm in any one month, August is markedly dry, and some soil water recharge is possible in the wettest months, but no true water surplus is achieved (Lomé). The north now becomes the area with the most pronounced deficit (Atar: –170). These northern regions often remain north of the ITD even in July and August; the near zenithal sun produces temperatures well in excess of 40°C and PE is greater than 150 mm (Atar). Rainfall remains slight, however, so that AE = P even in these rainy season months.

In the south–east the rains significantly increase after February, exceeding 700 mm in July and August at Douala. By late March, in this cloudy, humid zone, soil moisture storage has reached maximum levels so that PE = AE for nearly nine months and the water surplus is pronounced until the precipitation again falls below the PE level in December.

The greatest water surplus, however, is achieved in West Africa's wettest region, the south–west. Here the summer monsoon is most dramatically felt. After three or four months with little rainfall and a marked water deficit, very high precipitation levels are experienced. Conakry normally has in excess of 1000 mm in July and in August. Thus the pattern for Douala is repeated, though the water surplus is even more significant. This seasonal contrast has in the past meant that this wettest region has nevertheless suffered from drought in the dry season. The plight for a growing city such as Freetown could only be alleviated by the construction of the Guma reservoir in 1966, in which is stored some of that excess monsoonal rain for use in the dry 'winter' months.

A knowledge of water budgets is of significant practical value. The water deficit provides a measure of drought and provides basic information relevant to irrigation. The surplus will

evaporation is reduced despite clear skies (though dust may reduce insolation levels). The arid conditions mean that a minimal vegetation cover exists, so limiting transpiration. Thus, at this time, both PE and AE are very low. What slight precipitation may occur is immediately lost by evaporation, so that AE = P (see Figure 78, Tamanrasset).

It may seem even more anomalous, however, that the greatest water deficits are recorded in the south and along the Guinea coast, but the Togo

indicate whether water is available for this purpose, and soil moisture storage data will supply an insight into appropriate irrigation amounts. Water surplus data, furthermore, will provide essential information relating to ground water levels and stream runoff. The total annual water surplus in Nigeria is estimated to be in excess of 320 billion cubic metres; this is the volume of water that drains from Nigeria into the Atlantic (Obasi, 1972). The volume of 'fresh' water that flows into the Bight of Biafra, as into any oceanic area, helps to determine the salinity and density of the sea water locally, which in turn can affect fisheries.

6 Thunderstorms and line squalls

Thunderstorms

At any one moment there at about 2000 thunderstorms taking place in the world, 80 per cent of them in the tropics. Often associated with heavy rain and strong winds, they are important weather phenomena. As Figure 79 shows they are common in most parts of West Africa.

Normally the frequency of thunderstorms is expressed in the number of days per month or year with storms, or the percentage of days in a month or year having storms. The days in question are those on which thunder is heard, no matter whether it be once or a hundred times. Registering the sound of thunder without the aid of instruments implies that a storm must have passed within 24 km of the recorder, as the sound cannot normally be heard beyond this distance. (Lightning, on the other hand, may be seen even at 200 km distance at night.)

Thunderstorm records

The recording of thunder, it can be appreciated, is notoriously unreliable, depending as it does on the quality of hearing of the observer, on his location indoors or outdoors when the thunder occurs, on noise interference, and on weather conditions, especially wind direction. Thus it is likely that many recorded values are underestimates.

Published data on thunderstorm days, and for a number of years, are scarce. Attempts to map the world picture, for example those of Brooks (1925) and Berry et al (1945), showed significant differences, but indicated that West Africa, especially the south-east south of latitude 10°N, had 70 to 80 thunderstorm days (TSD) per year, the number diminishing rapidly northwards to

about 10 TSD at 15°N, and only three to five TSD in the Sahara. Walker (1962) indicated the average number of *local* thunderstorms per year at Ghanaian stations to be as follows: Axim 45.1; Takoradi 30.5; Accra 13.5; Akuse 48.1; Kumasi 63.1; Kete-Krachi 28.4; Tamale 41.1; Navrongo 36.4.

Now these figures are significantly lower than those in Figure 79, derived mainly from a WMO publication (1953), which was also probably the basis of later American and Russian material (USAAF, 1968; Lebedev, 1967). The number of days with thunder can obviously fluctuate widely from year to year and Figure 79 is based mainly, but not entirely, on records for unspecified and not necessarily coinciding five-year periods. However, Mulero (1973) drawing on ten years of records (1962–71) for Nigeria alone, produced data for January and July that coincided very closely with those mapped in Figures 80a and b.

Using five-year records (1954–8) of *visual observations of lightning* (presumably of 'local' distribution otherwise values would be higher still), storms were recorded in Ghana on up to 156 days per year at Tamale and 152 days at Accra (Ghana Meteorological Dept, 1959). In a study during the years 1963 to 1964 in Sierra Leone (Nicholl, 1965), 171 TSD per year were recorded at Freetown, 205 at Bo, 125 at Kenema and 180 at Daru. And Sansom (1966) published a small map of mean annual frequency of TSD for West Africa which showed south-eastern Nigeria and south-west Cameroon, central Liberia and eastern Ivory Coast with over 150 TSD, frequencies then falling off regularly northwards to 10 TSD at about 20°N.

In Figures 79 and 80 the values for five of the six Sierra Leonean stations used are drawn from 10 years of records, for 10 of the 17 Cameroonian stations the records are for nine to 14 years, for each of the four Togolese stations the records are

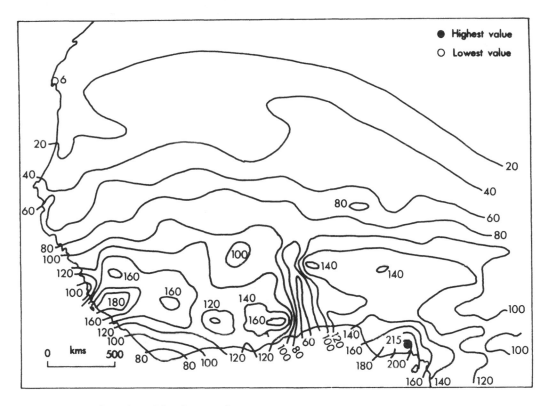

Figure 79 Mean annual number of thunderstorm days

for a minimum of 16 years, for the two Guinea-Bissau stations for 20 and 30 years, and for Tamanrasset 15 years. It is thus possible, therefore, that these maps present a more realistic picture of the distribution of thunderstorms in West Africa than has previously been published. The discrepancies between Walker's data and those used to produce Figure 79 are probably explained by his use of the term 'local'. The thunderstorms he recorded probably actually passed over the towns named, and others that were observable but passed by the stations were not counted.

Distribution of storms in West Africa

Most storms are recorded in south-eastern Nigeria and coastal Cameroon and in Sierra Leöne, closely followed by central Ghana, southern Guinea, western Ivory Coast and central Nigeria. North of 10°N the number of TSD rapidly diminishes so that the Sahelian states generally have fewer than 100 TSD per year. The more humid the region the more the TSD. The Togo Gap stands out strongly as an area of rela-

tively few storms. The apparent anomalies in south-eastern Ghana and northern Benin, where low and high values are in close juxtaposition, are probably explained by topography. Akuse (171 TSD per year) is by the Volta River, but only 10 km from the scarp front of the 560 m high Akwapim Hills which encourage the uplift of conditionally unstable air. Natitingou (143 TSD per year) in northern Benin is at 500 m altitude in the Atacora Mountains, whereas the 'nearby' Sansanne-Mango in Togo is low in the Oti valley and actually 60 km from the same highlands. The Jos Plateau in Nigeria, the Fouta Djalon in Guinea, the Freetown Peninsula mountains and Mt Cameroon seem to have less impact on thunderstorm incidence, although in Mulero's study (1973) the highlands of central Nigeria were shown to experience more storms in April and July at least.

The frequency of thunderstorms

Figure 81 shows the months of the year in which most or fewest thunderstorms occur.

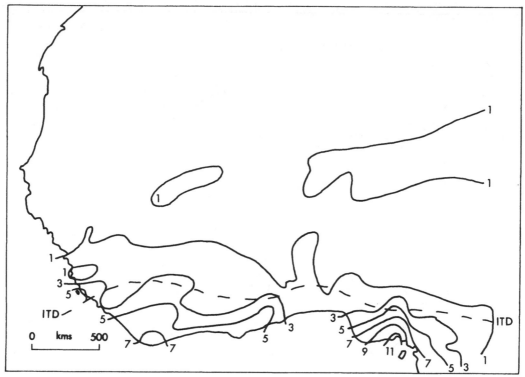

Figure 80a Mean number of thunderstorm days in January

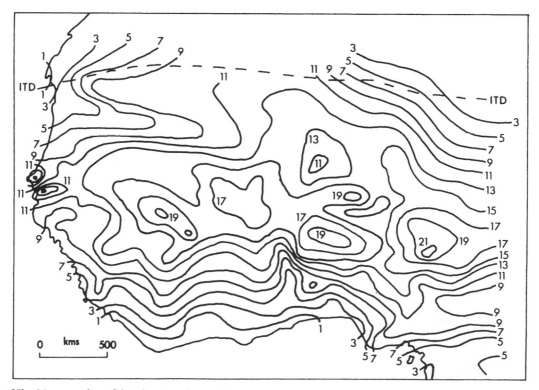

Figure 80b Mean number of thunderstorm days in August

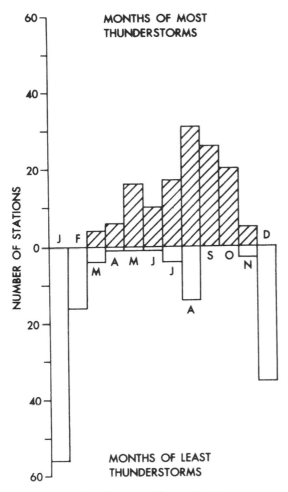

Figure 81   Months of most and fewest thunderstorms

south. In fact, almost without exception, all stations south of a line from Banjul to Maroua (northern Cameroon) show this reduction in thunderstorm activity in the July to August period. (North of this line only Sokoto seems to demonstrate this.) Furthermore, not only is the reduction more pro- nounced the further south the station, but it is also there longer in duration. It is in part related to the lowering of surface temperature and increase in atmospheric pressure that is brought about by the clouds and rain of the wet season.

Thus there are two clear thunderstorm seasons in the south, one preceding and one following the main rains of June to September, but only one peak storm period in the north. This pattern is clearly related to the movement of the ITD across the region. It is noticeable from Figure 82 that the autumn storm period is shorter than the pre-rains one, as the retreat of the ITD southwards is faster than its advance northwards.

Conditions in a thunderstorm

The thunderstorms of West Africa are essentially of two types, one the localized, convectional storm, which travels little but may drift with the prevailing winds from the south-west, and the sëond being the line squall, comprising a bow-shaped line of storms convex to the west and travelling sometimes right across the West African region from east to west.

The former would be classified as an airmass type of storm, produced by strong, local heating and convection of sufficiently deep moist air over a warm land surface, uplift often being further encouraged by topographic barriers. An individual convective cell may grow within a few minutes both laterally, from 2 to 3 kilometres wide to perhaps 10, and vertically, to perhaps 10,000 metres. Adjacent cells will normally develop and join together until a mature storm may reach 40 km across and 18,000 m in height, the cloud top glaciating and being stretched out into an 'anvil' by the high-level winds and on reaching the tropopause. Air is entrained into the developing cloud in which vertical air ascent can reach 30 ms^{-1}. In the cloud turbulence raindrops, snowflakes and ice crystals are swept up and down and the charge separation of perhaps a billion volts that ultimately leads to lightning flashes is created.

The atmosphere contains ions (molecules electrically charged as a result of gaining or losing

Overwhelmingly the months of minimum humidity levels and precipitation are also the months of fewest storms, and August stands out as the month with most thunderstorms. However, in this same month 10 per cent of stations record minimum levels of thunderstorm activity. These 10 per cent are all in the Niger delta and coastal Cameroon or along the Guinea coast between Bonthe and the Volta, extending inland for no more than 150 km. They thus lie in the west in the same area that most markedly experiences the August rainfall hiatus, the 'little dry season', but the coincidence does not fully apply to the Togo Gap or southern Nigeria and Cameroon. However, Figure 82 clearly indicates a number of stations experiencing a marked reduction in TSD in August, especially in the

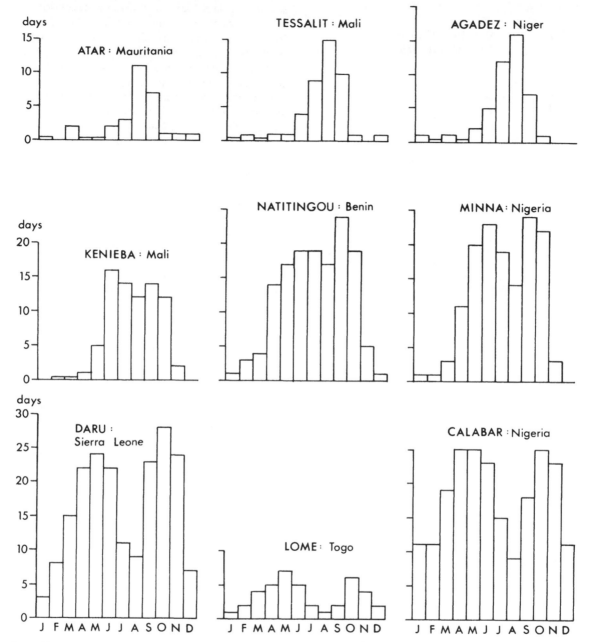

Figure 82 Mean number of thunderstorm days per month at nine stations

electrons). They are usually formed from oxygen or nitrogen molecules bombarded by cosmic radiation. When water drops are formed they attract negative ions and become negatively charged, surrounded then by air with a relative excess of positive ions (ions that have lost an electron). By collision ice crystals also gain electrons and so become negatively charged. As they grow and increase in weight so they sink to the lower areas of the cloud, in part explaining why the lower areas become dominantly negatively charged and the upper parts positively charged.

The lightning flash is a rapid discharge of electricity across the cloud, or from cloud to ground or vice versa, temporarily neutralizing the charge discrepancies. The length of a flash can exceed 5 kilometres in a large cloud, although the channel of the stroke may be only a few centimetres wide. In that narrow zone the air may be heated to 15,000°C in a few ten-millionths of a second. It is the shock waves caused by the explosive expansion of the heated air that produce the sound waves of thunder, which may be brief if the flash is near and short, or long (rumbling) if the flash is long and multi–channelled. As light travels at some 300,000 kilometres per second, but sound only at 330 metres per second, so the thunder follows the flash by about 5 seconds if the observer is 2 kilometres from the storm, 10 seconds if 4 kilometres away, and so on. A storm producing a flash at 20 second intervals may dissipate a million kilowatts equivalent to the energy involved in ten atomic bombs of the type used in the Second World War. It is perhaps worth noting therefore some of the lightning counts made by Nicholl (1965) in one year in Sierra Leone. At Freetown a lightning counter, effective to about 32 kilometres, recorded 35,541 strokes in 171 storms, an average of 208 flashes per storm. At Bo the count was 39,731 in 205 storms, or 194 flashes per storm, and at Daru 74,962 strokes were recorded in 180 storms, or 416 per storm. Why the storms near Daru should have been that much more vigorous is unclear.

Lightning, of course, can be very dangerous and destructive. Jeffreys (1952) records at least 47 persons killed by lightning in the then Bamenda Division in Cameroon between 1921 and 1944 (i.e. two deaths from this cause per 300,000 inhabitants, a rate eight times greater than in a cooler country like Great Britain), and at least 279 cattle, 96 in one flash. Lightning is no respecter of persons or property, including in its targets in the period noted a District Officer's house, a medical officer, a Roman Catholic school and a teacher in a mission school! The commonly held belief that lightning never strikes the same place twice is quite unfounded. Nevertheless it is possible that the percentage of lightning flashes in West Africa reaching the ground is only half that in temperate regions, storm clouds in the tropics commonly being higher (Chalmers, 1941). This is just as well, given the lightning counts observed by Nicholl.

After a short period of development, perhaps not exceeding 15 or 20 minutes, the storm clouds will normally produce heavy rain. The falling droplets will cause a downdrag of cold air that can reach the ground, where its outward divergence will then create the strong, cool surface wind that is commonly experienced during the height of the storm. After a further 30 minutes or

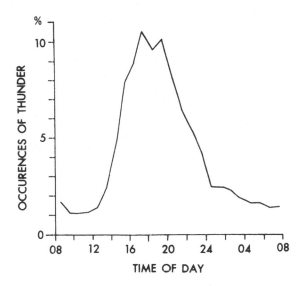

Figure 83 Percentage occurrences of thunderstorms by time of day, Sierra Leone, 1963–4 (using data in Nicholl, 1965)

so the accumulating downdraughts will subdue the convection, vertical motion ceases and over the next half-hour the storm will die away.

Because of their origin in the convectional uplift of the air, it is not surprising that the majority of airmass storms in West Africa occur in the late afternoon and early evening, when maximum heating of the earth has been attained. Figure 83 is again based on Nicholl's work in Sierra Leone, but suitably demonstrates the picture for all West Africa.

Line squalls

Of greater significance and interest in the meteorology and climatology of the region are the line squalls or disturbance lines. Of particular prominence in the Sahel (their mean latitude is 13°N) but recorded anywhere in West Africa, these lines of storms are not all equally active, of course, and any one storm is not consistently active during its passage. They may travel up to 3000 kilometres along westerly tracks with a mean speed of about 58 km hr^{-1}. Sometimes they fade before they have traversed the region, sometimes they peter out as they pass from land to sea; others may be rejuvenated as they move over the warm ocean.

Numbers of line squalls

Some interesting results came from a study of disturbance lines, identified from satellite photographs, in three periods: 28 June to 16 July, 28 July to 15 August and 30 August to 19 September 1974 (Aspliden et al, 1976). These are summarized in Table 20. It is immediately apparent that many more disturbance lines were generated over land than over the sea. Of the 176 storms 130 also decayed over land. The mean speed of east to west movement tended to be higher over land, so that greater distances were travelled by storms over the continent, though life-spans tended to be about the same. It may also be observed that the earlier period, that of the retreating zenithal sun, produced most storms over land, whilst it was the later period, when sea surface temperatures were at their highest at Sahelian latitudes off West Africa, that generated most disturbance lines over the ocean. (The origins and meteorological significance of line squalls are discussed in Part Two.)

Line squalls and associated weather

Rainfall is commonly between 40 and 90 mm during the passage of a storm, and as the word 'squall' implies they are often associated with the strongest winds experienced in West Africa (see Table 11). Line squalls often pass localities during a particular part of the day (for example the greater proportion of storms at Freetown occur between 22.00 and 05.00 hours) suggesting possible source regions for the storms. One such seems to be just east of Niamey, but many squalls originate even east of Lake Chad. Figure 84 is based on satellite imagery (Dugdale and Pearce, 1977) by plotting the location of the leading edge of the cloud cluster associated with a travelling storm. It illustrates the movement of one storm over a period of one and a half days. Its erratic progress will be observed. Between 0600 hours and 1200 hours on 9 September 1974, the line of storms travelled only some 100 km, but in the following six hours it advanced about 400 km. It then slowed again and moved northwards before covering 600 km in the next six hours, an exceptional movement. It finally dissipated as it crossed the coast.

These travelling storms are exciting phenomena to observe. The latent heat released may equate with 25 per cent of the total daily receipt

Table 20 *Line squalls in West Africa*

Periods	Storms generated over the ocean			Storms generated over the continent		
	I[a]	II[b]	III[c]	I[a]	II[b]	III[c]
Number of storms	1	5	8	63	43	56
Mean speed (kmhr⁻¹)	56	42	41	53	61	50
Mean distance travelled (km)	450	550	560	580	800	750
Mean duration (hours)	8	13	14	11	13	15

Notes: [a] 28 June to 16 July 1974; [b] 28 July to 15 August 1974; [c] 30 August to 19 September 1974.
Source: after Aspliden et al. 1976.

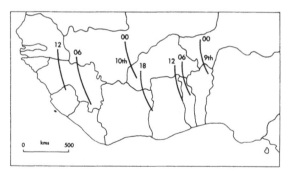

Figure 84 Passage of a line squall across West Africa

of insolation at the surface (Obasi, 1974a). Thus the storms play a major role in atmospheric energetics, and the partial conversion of this latent heat to eddy potential energy maintains the disturbances for the long periods noted. A romantic view of the fury of a passing line squall was quoted in the introduction to this book. A rather more scientific description was offered by a Nigerian meteorological observer in the early 1940s:

The day dawned quietly . . . landward visibility was moderate . . . (seawards) visibility was excellent. . . . The wind began to blow East, veering to North-east, Beaufort force one. . . . At about 08.54 G.M.T. the sky began to appear threatening . . . from the Eastern horizon and

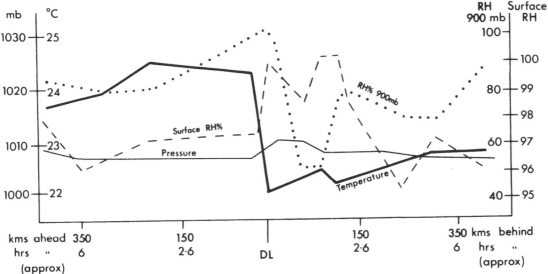

Figure 85 Atmospheric conditions associated with the passage of a line squall (after Obasi 1974a)

ominous rumblings proceeded from lumpy Cu–Nb below which scuds were observed. . . . The sunshine was pale. . . . The . . . rumblings approached nearer. There was a heavy downpour of rain at 09.45 G.M.T. followed by . . . lightning from the East and South. The visibility was reduced to 200 metres. The wind veered suddenly to North-east . . . force 7 . . . (which) tore down unripe coconut fruits . . . while some dilapidated . . . buildings were taken to pieces. . . . The sea lashed into foam and . . . fish were thrown on to the Marine Avenue . . . The storm reached its greatest intensity at 10.50 G.M.T. . . . At about 11.40 G.M.T. the flashes of lightning became less vivid. . . . The wind dropped to . . . force 3 . . . Precipitation was (intermittent) . . . (from) thin A–St. . . . The mountains and hills (became) clear of drifting mist. Rain gauged was 10.58 inches (269 mm) (Efiong, E. E. in Hamilton, 1943b).

Yet from this account are deleted comments upon the noise of the thunder, the severity of the rain and the beauties of the clearing sky.

The characteristic meteorological changes that occur at the surface with the passage of a line squall are neatly described by Obasi (1974a) and are summed up by him in Figure 85. It will be observed that in the example he studied the temperature rose from 23.7°C to 24.5° between six and three hours before the squall passed (assuming the speed of travel of the storm to have been the

mean of 58 km hr^{-1}). Just ahead of the storm the temperature was 24°C, but with the arrival of the squall it fell abruptly by 2°C. Falls of 5 to 6°C are not uncommon. With the drop in temperature came an associated and equally sudden rise in relative humidity, from 60 to 90 per cent followed by oscillations for some two and a half hours before it slumped to 40 per cent and then returned to the level experienced while the storm was still 350 kilometres away. Surface atmospheric pressure showed the least dramatic fluctuation although it still rose noticeably with the arrival of the storm. (In a severe squall at Kano on 12 June 1947 a wind increase from 32 to 129 km hr^{-1} and a temperature decrease from 32 to 19°C were noted (Dorrell, 1947)).

Line squalls and rainfall

Line squalls are particularly important in the climatology of the Sahel as they produce a high proportion of the rainfall of this area, but they are not insignificant in the south. Accra, for example, experiences some 22 line squalls per year, Navrongo 47, and these account for the greater part of the rainfall in April to May and September to October (Walker, 1962). Figure 86, which is self-explanatory, also illustrates this. Obasi (1974a) reports that 40 line squalls affected Lagos in 1962, producing 400 mm of rain, again the greater part just before and just after the main rainy season. Figure 87 illustrates this seasonality. The weather change from May to June and from

September to October and November is to be noted. In March, April and May the rainfall is essentially associated with easterly winds in excess of 40 km hr^{-1}, i.e. with line squalls, but in June, July and August most of the rainfall is associated with lower velocity winds from the west. However, as Ayoade and Akintola (1982) point out, there is little relationship between the number of rainstorms and total annual rainfall. They noted that at Ibadan in 1960 30 storms produced 1322 mm of rain, whereas in 1977 twice as many storms produced only 944 mm.

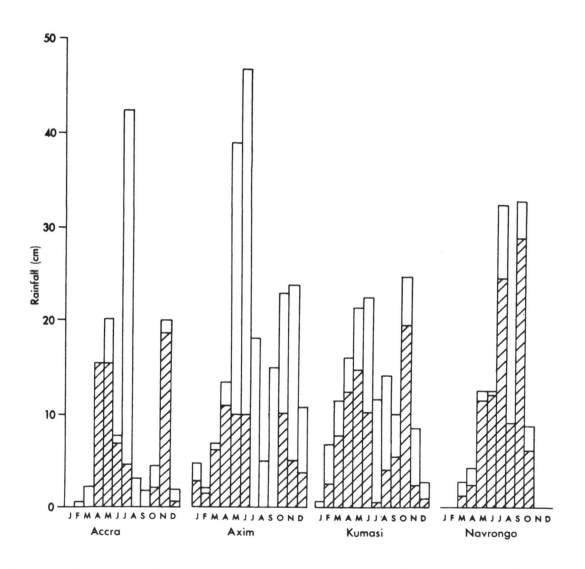

Figure 86   Monthly rainfall and the proportion of monthly rainfall due to line squalls (shaded) in 1955 at four stations in Ghana (after Eldridge 1957)

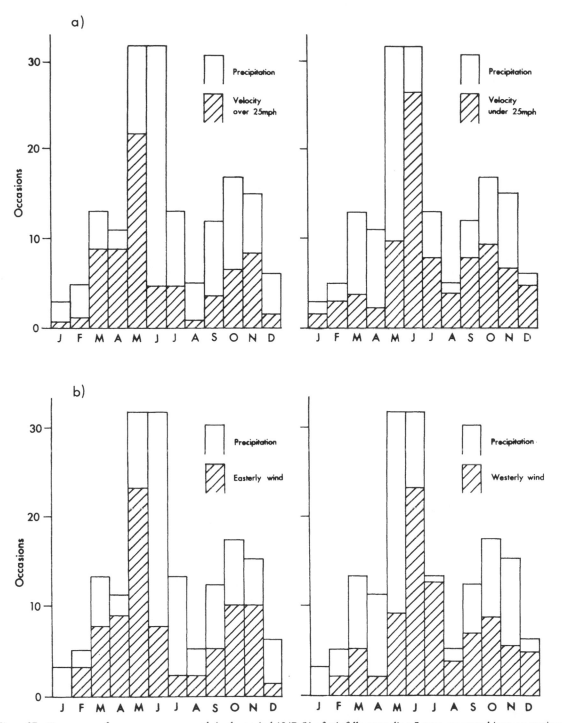

Figure 87 Frequency of occurrence per month in the period 1947–51 of rainfalls exceeding 5 mm, separated into *a* occasions when rainfall occurred in conjunction with winds greater or less than 40 km hr⁻¹; *b* occasions when rainfall occurred in conjunction with winds dominantly from east or west (after Swan 1958)

7   The seasonality of West African climate and climatic regions

Seasonality

The preceding chapters have quite clearly indicated marked seasonality in West Africa in terms of radiation and sunshine, temperature, pressure and wind patterns, in humidity and cloudiness, evaporation and precipitation, even the frequency of storms and in offshore marine conditions.

The basic seasonal variation in climate is summed up in Figure 88 (values rounded off), although this illustrates actual weather conditions at noon on two specific dates approximately six months apart. Contrasting situations are obviously indicated, perhaps seemingly thereby supporting the commonly held view that the climate of West Africa comprises two seasons, wet and dry. A close examination of the December and June maps and of the preceding six chapters will show that the seasonal pattern is, however, much less simple. At least four and sometimes five 'bands' or zones of distinctive weather are experienced over much of West Africa, and it is their passage across the region that determines the climatic seasons, which in some areas can number as many as eight.

The Inter-tropical Discontinuity (ITD)

The movement or migration of these weather areas is associated with northward and southward pulsations of a narrow zone of discontinuity between a dry, continental airmass to the north and a tropical, maritime airmass to the south. The range of movement was used to determine the northern limits of West Africa in the introduction to this book. The discontinuity is often no less abrupt than that at a typical cold front in temperate latitudes and has indeed been described in the past as the Inter-tropical Front (ITF). However, a lack of temperature contrasts on either side of the airmass boundary, especially in winter (see Figure 88a) and the lack of rain along the line of discontinuity, are sufficient to invalidate use of the term 'front' to describe the feature. Other common descriptive terms include Inter-tropical Convergence Zone (ITCZ) and Inter-tropical Discontinuity (ITD). Garnier (1967) expressed the view that none of these terms (ITF, ITCZ, ITD) is suitable, first, because of the lack of frontal activity, and second, because the discontinuity can separate airmasses that need not be of tropical origin. He suggested, therefore, the use of the term Surface Discontinuity (SD). Such a term, however, does not significantly distinguish this airmass boundary, which is usually within the tropics, from any other, and is only applicable to the boundary at the surface; yet the feature can be traced from the surface into the upper air. The term Tropical Discontinuity would seem to be adequate, but the common use of ITD and its apparent adoption by the World Meteorological Organization has encouraged the authors to apply Inter-tropical Discontinuity in this work.

The reasons for the north–south migration of the ITD are discussed in Part Two. Here its significance in relation to the patterns of weather characteristic of zones parallel to it is noted.

Weather zones in West Africa

Consider Figure 88a. The ITD lies far south in West Africa, so that the greater part of the region is under the influence of the northern, continental Saharan airmass. This being the northern hemisphere midwinter, temperatures north of the ITD

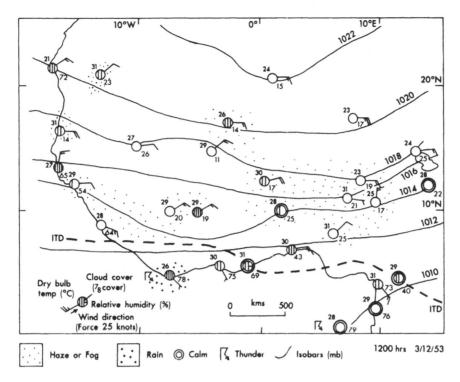

Figure 88a West African weather in December

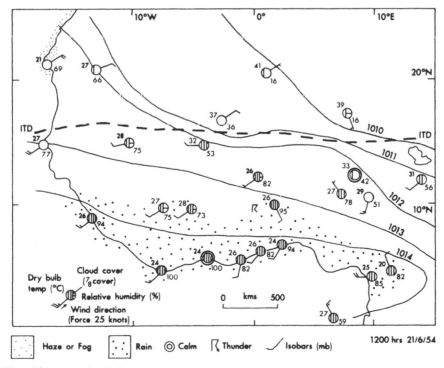

Figure 88b West African weather in June

range from about 30°C in the south to the low 20s in the north. More significant, perhaps, are the low relative humidity values; away from the coast these do not exceed 25 per cent and are as low as 11 per cent. Winds are light and from the east or north-east. There is little cloud, though dust haze is widespread.

South of the ITD the temperatures are much the same as to the immediate north, but humidity levels are in excess of 69 per cent, winds are from the south-east (they would more usually be from the south-west), skies are generally cloudy, but rain is only associated with a thunderstorm at one indicated locality. The sharp weather change at the ITD is obvious.

Six months later (Figure 88b) the maritime airmass from the Gulf of Guinea has pushed northwards so that its northern boundary, the ITD, lies at the surface at about 15°N. Nevertheless north of the ITD the conditions of humidity, cloud and wind remain little changed from the December situation. The temperatures, not surprisingly with the near zenithal sun and clear skies, are higher; the dust haze is less marked. Coastal fog is significant in the north-west.

Just south of the ITD and in a zone some 600 km wide, north to south, the weather is much the same as in December, though the winds are now markedly from the west. From 10°N southwards to the coast the weather is clearly different again, with temperatures in the mid 20s because of the overcast conditions and the widespread rain. Some of this precipitation, immediately south of the humid but rainless zone, is associated again with thunderstorms, but along the coast storms are absent. Humidity levels are particularly high. Towards equatorial latitudes, despite the cloud, rainfall again diminishes. Even this limited analysis of maps bearing limited information nevertheless suggests at least three and possibly four weather zones. These, effectively paralleling the ITD and the Guinea coast, have been identified by means of letters as follows.

Zone A: this is north of the ITD. There is little cloud, except possibly some high cirrus. Periods of dust haze occur. Vapour pressure is below 15 mb. The dry air plus low heat capacity of the ground create a large diurnal temperature range (days 35 to 40°C; nights 18 to 24°C). Winds are generally light in the morning, and east-north-easterly, with a common nocturnal temperature inversion near the ground, but strengthening towards midday with increasing convection.

Nevertheless, rainless conditions prevail.

Zone B: this is south of the ITD for some 320 km. Conditions are mainly rainless but humid (midday 50 to 60 per cent; nights even up to 100 per cent). Nocturnal temperatures are 20 to 24°C; daytime temperatures 29 to 35°C on the coast, 35 to 43° inland. Isolated thunderstorms in the late afternoon or evening can occur, but there is little rain. Cloud development is restricted. There may be some mist or stratus at night, but normally fair weather cumulus would be dominant. Overnight and morning winds are usually from the south-west, sometimes freshening during the day and becoming south-easterly. Sea breezes occur along driest coasts.

Zone C: this is south of Zone B in a belt perhaps 800 km wide, north to south. Temperatures and humidity are similar to those of Zone B, but daytime convectional activity is marked and local thunderstorms and westward moving line squalls are prominent. Rain therefore occurs in heavy showers. Overnight light south-westerly winds freshen during the day.

Zone D: this is south of Zone C for a further 300 km. Rain is an almost daily occurrence in showers commonly five or six hours long, but sometimes persisting all day. Mean winds are again invariably from the south-west and stronger than in Zones B or C. Thunderstorms are infrequent, especially in the west. Skies are cloudy and overcast, but two to three hours of sunshine may still be recorded on most days. Daytime temperatures may attain 26°C inland, 24°C on coasts; nights are 3 to 6°C cooler. Humidity is as in Zone C.

To these four zones may be added a fifth, *Zone E*, which affects only southernmost coastal areas in July and August. Conditions here are generally much less rainy than in Zone D, although cloud cover remains extensive. Temperature and humidity are as in Zone D, winds south-westerly and steady. The influence of this zone creates the 'little dry season' noted in Chapter 5.

Thus we have a five-zone model for West Africa. However, as the 'little dry season' is really only less wet than the 'deep' rains of Zone D, and as it is not experienced every year, even along the southernmost parts of West Africa, and rarely observed north of 8° latitude, it is perhaps preferable to adopt a four-zone model (A to D), regarding Zone E as a sub-zone of D.

The mean position of the ITD in December

determines that West Africa is virtually rainless at this time. The southernmost areas generally experience Zone B weather, whereas most of the region is under the influence of the harmattan, the air flow from the Sahara. From January to July there is a northward movement of the ITD, so that by late June Zone D is beginning to be felt in the south, which area has already perhaps experienced Zone A weather, certainly B and C weather in sequence. After August the pattern is reversed.

The movement of the ITD

The movement of the ITD is, of course, very variable within the overall repeating pattern. Clackson (1957, 1958, 1959) published diagrams for three consecutive years showing the day-by-day position of the ITD along longitude 3°E. In 1956 the southernmost position reached by the boundary was 5°N on 1 January, the northernmost about 23°N on 4 and 23 August. In 1958 the southernmost latitude reached was 6°N on 19

February, the northernmost 23° on 20 and 28 August.

The pattern for 1957 is reproduced as Figure 89. It shows the exceptional situation in mid-January when the ITD reached 2°N and on 1 August when it reached 25°N. The erratic day-by-day oscillations are a marked feature of the movement of the airmass boundary, as are some of the daily ranges. For example on 1 August from its northernmost position the ITD had within 24 hours moved south through 6° of latitude. Daily changes of twice this range can occur.

Figure 90 indicates the weather zones that were related to the migration of the ITD along 3°E in 1957. The ITD is drawn as a simple feature reaching to about 2500 metres altitude, but in reality the boundary is a more complex phenomenon. Clackson points out that as the north-easterly current overrides the airmass from the south, the latter wedging under the former, the southern air in the tip of the wedge is modified to some extent with regard to both temperature and humidity, by convectional mixing with the dry northern air

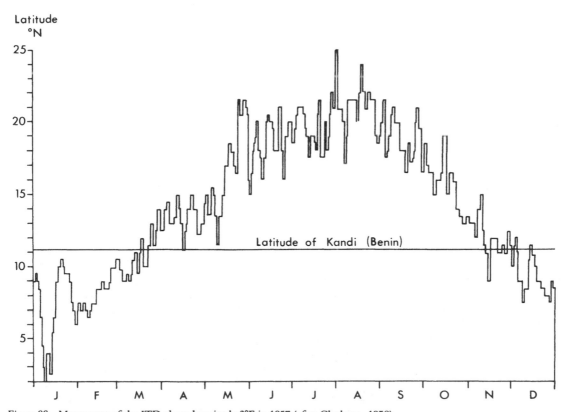

Figure 89 Movement of the ITD along longitude 3°E in 1957 (after Clackson, 1958)

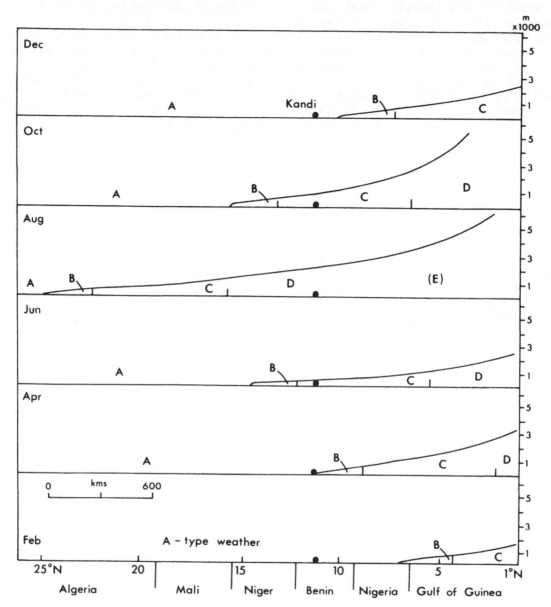

Figure 90 Location of the ITD along 3°E longitude, 1957

above. But the northern air at the surface just north of the ITD is modified much less. Clackson used the southern boundary of the air of northern origin, therefore, to delimit the ITD.

In Figure 90 the ITD is illustrated as a feature concave upwards. More commonly it is pictured as in Figure 38 (p. 58), the maritime tropical airmass wedging beneath the continental air with a convex front. The basis for this new presentation is research undertaken by the Bureau d'Etudes of the Agence de Securité de Naviga-

tion Aérienne en Afrique et Madagascar, Dakar, which published maps showing the mean position of the ITD at various altitudes expressed in millibars over West Africa in January and July (ASECNA, 1973). Sections drawn across these published maps along 3°E longitude, showed the ITD to have approximately the form illustrated in Figure 90.

Residents at Kandi (Northern Benin), for example, experienced only Zone A type weather in 1957 until about 16 March, when the first influences of the humid southern airmass were felt. By late May Zone B weather had given way to the thundery weather of Zone C. Throughout much of August the 'deep' rains of Zone D were present, but by mid-November the harmattan was again being experienced, and after 3 December, apart from one day, Kandi was permanently under the influence of the northern airmass. Thus seven seasons were encountered through the year: A, B, C, D, C, B, A at Kandi. Areas in south-western Nigeria probably experienced the 'little dry season' as well, so the sequence there was A, B, C, D, E, D, C, B.

An interesting feature of the seasonality of the West African climate that is brought out by Figure 90 is that the movement south of the ITD, at a mean rate of about $3\frac{1}{2}°$ latitude per month, is quicker than the movement northwards ($2\frac{1}{2}°$). Thus the storm season (Zone C), as noted above, is normally of shorter duration in the second half of the year than in the first half. If we regard Zone B as being approximately $2\frac{1}{2}°$ wide and Zone C $7°$ of latitude wide, and allowing for oscillations of the ITD, we can see that in 1957 at Kandi the storm season lasted approximately 77 days (11 May to 27 July) before the August rains, but only 57 days (26 August to 22 October) after the main rains.

Figure 62 (p. 85) shows the mean position of the ITD at its most southerly and most northerly locations. If a steady and regular movement of the ITD could be assumed (though Figure 89 above shows this to be a dangerous assumption) and if the B and C weather zones were of unvarying widths, then one could calculate for any locality in West Africa the percentage of each year in which the seasons would be experienced. So, for example, at Freetown, Sierra Leone, little A zone weather is actually encountered. The sequence of seasons is approximately as follows:

January to mid-February (i.e. 13 per cent of the year): mainly Zone B, occasional A.

Mid-February to early March (a further 5 per cent): Zone B.

Early March to early June (24 per cent): Zone C.

Early June to mid-October (30 per cent): Zone D.

Mid-October to mid-November (13 per cent): Zone C again.

Remainder of the year (14 per cent): Zone B repeated.

Thus for about 32 per cent of the year Freetown will experience the humid but dry weather typical of Zone B; 37 per cent of the year will have similar weather disturbed by 87 or so thunderstorm days in the period (see Chapter 6) and a further 30 per cent of the year, a single period from June to September, will produce about 78 per cent of the mean annual rainfall of 3207 mm.

The length of the rainy season

Despite the observations made at the start of this chapter, it is in terms of rainfall that seasonality is most frequently assessed, and numerous attempts have been made to define the rainy season, its commencement and end. A number of such attempts are noted below. Despite the day-to-day, north–south oscillations of the ITD, the fact that rainfall variability can be as pronounced as indicated in Chapter 5, and the failure of the rains in some years, as exemplified in the early 1970s in the Sahel, there is nevertheless a regularity of seasonal change so marked that it has encouraged many students to seek to pinpoint a particular date around which it can be safely assumed that the change in weather will occur at any locality, thus permitting the creation of maps showing isochrones of the onset and retreat of the rainy season across countries. Charre (1973) for example, claims that there is very little deviation from the dates indicating the start and end of the wet season at major towns in Niger noted in Table 21, even though rainfall amounts are very variable. If there is any validity in the table it shows interesting variation within one state, large as Niger is. The towns have been listed more or less in order from the most southwesterly located (Niamey) to the most northeasterly (Bilma), and thus the table shows the advance of the rains in that direction, though less clearly their retreat in the opposite direction. A very slow advance of the rains in the east is indicated. N'Guigmi is only 170 kilometres from Maine-Soroa but its rainy season commences 39 days later than at Maine.

Of course the terms 'wet season' and 'dry season' are highly subjective. What a citizen of Niger might describe as a wet period would probably seem like the opposite to a resident in Guinea or Liberia or southern Cameroon! The start and end of the rains means something different to Stern et al (1981), who quoted 21 June as the date for the commencement of the rains at

Tahoua, 10 June at Birni, 21 September for the end of the rains at Tahoua, 23 September at Birni. Clearly with the passage of the ITD (in one 24 hour period, 15 to 16 May 1956, it moved from about 12½°N to 20°N, a distance of 850 kilometres and with a mean speed of 35 km hr⁻¹) there can be very abrupt changes in weather characteristics (A to B type weather or vice versa), but the boundaries between B and C zones, or, more particularly, C and D, are much less obvious.

Stern et al noted that in the rainy season there is a mean interval of four days between successive rainy spells. Gregory (1965) defined the wet season as that sequence of months in which the mean rainfall is at least four inches (102 mm). Acheampong (1982), with reference to southern Ghana, assumed that a monthly rainfall of 102 mm is adequate to offset water loss by evapotranspiration. Any month with more than 102 mm is therefore 'wet', and the first and last months in the year with this amount of rainfall mark the limits of the rainy season. Thus, it is claimed, Axim has two wet seasons: March to July and October to November, whereas Accra has only one, from March to June.

Walter (1959b) chose the start of the rains to be when the accumulated total of two inches (51 mm) of rain had been reached, and the cessation of the rainy season to be when the reverse pattern

was identifiable. Oshodi (1971) noted that daily rainfall amounts are too variable and the pattern of rainfall too chaotic to permit the use of daily values to determine the start and end of the rains. On the other hand within a period as long as a month, far too many details of the rainfall pattern are lost. He therefore chose to use pentads, totalling rainfall for five-day periods in 30-day months; and 0.5 inches (12.7 mm) of rain within a pentad, when followed by two or more consecutive pentads also recording 0.5 ins, indicated the end of the dry season. A total of 1 inch (25.4 mm) of rain in similar periods indicated the start of the rainy season proper (the 'deep' rains). The reverse pattern was again used to determine the end of the main rains.

Griffiths (1960) required three pentads together totalling 1.5 inches (38.1 mm) of rain, with no one pentad having less than 0.3 inches (7.6 mm), to indicate the onset of the rains. Ilesanmi (1972a, 1972b) plotted cumulative percentages at five-day intervals and took the commencement of the rains to be the first point of maximum curvature on his graph, with the reverse situation denoting the end of the rainy season. Alternatively he expressed the onset of the rains with an accumulated 7 to 8 per cent of the annual rainfall, the rains ending after the accumulation of 90 per cent of the annual rainfall. Stern et al (1981) defined the start of the rains as the first occurrence of 10, 20 or 30 mm of rainfall within two successive days.

Applied to Figure 91 the rains started on 26 March if Walter's method is adopted, on 24 April after Oshodi, on 24 March (Griffiths), on 26 May (Ilesanmi) and on 24 March (Stern et al). At least three of the dates are effectively the same. The rains ended on 6 November (Walter), 6 November (Oshodi), 22 October (Ilesanmi). Only Oshodi distinguished between the 'rainy season' and the 'deep' rains, the latter commencing on 26 June and ending on 22 October.

The inadequacies of these methods, or indeed of others that might seem more precise and sophisticated because they incorporate the determination of pluviometric coefficients, are noted by Ayoade (1970). Pluviometric coefficients are obtained by expressing the mean monthly rainfall as a percentage of the expected rainfall if the annual precipitation were uniformly distributed throughout the year. Thus they are measures of departures from expected values if uniform distribution occurred. However the variations shown are still only departures from arbitrary

Table 21 *Start and end of the wet season in Niger*

	Date of start of wet season	Date of end of wet season	Length of season (days)
Niamey	3 May	9 October	161
Tillaberi	11 May	7 October	149
Birni	7 May	5 October	152
Maradi	7 May	3 October	149
Tahoua	7 May	6 October	154
Zinder	16 May	2 October	140
Maine-Soroa	20 May	4 October	137
N'Guigmi	28 June	15 September	80
Agadez	2 June	16 September	107
Bilma	May have no humid season at all		

Source: after Charre 1973.

values. Thus Ayoade proposed a new seasonality index:

$$SI = \frac{100}{22 \, Ry} \sum_{i=1}^{i=12} \left| 12 \, Ri - Ry \right| \text{ per cent}$$

where SI is the index, Ri is the mean monthly rainfall and Ry is the mean annual rainfall for the station. In this equation are summed the absolute deviations of the mean rainfall for each month from the expected monthly rainfall if the annual fall were uniformly distributed through the year. SI must range between 1 (maximum seasonality with all the rain in one month) to 0 (no seasonality, the rain being evenly distributed in all months).

Acquiring the data and calculating the index for many stations for a large area would be a major task. It is probably adequate simply to note the approximate periods in particular months as the times when seasonal changes occur, as has been attempted above for Freetown!

A year of weather at Freetown

A careful analysis of Figure 91 suggests that in 1969 the dry season (Zone B) was extended from early to late March, and the main rains (Zone D) commenced in mid-June, ending in late October. The writers have found for Sierra Leone at least that it is not unreasonable to define the 'deep' rains as that period within which there is no spell longer than two days without rain. If we are to select actual dates for the change of seasons at Freetown in 1969, based on the rainfall record alone, they would be: 23 March (Zones B and A giving way to C); 15 May (Zone C to D); 31 October (Zone D back to C); 20 November (Zone C back to B).

However these dates and the basis for choosing them hide the fact that there were seven line squalls in May and nine in June, the stormiest month. Thus the break between Zones C and D is far from precise. 'Deep' rain conditions had set in by the end of May (41 per cent of the month's rain came from line squalls), but in June storms still produced 23 per cent of the monthly rainfall total.

The seasonal changes are not so obvious from the humidity, temperature and sunshine records, though from about 24 March one notes a progressive tendency towards persistently high humidity levels, significantly lower temperatures and temperature ranges, and much less sunshine, until mid-August. Thereafter there is a steady return to dry season conditions.

A number of interesting features are revealed by the details of Figure 91. There was a marked relatively sunless period from 12 to 18 January. This was associated with winds from an easterly direction and heavy dust concentrations to ground level. Yet the humidity levels remained characteristic of Zone B weather, not of Zone A. There was no incursion of true harmattan conditions longer than five hours in the month, bringing the lower humidities noted on a few days. The humidity record for February might suggest that the harmattan was particularly strong in that month, especially from the 13th to the 19th. But again it should be noted that on those same days, and indeed throughout the entire month, at some time in the day the relative humidity levels were in excess of 90 per cent. In fact in the month easterly winds were recorded for only 10 hours. For only two hours in the month were relative humidity levels below 50 per cent associated with winds from the east. There were no periods in the month with intense dust haze. In fact February of 1969 was the windiest month in that year, with a mean speed of 3.6 knots. The astonishingly low humidity values and remarkable daily ranges, were not then associated with the harmattan but with overturning and turbulence along the ITD, rather than a clear movement of the boundary to a position south of Freetown. The abruptness with which the weather can change, however, is also brought out by the humidity patterns in the graph. Oscillations in the location of the ITD are suggested by the sudden return to a period of heavy rain in late October, just when the indications were that the rains were diminishing.

Another interesting feature indicated by Figure 91 is the level of sunshine, even on wet days, denoting the showery nature of the rainfall. Even on 11 July when 260 mm of rain were recorded, there were also nearly four hours of sunshine. July 1969, with 1295 mm was an exceptionally wet month, and it was unusual to have as many as 11 days without sunshine in the one month. The longest unbroken period of rainfall was 17 hours 30 minutes on 15 to 16 July.

Announcing with hindsight when the rains commenced is of little value to the farmer. It is obviously of much greater importance that the meteorological offices of the various West

African states are able to *forecast* rainfall distribution and quantities for local areas. To be able to do this, and bring the information to remotely located villages, will require a much larger investment in meteorological services in many states, but it would seem to be an appropriate investment in nations whose fundamental well-being rests in agriculture.

Summary

Seasonality in temporal terms has been strongly demonstrated in the preceding chapters in a number of histograms indicating the months of maximum and minimum values (Figures 19, 22, 47, 48, 53, 67, 68, 76, 81). In regional terms seasonality has also been expressed by noting the annual ranges of various climatic values, and is illustrated in Figures 13, 20, 21, 26, 45, 49 and 62. In summary they indicate the following:

1 The far north experiences the least seasonality in terms of rainfall, being dry throughout the year.
2 The south-west coastal region experiences the most marked seasonality in terms of rainfall, but also of sunshine.

3 A small area around Douala also experiences marked rainfall seasonality but very low seasonality in terms of sunshine, being the most persistently cloudy part of West Africa.
4 The Sahel experiences the greatest seasonal ranges of vapour pressure and relative humidity, but the lowest ranges in terms of sunshine.

The climatic regions of West Africa

Previous classifications of the West African climate

A number of attempts have been made to delimit climatic regions in West Africa. Some regions have been defined at very small scales within broad, global classification schemes such as the well-known systems of Köppen and Thornthwaite. (These and other less familiar systems have been amply described elsewhere, for example, by Gentilli (1958).) The complexities of such schemes, utilizing climatic data, often involving computation and the production of indices, make them cumbersome to use and apply, and by their results often imply the existence of discrete climatic regions that just cannot be so precisely defined, especially at the small scale.

Such climatic classification techniques have

(*Figure 91* continues on the next page)

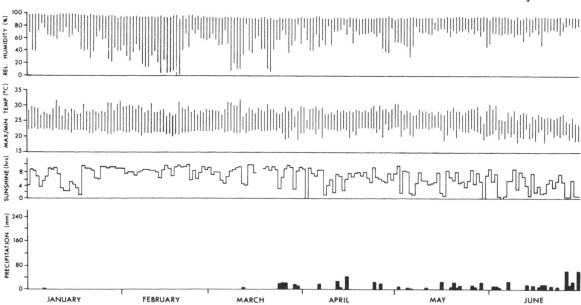

been more readily applied to countries within West Africa (see, for example, Ojo (1977) and his discussions of the climatic regions of Nigeria). For West Africa as a whole rather more subjective schemes have been applied, their basis being the grouping together of places having the same general climatic conditions at the same time period. (It is almost inevitable that seasonality is incorporated into climatic classifications.) Nevertheless there is usually a concentration on a particular climatic element. Thus Hubert in Grandidier (1934) gave priority to temperature, especially to seasons of maximum and minimum temperature, dividing West Africa into 14 regions, eight of which occupied neat, parallel east-to-west trending bands extending from the Sahara to the south coast, where the regions around Douala and the Togo Gap were separately delimited, as were also the north-west coast (north and south of Nouadhibou), the south-west facing coast and the Fouta Djalon and Guinea Highlands.

In the map of climatic regions compiled by Aubréville (1949), levels of precipitation are also considered along with the number of months with rain, the number of dry months, levels of water vapour, saturation deficit, mean annual temperature and temperature ranges; in consequence his map is complex. He has four main divisions: Saharien, Sahélien (with three subdivisions), Soudanien and Guinéen (with 11 subdivisions), plus the côte sénégalaise and the côte togolaise et dahoméenne.

Harrison Church (1961) utilized seasonal patterns of precipitation as the basis of his map of climatic regions in West Africa. Once again the pattern north of 8° latitude is a simple one of six parallel zones, and a very narrow north-west coastal strip, the south-west facing coast, the Fouta Djalon and the Togo Gap are again delimited. The remaining area, however, is subdivided in a more complex fashion into ten more regions.

The most recent attempt to produce a map of the climatic regions of West Africa is that of Leroux (1983). Based on six ranges of humidity, seven of temperature and seven of rainfall, he delimits three coastal zones: the north-west, the Libero–Guinean and the Guinea coast (described as the Domain of Permanent Atlantic Monsoon), inland of which are four west-to-east trending regions: Sahara tropical, Sahelian domain, North Sudanian and South Sudanian domains, with minor sub-regions such as the Fouta Djalon mountains.

All four authors cited seem to be faced with the problem of choosing names for their delimited regions. Some are reasonably precise: Guinea Foothills or Mauritanian coastal (Harrison Church); others are confusing: Baouléen or Béninien (Hubert), Equatorial (Harrison Church); others are vague: Savannah (Harrison Church), Soudanien (Aubréville, Leroux).

Figure 91 The weather at Kortright, Freetown, Sierra Leone, 1969

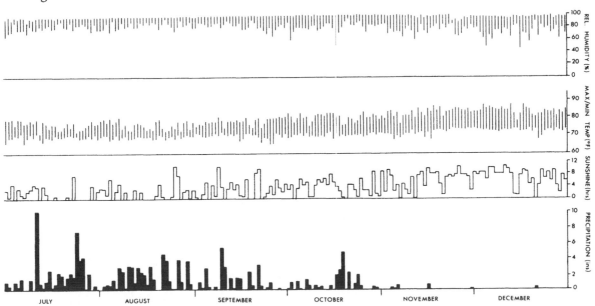

A new classification and map of climatic regions

This problem is avoided in the new classification outlined below, in which 10 major climatic regions are delimited, identified simply by numbers, subdivisions denoted by letters. In the system priority is given to rainfall, as this undoubtedly is the most significant element determining regional differences in West Africa, and not only in terms of climate but also in terms of flora and fauna, agriculture, even local architecture and basic ways of living. However seasonal patterns of temperature are not ignored.

The most fundamental fact about the regional climatology of West Africa is that the south is wetter than the north. As precipitation is related to the advance northwards and retreat southwards of the ITD across the region, it follows that the shorter the period in which an area lies south of the ITD, the less will be its rainfall. Thus West Africa is divided into regions primarily on the basis of length and dates of the rainy season, with further subdivisions based on mean annual rainfall using arbitrarily chosen values. Figure 92 illustrates the regional pattern, which is described below.

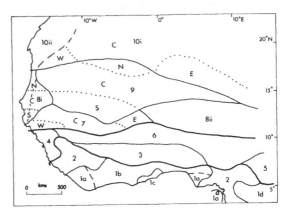

Figure 92 The climatic regions of West Africa

Regions 1 and 2
Rain (10 mm or more) in all months, i.e. no dry season proper.
Region 1
Delimited on the basis of a reduction in rainfall around August ('little dry season'), so that two main rainy periods are identified two months on either side of August. Nine to twelve months of the year are humid (Figure 50, p. 70).
Sub-region
  1a generally having over 2000 mm per annum.
  1b with 1200 to 2000 mm per year.
  1c anomalous drier zone (Togo Gap) having less than 1000 mm.
  1d wetter southern Cameroonian zone, 1500 to 2000 mm, with wettest months September to October.
Region 2
No 'little dry season'. Rainfall 1700 mm per year increasing towards the coast where it exceeds 3000 mm. Most stations with September or October as the wettest months.
Regions 3 to 10
Less than 10 mm of rain occurs in one or more months, i.e. there is a clear dry season. The

lack of an August hiatus means that there is a single clear rainy season.
Region 3
Rain in 8 to 9 months, March to October. 1000 to 1500 mm per year. Peak period July to September.
Region 4
Rain in 7 to 8 months, April to November, 2000 to over 4000 mm. Peak: August. Humid 8 to 12 months.
Region 5
Rain in 7 to 9 months, March to September, October or November. 1250 to 1600 mm. Peak July to August. Humid 6 to 9 months.
Region 6
7 months with rain, April to October. 1000 to 1500 mm. Peak clearly August.
Region 7
Rain in 6 months. May to October except on the coast where the period runs from June to November. Peak throughout in August. Highest mean temperatures in April to May.
Sub-region
7 *West*
  (Guinea-Bissau). Rains in June to November, 1500 to 2000 mm.
7 *Centre*
  1000 to 1500 mm.
7 *East*
  700 to 100 mm.
Region 8
Rain in 5 months. Peak is August.
Sub-region 8i
  Rain months are June to October.
8i South
  2000 mm in south diminishing to 1000 mm

in a distance of only 180 km northwards.

8i Centre

(Central Senegal) 1000 mm in the south declining to 500 mm over 170 km northwards.

8i North

500 to 100 mm only.

8i Coast

A coastal zone, 100 to 150 km wide, can be delimited where highest mean temperatures are attained a month later than in the east of this region, due to the cooling effect of northerly winds off a relatively cold ocean.

Sub-region 8ii

Rainy months are May to September. Rainfall ranges from 1000 mm in the south to 250 mm in the north. Highest mean temperatures are experienced in April to May.

Region 9

Rain in June to September. Peak in August. Highest mean temperatures in May to June.

Sub-region

9 South

1000 to 500 mm.

9 Centre

500 to 100 mm.

9 North

Less than 100 mm.

Region 10

Rain in 2 to 3 months only, July to September. Peak August (but September at Tamanrasset). Highest mean temperatures in July to October.

Sub-region 10i East

200 mm in south, 100 mm in north.

10i Centre

Under 100 mm.

10i West

Over 100 mm.

Sub-region 10ii

(north-west coast). August start to slight rains. Peak September in the south, progressively later northwards. Less than 100 mm yr^{-1}.

Regions 8i and 10ii are also set apart as the windiest in West Africa and by the dominance of northerly winds.

The northern boundary of Region 6 approximately coincides with the location of the line used in Figures 20 and 21 (p. 39) to delimit northern from southern West Africa in terms of seasons of maximum and minimum temperatures. Thus climatic regions 1 to 6 have their highest mean temperatures before the main rains and the mid-year period (the rains) is the coolest. Greatest seasonality in terms of radiation, sunshine and rainfall is also experienced in this southern zone, as noted above. Regions 7 to 10 comprise northern West Africa and, in contrast to the south, have their lowest mean temperatures in the northern hemisphere winter months, their highest temperatures during the wetter part of the year. Regions 8ii and 9 have the greatest seasonality in terms of vapour pressure and relative humidity, least seasonality in terms of sunshine. Region 9 experiences the highest temperatures in West Africa and region 10 has the least seasonality in terms of rainfall.

PART TWO

Explanation *West African Climatology and Meteorology*

8   The atmospheric circulation over West Africa

The surface wind flow

The essential pattern of surface air motion over West Africa has been illustrated in Figures 28 and 29. (pp. 46–48). Explanations for this pattern are now considered with reference to surface pressure gradients, vorticity and horizontal motion.

Surface pressure gradients

The atmosphere is a fluid medium under the two basic forces of gravity and friction. The former accelerates the atmospheric gases towards the earth's surface, so that they exert a pressure upon the surface. If that pressure is not uniformly distributed there will be an air flow from the regions of higher pressure to those of lower. There are thus horizontal as well as vertical pressure gradients.

The most common cause of a horizontal pressure gradient is differential heating of the global surface and, therefore, of the air immediately above it. On a non-rotating world of uniform surface there would tend to be a simple flow of surface air from cooler, higher, latitudes towards the warmer lower latitudes, where the convergent flow from the two hemispheres would have to rise and circulate back at high altitudes towards the polar regions, where it would sink to maintain two simple hemispherical cells of flow. Such patterns are known as Hadley cells after the eighteenth century scientist who first postulated their existence in the tropics (Figure 93).

There are two basic reasons why such a simple, global, atmospheric circulation does not exist. One is the variation in, and thus differential heating of, the earth's surface. Even the temperatures, densities, colours and albedos of the oceans

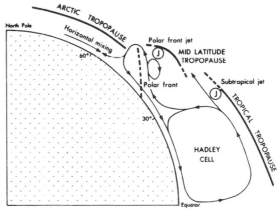

Figure 93 Latitudinal air movement and the Hadley cell (after Palmen 1951)

are not uniform, and variation on land and marine-terrestrial contrasts are obvious. Over warmer surfaces air will tend to rise, over colder to sink.

Over West Africa in the northern hemisphere winter net radiation levels are generally below 225 langleys and albedos are high (Figures 10a, p. 25 and 7b, p. 21). So, although sunshine amounts are also high, mean monthly temperatures across the region tend not to exceed 26°C in the south and are as low as 18°C in the north (Figure 17a, p. 35). On the other hand sea surface temperatures in the Gulf of Guinea, which tend not to fluctuate beyond 2 or 3°C round the year, are generally in excess of 27°C. Thus pressure tends to be lower over the Gulf (under 1010 mb) than over West Africa, where it increases northwards from 1010 mb at the coast to 1016 mb and higher in the north (Figure 25a, p. 42). Thus surface air flow is out from the North African high pressure area towards the lower pressure areas prevailing over the ocean. (Commonly, however, this northerly air flow fails to reach West Africa south of 7 or 8°N latitude as the ITD does not generally move farther south than this in the region as seen, because of the north to south land – sea temperature and pressure contrasts. From central Africa eastwards to the central Pacific Ocean, however, the ITD generally lies from 10°S to 20°S in January.)

In the northern hemisphere summer the pattern is reversed. By 21 March each year the zenithal sun, on its apparent northward journey from its December location over the Tropic of Capricorn, reaches the equator. By 9 April (Figure 21, p. 39) it reaches 5°N and thus West Africa. By 29 April it is overhead at noon along 10°N; by 19 May at 15°N; by 21 June attaining its most northerly position at 23½°N. With the advancing sun comes a rise in net radiation and surface air temperature in northern West Africa. Mean temperatures at Arouane in northern Mali that are only about 18°C in January increase to 37°C by June. South of about 12°N the highest mean temperatures occur about one month before the date of the overhead sun because of the mid-year increase in cloud cover.

The rising temperatures inland in West Africa bring about a replacement of the high pressure zone of the dry season by one of low pressure relative to that over the Gulf of Guinea (Figures 25c, p. 42 and 29b, p. 48). Thus the air flow progressively strengthens from the south and spreads across the region. However the movement of air at the surface is not directly from north to south in 'winter' or the reverse in 'summer'. The hemispherical Hadley cell described above is not replaced by a local West African version. Figures 28a and c (pp. 46 and 47) indicate that winds tend to be dominantly from the north-east in 'winter' and from the south-west in 'summer', blowing obliquely across the isobars. Hence the second basic reason for the absence of the simple pattern.

Vorticity and horizontal motion

The Hadley cell, as described, is appropriate to a non-rotating globe. However, the real world does rotate and the atmosphere with it. The horizontal component of the rotation is of great importance in meteorology and climatology. Imagine a small parcel of air resting precisely on The North Pole (Figure 94). The pole is the end point of the axis of the rotating earth. Thus the air parcel will also rotate, once every 24 hours.

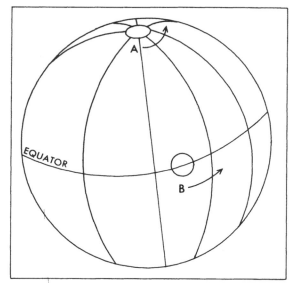

Figure 94 Vorticity and horizontal air motion

It possesses vorticity; a similar parcel on the equator, however, will have no vorticity at all. The rate of rotation varies with the sine of the latitude, progressively increasing from equator to pole.

If a parcel or column of air is also rotating independently it will possess positive vorticity if rotating in the same direction as the earth and negative vorticity if rotating in the opposite direction.

On the other hand, the equatorially sited parcel of air, although not rotating, moves around the axis with a velocity of some 1670 km hr⁻¹, the earth rotating once in 24 hours and its equatorial circumference being 40,075 km, whereas the polar sited air does not have this horizontal motion.

Consider now the low-latitude parcel moving from equatorial latitudes along a course due north. Because it is freely moving it will now have two velocity components, that towards the pole and that at right angles to the northward track with the original velocity of 1670 km hr⁻¹. As the air moves northwards it will pass into latitudes where the surface beneath is moving around the polar axis at speeds lower than 1670 km hr⁻¹. In accordance with the principle of the conservation of angular momentum the air parcel will appear to move steadily sideways, that is to the right in the northern hemisphere viewed from above. A particle, or the hypothetical parcel of air, moving from polar latitudes southwards, will possess little circumpolar velocity. Therefore as it moves equatorwards into latitudes where the speed of motion of the earth's surface around the axis is considerably increasing, the parcel will appear to be 'left behind' as the surface moves quickly eastwards beneath it. The impression to an earthbound observer is again, therefore, that the airflow is being diverted to the right and the wind that is actually northerly is described as north-easterly.

In the southern hemisphere the apparent deflection of the airflow is to the left. This deviation is known as the Coriolis effect. Airflow (wind) on a non-rotating globe would be along the pressure gradient, i.e. at right angles to the surface isobars. The Coriolis effect causes the winds to blow at the surface, where frictional forces also operate, obliquely across the isobars, commonly at an angle of 10 to 20° over the oceans, 25 to 35° over land.

The seasonal climatic changes in West Africa can obviously be related to this seasonal pattern of surface winds, but only in part. Much of the explanation for West Africa's climate and weather is to be found in the upper air.

The upper air over West Africa

The frictional effects of the earth's surface on air flow are often restricted to the lower 1000 metres of the atmosphere, though sometimes observed up to 2 km. With increasing altitude and diminishing friction, both wind speed and the Coriolis effect may increase. The point may be reached where the horizontal component of the Coriolis effect balances the pressure gradient force, so that the wind no longer flows across the isobars but actually parallel to them. In these conditions it can be appreciated that, looking downstream in the northern hemisphere, the lower pressure will be to the left of the wind flow, the higher pressure to the right. Such a wind is termed geostrophic and its velocity is provided by the formula:

$$V = \frac{1}{2\upsilon \sin\phi} \cdot \frac{dp}{dn}$$

where dp/dn is the pressure gradient, $2\upsilon \sin\phi$ is the Coriolis effect (υ being the angular velocity of spin and ϕ the latitude).

The winter situation

In the northern hemisphere winter surface north-easterlies are dominant over most of West Africa. Flow is essentially *from* a high pressure cell, or anticyclone, over northern Africa. It is therefore divergent. Deflection of the air movement to the right, for reasons noted above, produces a clockwise circulation of the air out of the anticyclone. The surface air that moves from the high pressure zone must be replaced by an inward flow aloft, a flow with an anticlockwise rotation. It may help, perhaps, to envisage the atmosphere over northern West Africa as a vast, vertically-standing cylinder, into which the air is entrained with an anticlockwise flow at the top and sides, to sink downwards with cyclonic curvature, only to reverse direction at about 4000 to 5000 metres altitude and spiral outwards with a clockwise flow near or at the surface.

This simple picture of air movement over West Africa in January cannot, of course, be isolated from the general atmospheric circulation of the earth, especially of low latitudes. Over the tropical oceans, where the land-marine contrasts associated with West Africa do not occur, the thermal equator (the ITD) is a zone of convergent air from north and south, very similar not only in temperature but also in humidity. There is thus no tendency for either airmass, from north or south, to override the other. (Over the Pacific Ocean there are often two ITDs, one 5 to 12° north of the equator, the other 6° south, separated by a rainless zone along the equator where the Peruvian Current and upwelling cool waters produce lower air temperatures and thus diminish cloud development.) The convergent surface airflow tends to rise in a marked convective layer, perhaps 12,000 m high, from the top of which it then drifts out to approximately 30°N and S, where subsidence then occurs to produce the sub-tropical oceanic anticyclones. It is this subsiding air, warming adiabatically as it sinks, that prevents significant convectional uplift from the seas below, thus prohibiting precipitation in these areas.

The pattern described here is the classic Hadley cell and within it, were it not for the Coriolis effect, the circulation would consist in the northern hemisphere, of surface winds blowing due south from about 30°N and upper winds blowing in the reverse direction. The Coriolis effect, however, will make the northerly surface winds north-east trades and the counter trades aloft *westerlies*. Because of the diminishing effect of friction with altitude, so the high-level winds flow geostrophically as westerlies, not south-westerlies.

The sub-tropical jet stream (STJ)

At about 12,000 m altitude and at 20 to 30°N latitude the winds also travel with significantly greater velocity than at lower levels where friction is operative. They form the *sub-tropical jet stream* (STJ). These high-level westerlies and the STJ at the 200 mb level (12,000 m) pass from the Atlantic in the northern hemisphere winter over

Figure 95 Mean meridional cross-section of actual temperature for 1 January 1956 (°C) (after Defant and Taba 1957)

Note: TPT – Tropical Tropopause; MT – Middle Tropopause; J – Jet cores.

West Africa, attaining speeds of 35 ms⁻¹ (126 km hr⁻¹; 68 knots) over the coast of Mauritania, increasing to 40 ms⁻¹ (144 km hr⁻¹; 78 knots) over Mali, with convergence into the high-level low-pressure area lying above the surface anti-cyclone. The surface high-pressure cell is fed from above and spreads its influence almost to the Guinea coast as the harmattan.

There are two main reasons why the winds at 200 mb and 20 to 30° latitude blow with greater velocity than the winds around them. One of these relates to the principle of the conservation of angular momentum mentioned above. The northward drifting airflow associated with the upper part of the Hadley cell circulation progressively moves into latitudes where its velocity increases relative to that of the rotating earth's surface. Thus towards the northernmost limits of the Hadley cell aloft, westerly wind velocities are

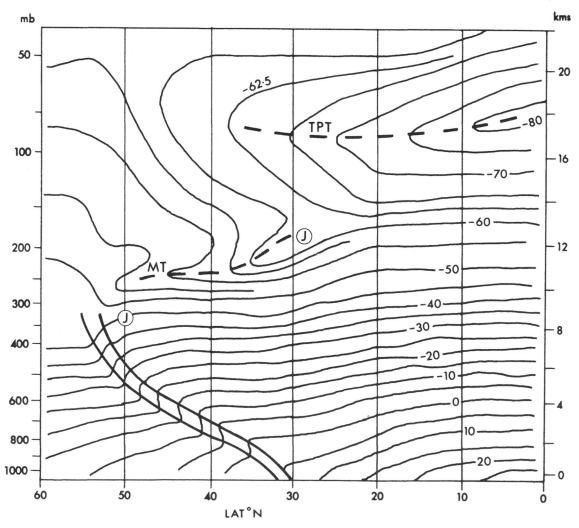

increased relative to the flow beneath where sinking and divergence is taking place.

The second reason relates to temperature and pressure gradients aloft. Readers will be aware that the earth's atmosphere is thermally stratified. The mean situation is that air temperatures tend to decrease with altitude up to about 8 to 11 km in high- and mid-latitudes and about 14 to 18 km in the tropics, within a layer known as the troposphere. At these altitudes temperatures will tend to be in the vicinity of −55°C. (Thereafter in the stratosphere they remain steady to about 35 km, then rise to nearly +80°C at 55 km, falling again through the mesosphere to near −40°C at 80 km, then rising beyond in the thermosphere even to levels in excess of 1000°C.) About 75 per cent of the earth's atmosphere is compressed into the troposphere, the upper thermal boundary of which is termed the tropopause. The tropical tropopause, at some 16 km over equatorial latitudes, is thus separated at the northerly limits of the Hadley cell from the mid-latitude tropopause at about 12 km by a sharp break in height (Figure 95). There is thus a marked south-to-north temperature gradient at about 12 km altitude just south of the mid-latitude tropopause. Such a gradient will also create a significant horizontal pressure gradient across which winds would tend to flow more strongly. However these winds, for reasons already noted, become geostrophic, forming a prominent flow from west to east, the STJ, in the northern hemisphere winter.

This jet characteristically passes around the earth in a series of three long and rather symmetrical waves, the ridges (northward bends) extending to 35°N, even 40°N, over south-east Asia, eastern North America and north-east Africa, and having the strongest winds of up to 77 ms⁻¹ (278 km hr⁻¹; 150 knots; 173 mph), with a mean value over Africa of about 45 ms⁻¹ (160 km hr⁻¹; 87 knots; 100 mph). The troughs (or southward bends) may reach 20 to 15°N over the Pacific Ocean, the Atlantic and the Arabian Sea and India, with maximum velocities perhaps not exceeding 51 ms⁻¹ (185 km hr⁻¹; 100 knots). The jet stream is part of the dynamic circulation of the atmosphere and must occur on a rotating earth. However, it is not unrelated to conditions beneath, as seen, and it plays a major role in determining surface climate, as the next chapter will confirm.

The jet velocity will increase where contrasting airflows are most prominently brought together, that is, where baroclinicity is therefore strongest. These areas tend to be especially on the north-west flanks of the upper-air high pressure cells where the carriage of warmer air out of the tropics to higher latitudes is most marked. One such area, as noted, is over north-eastern Africa. Furthermore, the pattern of upper air isotherms is such that the westerly flow tends to be directed slightly northwards in the same area.

In the light of the comments above concerning vorticity, consider what happens to the air as it flows away from the equator. Not only will it find itself passing into latitudes where the earth beneath rotates about its axis more slowly than at the equator, and accelerating relative to that diminished earth surface rotation, but also this northward tracking air is moving into latitudes where the surrounding air possesses more vorticity. The principle of the conservation of vorticity applies as does that of the conservation of angular momentum. The tendency, therefore, is for the northward moving air to swing back southwards. Having gained some vorticity, however, on its passage north, on coming again to lower latitudes the air finds itself possessing more vorticity than it should have at those latitudes and so it swings northwards again. Hence the north-to-south wave flow of the STJ about the earth. Unlike jets in higher latitudes, however, because of the general lack of vorticity in the tropics, the waves in the STJ are long and gentle and the flow steady.

The northward and southward swings of the jet have significance for the weather and climate beneath. It will be appreciated that at the region where the jet stream approaches its most southerly position, the flow of the various strands which make up the stream will be convergent especially to the right, lower-latitude side of the jet, and divergent to the left (see streamlines in Figure 96), and that the convergence will be accompanied by a deceleration in stream flow. Convergence and deceleration aloft will produce downflow, sinking of air and thus outflow, that is anticyclonic high pressure conditions, at the surface below. Such a situation prevails over north-west Africa (Figures 25a, p. 42 and 29a, p. 48).

As the jet stream swings northwards, however, it starts to accelerate, conserving angular momentum. Greatest velocities will be reached as the jet commences to swing south again. At the approach area to maximum velocity (the entrance zone of the jet) convergence will be strong on the left, north, side of the flow, but, with an ever increasing velocity, the air particles ahead

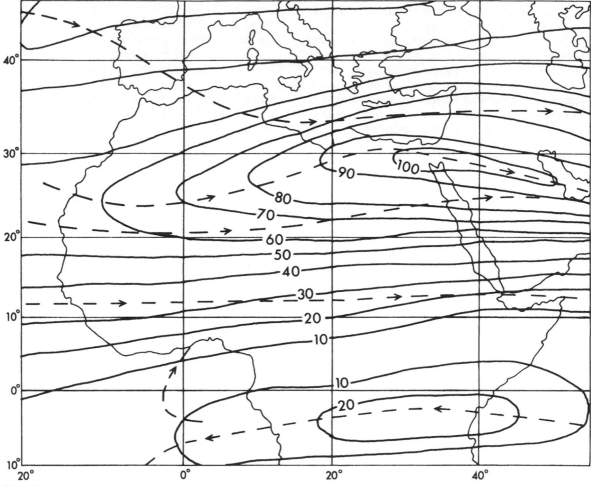

Figure 96   Mean winds at 200 mb (12,000 m) over North Africa in January (after Air Ministry 1962)

Note: Isotachs at 10 kt intervals; streamlines shown as pecked lines.

pulling away from those behind, there will occur increasing divergence in the jet flow. Divergence aloft will mean associated uplift from below, meaning relative low pressure there. Such conditions prevail over the eastern Mediterranean in winter.

The summer situation

With rising air temperatures in West Africa comes a replacement of surface high pressure conditions by low pressure relative to that over the Gulf of Guinea. At the same time there is a slight increase in mean surface pressure in the region of the St Helena anticyclone in the south Atlantic. Inexorably, south-westerly winds are drawn across West Africa as the low

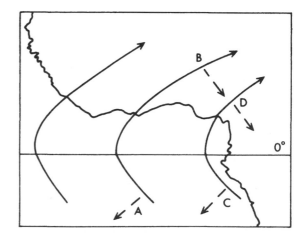

Figure 97   Convergent airflow over West Africa from the south-west

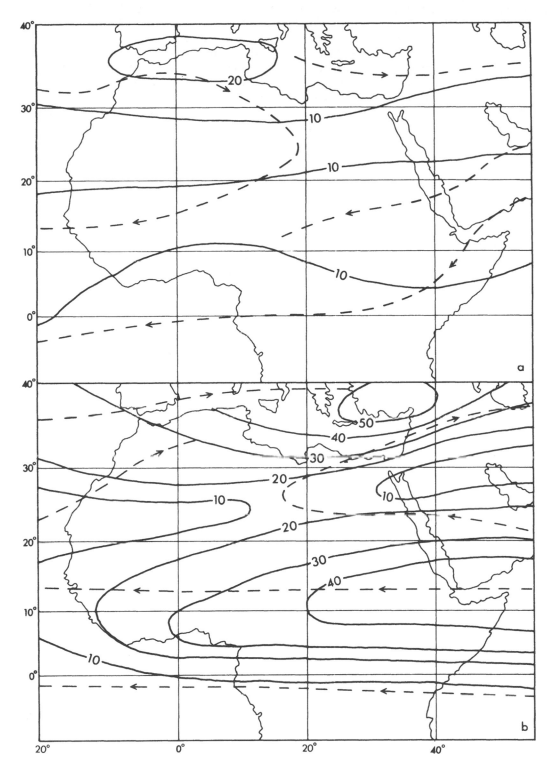

Figure 98 Mean winds in July at *a* 500 mb (5500 m); *b* 200 mb (12,000 m) (streamlines pecked; isotachs at 10 kt intervals) (after Air Ministry 1962)

there intensifies. The leading edge of the maritime airmass, the ITD, advances northwards, following the sun but with a time-lag of some six weeks and with typical diurnal oscillations, moving north in the afternoons, south in the mornings, perhaps by as much as 200 km, and with longer term surges. As the north African low pressure intensifies, relative high pressure may build over the Mediterranean to the north and the Azores anti-cyclone also becomes more clearly developed.

There is thus a tendency to convergent airflow from south-west and north-east into northern Africa near the surface, this enhanced by two further factors. One is the increased friction on the surface airflow as it passes from ocean to continent, a diminished flow, as noted above, leading to convergence. Second, as the air from the Gulf of Guinea moves north-eastwards it will be subject to an increased Coriolis effect. In a south-westerly airstream the left side of the flow will be further north than the parallel flowing right side, and will thus turn further to the right (east) to converge on the right side (Figure 97).

Convergence at the surface should lead to ascent and divergence aloft. The low pressure centre over the Sahara is overlain by a high-level anticyclone from which, therefore, the air flows with a clockwise spiral. On the south, West African, side of the anticyclone the winds will thus flow from east to west, part of a circumglobal easterly flow in the northern hemisphere at this time.

Towards the middle of the year with the continuing advance of the maritime airmass from the south, so the baroclinic and barotropic gradients from south to north across West Africa increase and the upper easterlies become the Easterly Tropical Jet (ETJ) (Figure 98). At the 100 mb level (15,000 to 20,000 m) wind velocities can be in excess of 20 ms^{-1} (72 km hr^{-1}; 39 knots; 45 mph). Near the 500 mb level (c 5000 m) in a rather broader band from about 12 or 13°N to 20°N, easterly wind speeds are also commonly more than 10 ms^{-1}. Though the weaker of the two flows, it is the lower, the African Easterly Jet (AEJ), that is particularly important in West African climatology.

9 Perturbations in the West African atmosphere and the origin of line squalls

Waves in the south-westerlies

The various positions of the Inter-tropical Discontinuity over West Africa are related not only to pressure changes within the region but also, as noted, to those further afield, even in the southern hemisphere. In turn fluctuations in such areas as the Azores, the Mediterranean, East Africa and St Helena, reflect atmospheric activity along the polar fronts. Thus developments in West Africa cannot be disassociated from conditions 'outside'. It is not surprising therefore that the advance and retreat of the ITD across West Africa is frequently erratic. At times it may surge forwards many degrees of latitude in a few hours; some years it may hold to a southerly location and the rains associated with its advance then fail across much of the region with disastrous consequences.

For reasons that are still little understood, meridional oscillations of the monsoon trough, which is the low-pressure zone associated with the ITD, seem to occur with periodicities of four to six days, and during poleward surges within deeper layers of humid air (Zone D) perturbations (waves) originate at lower levels (850 mb) and spread to 700 mb, over a six to twelve hour period (Adefolalu, 1977). Bands of often heavy and prolonged rain, some 160 km long west to east, 50 to 80 km north to south, associated with these perturbations, travel with the south-westerly airflow across West Africa from the Guinea coast, diminishing in significance as the depth of humid air decreases. They are the 'summer monsoon' of the region. Surface convergent airflow and topographic barriers, such as the Freetown Peninsula mountains, can help to produce in excess of 250 mm of rain in a day in extreme conditions and commonly 120 mm may be recorded.

Waves in the upper easterlies

Of even greater significance for the climate of West Africa, as they affect possibly the whole of the region, are perturbations most marked in the easterlies overlying Zone C. Northward advances of the maritime tropical (mT) air from the Atlantic Ocean lead, as we have seen, to an increase in baroclinicity and barotropicity across the region, and to the development of the AEJ. The baroclinic exchange entails conversion of eddy potential energy to eddy kinetic energy through the jet. The shearing that takes place at about the 700 mb level reaches a maximum when the AEJ is most developed and when the ITD has advanced sufficiently far northwards to introduce a level of vorticity into the surging airflow. The shearing effect on the boundary of contrasting fluid mediums moving by each other, as can be observed as the wind blows over water surfaces, creates wave motions. The release of latent heat in the development of such waves in the atmosphere may aid their growth. The waves in turn transport released heat and feed further atmospheric disturbances.

Characteristically, therefore, in the area between Khartoum and N'Djamena, where the upper easterly airflow commences to accelerate to become the AEJ over West Africa, waves develop south of the core of the jet at about 700 mb where the mean zonal flow is least stable; in effect in the base of the dry easterlies where they shear against the underlying humid south-westerlies, where horizontal and vertical shears are most marked, providing energy for the waves. These waves, or perturbations, with lengths between 1500 and 4000 km, typically cross West Africa from mid-June to early October, occurring at intervals of three to five days. They travel with a mean speed of 5 to 10 ms^{-1} (Burpee, 1972, 1974). Associated particularly with Zone C, the waves and the disturbed weather associated with them affect southern latitudes in West Africa early and late in the year and northern latitudes, even as far north as 30°N, in the middle of the year. They are especially significant for regional climate near 14°N.

Figure 99 (a to f) shows how these waves appear as north-to-south undulations in isotherms and streamlines, and shows the passage of waves right across West Africa. They travel some 5 to 10° longitude per day, slowing near the west coast, but often taking on cyclonic and closed circulations. Some 50 such disturbances per year pass over Dakar, and on average four may continue to intensify into major circular storms over the Atlantic. In fact the centres D and E in Figure 99 developed into the hurricanes codenamed Debbie and Esther in the Caribbean in September 1961; 60 per cent of West Indian cyclones originate in West African perturbations.

The dramatic consequences for surface weather of these disturbances are discussed below, but it may be noted now that to the west of the wave axes, ahead of the downward bulges in the base of the undulating easterlies, strong updraughts of 1 to 2 ms^{-1} can occur. Large vertical and lateral shears along the jet flow, well-marked troughs and vortices in the lower air layers, an ample moisture supply in a sufficiently deep humid airmass and the updraughts are all guarantees of precipitation.

The development of the West African line squall

It is probably true to say that waves can occur in the easterlies without the development of storms or even much cloud, but that there are no line squalls without the perturbations brought about by baroclinic and barotropic instability. One theory for the formation of the line squall is suggested in Figure 100 (after Leroux, 1976). In phase 1 the upper, dry easterlies ride smoothly over the advancing humid south-westerlies. Not surprisingly, phase 2, shearing along the humidity boundary that is the ITD, leads to the creation of an asymmetric wave, to compression therefore of the mT airmass at the point marked X, and to increased airflow above and below the developing wave. This in turn will exacerbate the wave development.

The wave may sometimes not grow beyond this stage. Given an active jet flow aloft, however, phase 4, the 'phase of blocking' (Leroux, 1976), may be reached. The temporary blocking of the advancing easterlies at Y will cause them to spread north and south of their central east-to-west axis, thus leading to the ultimate development of a curved north–south line of thunderstorms when phase 5 is established. Here the easterly flow has now broken through at the surface to create the squall that is associated with the arrival of the storm and which follows the calm that will occur at Z. The ascending westerlies ahead of the advancing easterlies will produce the heavy rain also characteristic of these travelling storms.

Phase 5 illustrates the fully-mature thunderstorm, cumulus and cumulo-nimbus cloud growing in the ascending mT air, with a downflow at the rear of the storm. The storm will advance or grow ahead of the frontal situation now existing, not only because the perturbation moves westwards in the base of the upper easterlies, but also because the ascent of air ahead of the 'front' produces lower pressure conditions there into which the descending air at the back of the storm will move. The passage of such a storm will produce the local weather phenomena described in Chapter 6.

The main sources of line squalls are at the most marked confluence zones of advancing air from the Gulf of Guinea and of cT air circulating around the North African upper-air anticyclone. This situation probably occurs most significantly in the vicinity of Lake Chad where the lower levels of the easterlies are channelled between the Tibesti and Darfur mountains, and north-west of the inland Delta of the Niger, where the easterly overland limits of the air moving round the Azores anticyclone meet the humid south-westerlies. As noted above (and in Chapter 6) the lines of squalls generated in the east may traverse the whole West African region. Some may dissipate over the cooler ocean as they reach the Atlantic. Yet others, for reasons beyond the scope of this book, may grow to become the hurricanes of the Caribbean and the Gulf of Mexico, as noted above with reference to Figure 99. Furthermore, dying remnants of some of these violent tropical storms, moving progressively northwards and then eastwards in the higher latitudes, may bring severe weather to north–eastern USA and even, back again across the Atlantic Ocean, to north-west Europe.

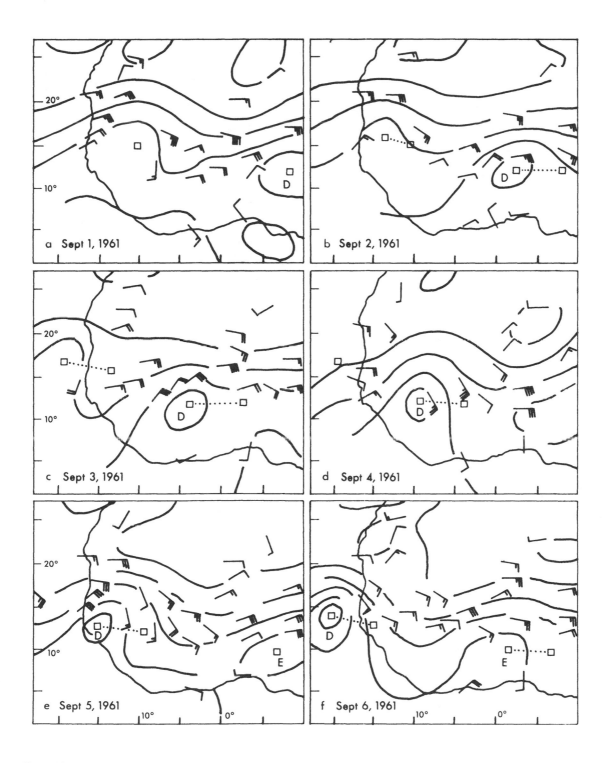

Figure 99 Waves in the easterlies at 700 mb, 1–6 September 1961 (after Erickson, 1963)

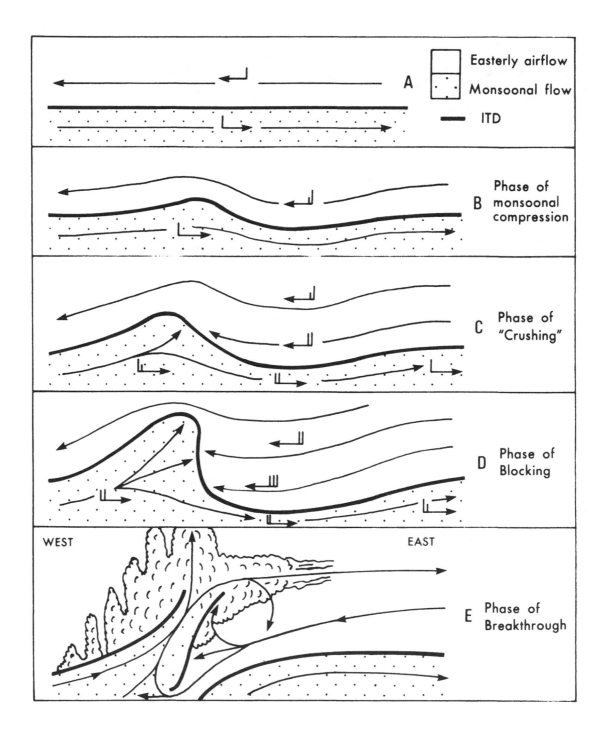

Figure 100   Formation of a line squall (after Leroux 1976)

Summary: the origins of the West African seasonal climate

In the period December to February the overhead sun is in the southern hemisphere and the equatorial trough (the ITD) lies approximately along the Guinea coast in West Africa. Sea surface temperatures in the Gulf of Guinea are about 27°C, but inland surface shade temperatures are only 16 to 18°C. In consequence, mean sea level pressures ranges from about 1010 mb over the Gulf to 1016 and 1018 mb in northern West Africa. The surface high pressure in north-west Africa results in a clockwise outflow of surface air from that region, producing, as a consequence of the Coriolis effect, north-easterly winds over most of West Africa. The area under the influence of the harmattan is designated as having Zone A type weather, which is normally rainless. The surface high pressure is further strengthened by the convergent air flow that must occur above it and the subsidence from above that must then take place. The location of the ITD just north of the Guinea coast means that humid south-westerlies affect only the extreme south of West Africa. The wedge of moist air is too thin and overlain by dry easterlies to permit significant vertical cloud development. Humid but generally dry conditions, Zone B weather, are the norm along the Guinea coastlands.

As the year progresses and the overhead sun advances northwards, so the St Helena anticyclone in the south Atlantic expands and pushes its influence north. At the same time net radiation levels over West Africa rise, the anticyclone over the north starts to break down as pressure falls inland, and the ITD advances inland with convergent streamlines. As the wedge of mT air advances, thickening towards the south, it is 'opposed' by strengthening upper easterlies, which develop on the southern side of an upper air anticyclone that replaces the relative low pressure that prevailed through the winter at high altitudes. Shearing at the boundary between the lower, humid and upper, dry airmasses, leads to the development of waves in the base of the easterlies. From these waves may originate line squalls which may travel across West Africa from east to west with the waves as they 'ripple' along the airmass boundary. The season of thunderstorms, Zone C weather, will be felt first in southern West Africa from March to May, then moves progressively further north to affect the Sahel through May to August. As the ITD then moves back southwards behind the zenithal sun, so a second season of thunderstorms is experienced in southern states. South of Zone C the wedge of advancing mT air is at its deepest. Perturbations within the humid airmass, the destabilizing effect of the near jet flow aloft, the very depth of humid air within which convection is possible, all aid the growth of cloud and more persistent rainfall than normally occurs in Zone C. In areas where the coastline runs at right angles to the main wind direction and high land also occurs (as in Guinea, in the Freetown Peninsula, at Mount Cameroon and on Bioko), Zone D conditions may lead to rainfall in excess of 3000 mm in the four months May to August. The increased cloud that is associated with Zone D serves to lower air temperatures in the south, this situation being exacerbated by a fall in sea surface temperatures between Tabou and Lagos. With these lower temperature conditions comes a rise in surface atmospheric pressure to some 1014 mb, further encouraging the development of the sequence of events described in outline above.

As the year progresses (September to November) the overhead sun moves away again into the southern hemisphere, net radiation levels fall over northern Africa, the low pressure centre of the mid-year gives way steadily to the return of the winter anticyclone. The ITD moves south with its related weather zones. The easterly jetstream weakens to be replaced by the high-level westerlies. Their southward, convergent flow over north-west Africa strengthens the development there of the surface anticyclone. The harmattan once again spreads its influence across West Africa.

10 The Togo Gap, the 'little dry season' and recent climatic fluctuations in West Africa

It remains in this section of the book to seek to explain the seemingly anomalous climatic phenomena that are commonly referred to as the Togo Gap and the 'little dry season' and to look for possible causes for climatic variability in West Africa, and particularly to note the reasons for the failure of the rains, for the drought, for instance, that has persisted in the region, especially in the Sahel, since the late 1960s.

The Togo Gap

Numerous maps in Part One of this book have shown the existence along the Guinea coast of an anomalous climatic region in the south-eastern corner of Ghana and in southern Togo (see Figure 92, p. 124). Figure 101 illustrates the presence of this region with reference to rainfall (see also Chapter 5) and helps to confirm how localized and coastal the Togo Gap is.

If the main agents of precipitation in West Africa are seen to be perturbations in the base of the upper easterlies overriding a humid airmass with a depth of at least 2000 m, and waves in the mT airmass itself, it is not easy to understand why these agents seem to bypass Accra and Lomé! The assertion (Navarro, 1950) that line squalls travelling east to west across southern parts of West Africa have characteristic tracks that terminate along the Guinea coast, most commonly between Mt Cameroon and Cotonou and between Abidjan and Tabou, thus avoiding the Accra–Lomé area, might seem to offer a partial explanation for the relative absence of precipitation in the Togo Gap. However it has been noted (Walker, 1962) that Accra experiences the passage of nearly 22 line squalls per year on average, whereas Axim in south-western Ghana

has only 12. Yet, as Figure 101 shows, Axim receives almost 2000 mm of rain per annum, Accra only 732 mm. Furthermore (Figure 102), data gathered by the authors indicate that thunder, although registered significantly less frequently in the Togo Gap than immediately east or west of it, also diminishes in occurrence west of Abidjan. Also Axim experiences some 45 *local* thunderstorms per year, but Accra only 13.5. This suggests that meteorological conditions in the Togo Gap somehow depress convectional tendencies despite the upper air situation.

In the one year 1955 (Figure 86, p. 112), of the 124 cm of rain that fell at Accra, 56 per cent was derived from line squalls, 61 per cent of those storms occurring in the three months March to May and another 14 per cent in November, following the little season of August. At Axim in 1955, of the 206 cm of rain recorded, 60 per cent came from line squalls, 67.5 per cent of these being in the four months March to June, a further 18 per cent in October and November. Thus, if it is a fact that on average Accra experiences more line squalls than Axim, and perhaps therefore did so in 1955, and the proportion of rainfall received from the storms differs little in the two centres, then Accra should be the wetter place. However it is precisely in the months of greatest storm activity, March to May and October to November, that the difference in precipitation levels between Axim and Accra is most pronounced (Figure 103).

Explanations for the unique climate of the Togo Gap have been sought for many years and the main hypotheses have been neatly summed up by Trewartha (1962). There are essentially two of them. The first points to the fact that the overwhelmingly dominant wind of the coastal area in question, from the Niger delta in the east to Tabou in south-west Ivory Coast, is a south-westerly. In the rainy months a wind from this

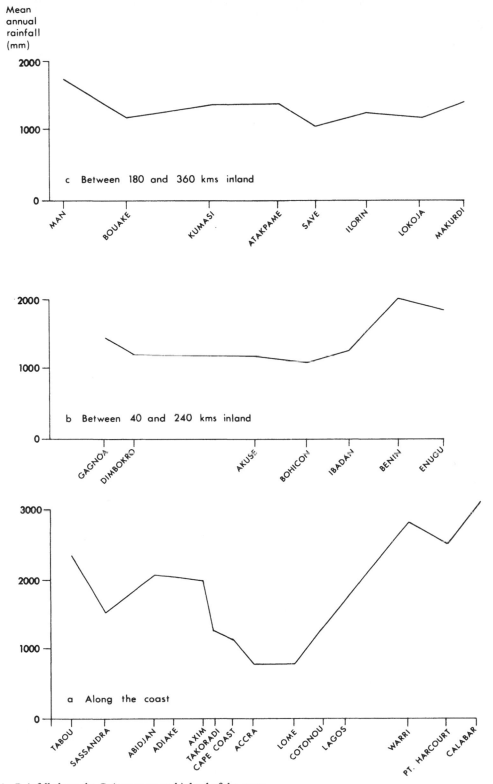

Figure 101   Rainfall along the Guinea coast and inland of the coast

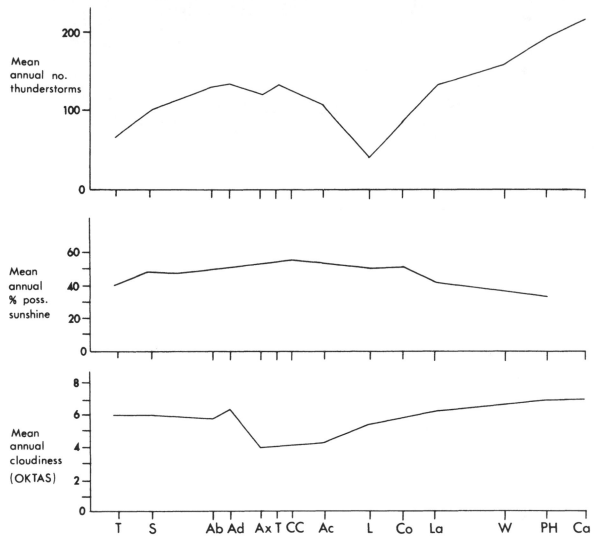

Figure 102 Cloudiness, sunshine and thunderstorms along the Guinea coast

quarter is also dominant along the coasts of Liberia, Sierra Leone and Guinea (Figure 28, pp. 46–7). Where the advance of the humid air of sufficient depth is more directly onshore, as between Bissau and Cape Palmas, so it will suffer more pronounced drag due to friction and, as noted above, a lessening of the air flow will lead to convergence within the flow. Where such convergence is overlain by an active and often divergent African Easterly Jet, so most marked uplift will occur and significant rainfall. On the other hand, where the wind direction tends to parallel the coastline, as between Lomé and Takoradi, Sassandra and Tabou, this tendency to convergence and uplift will be lessened.

With reference to Figure 101 it is noticeable that between Abidjan and Axim, and along the south-west facing coastline of the Niger delta (data for Warri), where the prevailing wind tends to be more directly onshore, mean annual rainfall is greater than in south-western Ivory Coast or the Togo Gap. The situation at Tabou is the exception, but here, as at Takoradi, some of the effects of the onshore winds a few kilometres to the west are carried forward.

However, if the alignment of the coastline was to be the sole cause of the lower rainfall in the areas noted, the area about Sassandra in Ivory Coast should be as dry as that around Accra and Lomé, but it is not. Hence to the first hypothesis is added the second, which relates to sea surface temperatures off the Guinea coast.

Table 10 and Figures 14 to 16 (in Chapter 1)

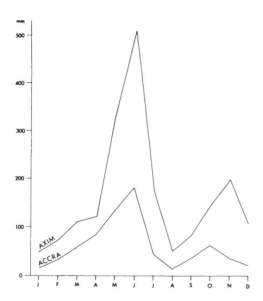

Figure 103 Mean monthly rainfall at Accra and Axim

temperature, he obtained a correlation coefficient (r) = +0.72 for the relationship between sea–air temperature difference and rainfall, and r = + 0.71 for rainfall plotted against the coastline direction. Of course these values do not prove that the variables are causally related, but they do suggest a relationship. Gallardo also noted that mean sea salinity levels off Accra and Lomé tended to be higher than elsewhere between Monrovia and Lagos, even allowing for the inflow of fresh water into the Gulf of Guinea from the River Volta (see also Chapter 1). However, the correlation of

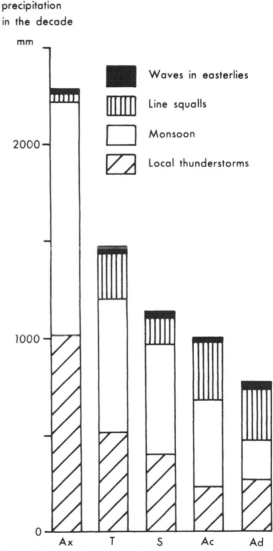

Figure 104 Contributions of four origins of rainfall to precipitation totals in southern Ghana (after Acheampong 1982)

have all clearly indicated a lowering of sea surface temperature between Cape Palmas and Lagos in the period July to October, that is when rainfall amounts along the coast also decrease dramatically. A cooler ocean surface, over which the prevailing winds pass before moving onshore, should lead to a lowering of surface air temperatures and reduce the tendency for uplift and rain. However the period of minimum temperatures in the Togo Gap is also the period of lowest temperatures throughout southern West Africa as a whole, and the mean minimum values for coastal stations in the Togo Gap are not significantly lower than elsewhere along the Guinea coast. The reduced temperatures throughout the region reflect increased cloudiness in mid-year. However, a marine influence may still be indicated in the Togo Gap if only because the gap is so very coastal in location and because mean cloudiness is rather less in that area (Figure 102), even though mean temperatures are as low as elsewhere along the coast. Nevertheless, cool surface waters are prevalent off Accra and Lomé only in the period July to October; even so, throughout the year, and especially in the wettest period May to July, the Togo Gap experiences less rainfall than its surrounding areas.

Thus a convincing explanation in these terms for the anomalous Togo Gap would seem still to be lacking. Yet when Gallardo (1978) sought to quantify the level of correlation between precipitation and coastline direction and sea surface

145

salinity with precipitation produced a value for r of only −0.63.

Acheampong (1982) looked elsewhere to seek to explain the existence of the Togo Gap. He noted what has been indicated in the preceding chapter, that rainfall in West Africa comes from four sources: the monsoon (waves in the westerlies), line squalls, local thunderstorms and waves in the easterlies. The contributions of these four to the mean annual precipitation in the decade 1961–70 at Axim, Takoradi, Saltpond, Accra and Ada in southern Ghana is illustrated in Figure 104. It is noticeable that the greater proportion of the rainfall comes from the monsoon, especially in the west. In the east (Accra and Ada) line squalls become relatively more important. Acheampong suggests that these variations reflect the location of the ITD which is characteristically nearer the Guinea coast in the vicinity of the Togo Gap in the early months of the year (see Figures 28a and 29a, p. 46 and p. 48). Thus the drier Zone B and C type climates may still be affecting the Togo Gap area even as monsoonal rains (type D) are beginning to make themselves felt further west and east, as the ITD moves north.

The question is also asked by Acheampong: to what extent does the absence of forest in the Togo Gap further contribute to the lack of rainfall, rather than the reverse? Perhaps, for all the reasons noted above, the region has been drier than those to the west and east in the last few thousand years, yet was forested. Human destruction of the forest has led to the development of savanna grassland in the Togo Gap, whereas to the wetter east and west 'bush' or secondary forest, with their greater levels of evapotranspiration, have replaced the original rain forest.

The 'little dry season'

In Chapter 5 reference was made to a decrease in rainfall that normally occurs in August along the Guinea coast between Sulima in Sierra Leone in the west, to Warri, or even Port Harcourt, in Nigeria in the east. Figure 71 (p. 92) illustrated the distribution of this climatic phenomenon, popularly called the 'little dry season'. East of Nigeria the island of Bioko also experiences

Table 22 *Mean rainfall (mm), June to September, at seven stations experiencing the 'little dry season'*

	June	July	Aug	Sept
Robertsfield	630	698	508	715
Gagnoa (Ivory Coast)	213	97	61	172
Kumasi	234	126	74	176
Atakpame (Togo)	182	194	164	189
Ibadan	188	155	84	178
Enugu	290	193	170	325
Meiganga (Cameroon)	178	260	228	253

this August hiatus in rainfall. Inland from the southern coastline of West Africa the little dry season is registered for some 300 km. At 6½°N in Cameroon, Meiganga's mean monthly rainfall records also indicate that it is experienced there. In the far south of Cameroon, however (records of Kribi and Lomié at 3°N), the drier month in the middle of the rains is July not August.

As noted in Chapter 7, some writers have designated the little dry season as weather Zone E, and its spatial and temporal distribution clearly suggests that its origins are related to the advance and retreat across West Africa of the ITD. However, in seeking to explain the causes of the main rains in West Africa, the depth of the advancing mT air from the Atlantic was seen to be important. If the advance is like a wedge (Figure 90, p. 118), the depth in Zone E should be as great as, if not greater than, that in Zone D. Yet by July in the far south-east of the region, and by August south of 8°N latitude elsewhere, the precipitation markedly diminishes (Table 22).

Explanations for this August hiatus offered by various writers, and all inadequate by themselves, have been summarized by Ilesanmi (1972c). They include subsidence inversion at the 800 and 850 mb level, or refer to the influence of coastal alignment or the summer upwelling of cool waters off the Guinea coast. Obasi (1965a) suggested that wind and moisture divergence occurs along the coast with the northward shift of the St Helena anticyclone. This hypothesis, however, fails to explain the absence of the little dry summer in August east of about 5°E in southern West Africa.

As seen in Chapter 5, for rainfall to occur, four basic mechanisms are needed:

1 a lifting mechanism to cool the air;
2 condensation of water vapour and the formation of water droplets;
3 a mechanism to merge the droplets to make them large enough to overcome 1;
4 the inflow of moist air to recharge losses due to rainfall and to supply sufficient water vapour to permit the rainfall amounts recorded.

These mechanisms are active in West Africa in mid-year, with the AEJ being of especial importance in destabilizing the atmosphere.

The mechanisms all require the use of energy. The total energy of the atmosphere comprises: kinetic energy (that possessed by virtue of motion) which accounts for only about 0.5 per cent of the total energy even in the vicinity of jet flow, plus the latent heat content, plus the sensible heat content, plus the potential energy (that measured as the amount of work required to bring a body from one position to another). Excluding kinetic energy, the remainder form the static energy content of the atmosphere, and it can be appreciated that not all parts of the tropics need have similar levels of static energy. Much energy is stored as latent heat in water vapour. North of the ITD the latent heat content may be less than 5 per cent of the total energy content, but south of the ITD it may comprise 15 per cent.

The sensible heat will reflect levels of incoming radiation. The atmosphere is perpetually losing sensible heat by radiation to space, but also gaining it from the radiating surfaces beneath and from the latent heat released by the condensation of water vapour. Potential energy is utilized in areas of sinking or rising air, decreasing in the former to reappear, as it were, as sensible heat and vice versa. By far the greatest proportion of atmospheric energy is in the form of sensible and potential energy, known together as the total potential energy. Where atmospheric processes are particularly active, for example where jet streams are prominent and precipitation mechanisms marked, energy levels will be greatest.

The distribution of total energy content of the atmosphere can be indicated by plotting equivalent wet-bulb potential temperatures. (An equivalent temperature is the sum of the actual air temperature and an addition due to the latent heat of the water vapour it contains. The potential temperature is the temperature that the air would take if brought adiabatically to a standard pressure, usually 1000 mb.) Wet-bulb potential temperatures by height and latitude have been plotted for West Africa as shown in Figure 105 (after Bernet, 1968), and, as Adefolalu (1972) has pointed out, low values are found north and south of the AEJ between 700 and 500 mb. The lower the values the less the precipitation.

As Figure 106 helps to demonstrate (after Leroux, 1976), the levels of energy at the 700 mb level reflect the transfers of energy associated with the horizontal and vertical wind shears, related in turn to the location of the AEJ. When the axis of the jet is north of 10°N the zone of upper air potential wet-bulb temperature that exists south of the jet impinges onto the Guinea coast with its related decrease in rainfall.

In other words the zones of heaviest rainfall in West Africa (Zones C and D) lie effectively beneath or within 750 km south of the core area of the AEJ. Any part of West Africa that lies beyond about 750 km from the jet experiences less precipitation. In July the AEJ is sufficiently far north of southern Cameroon to cause the little dry season to exist there in that month. By August the ITD and related AEJ have moved to their most northerly positions leaving the Guinea coastal area, south of about 8°N, beyond the influence of the jet.

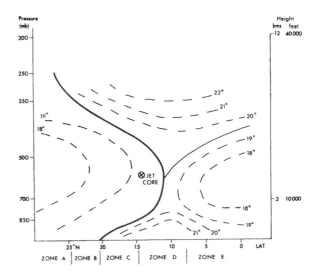

Figure 105 Height–latitude cross-section of wet-bulb potential temperature (°C) (after Bernet 1968)

147

Climatic fluctuations in West Africa

This book is concerned with the present climate of West Africa and so does not include a discussion of past climates and climatic change other than the brief considerations below and in Chapter 11. However, even within the short term, such as the lifetime of an individual person, fluctuations in climate may be experienced and in the West African context this essentially means significant short-term variations in precipitation levels.

Since the turn of the century there seem to have been periods of serious, below-average rainfall in the Sahel, when the rains have failed over three, four or more consecutive years, and these periods have recurred, for unknown reasons and possibly by chance, at approximately 30 year intervals. Drought conditions occurred at the turn of the century, in the 1930s and, infamously, in the present period since 1969 (Figure 107). Sadly the poor rains of the 1970s, continuing into the 1980s, have affected even parts of the Guinea coastlands. In southern Ghana in 1982 and 1983 the rains were only half mean levels and arrived up to two months late. So prolonged and devastating has the most recent drought been that the fear now exists that permanent aridity may have to be contended with, that the Sahel is to become part of the Sahara. In Mauritania the isohyet for 250 mm has shifted some 200 km to the south since 1960. Rainfall in the country generally in 1983 was only 27 per cent of the 1941–70 mean. Nouadhibou had no rainfall at all in 1982 and 1983. In northern Mali rainfall averaged only 25 to 40 per cent of the mean in these same years. In effect the Sahara desert advanced south by 200 km. In 1983 Nouakchott experienced over 200 days with sandstorms.

The tendency to desertification has been exacerbated, of course, by overpopulation, overgrazing, land mismanagement and policy failures (see Chapter 15), but the fact still remains that for some reason the rains have not advanced across the region in the 1970s as they did in the 1950s, that is, the ITD did not penetrate inland in its customary fashion. During the drought period the ITD has tended to fall short of its normal August advance by up to 300 km, which is 3 or 4° latitude. Thus the related easterly jet activity has also tended to remain south of its normal location and, indeed, has also been less promi-

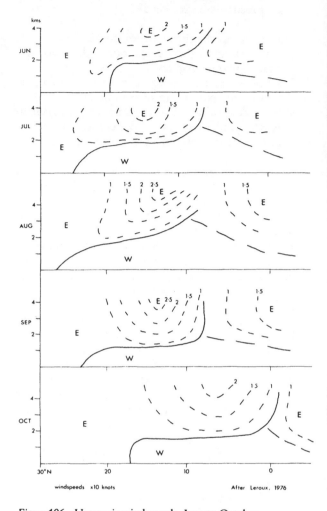

Figure 106 Upper air wind speeds, June to October

nent, and the Atlantic subtropical anticyclone has held a more dominant role over north-west Africa. Why these changes should come about and persist for a number of years is less easy to explain.

A number of writers have pointed to apparent related conditions in and outside West Africa, although such observations do not, in themselves, account for the variability in rainfall levels over a period of years in the region. For example, since the mid-1940s the mean annual sea surface temperature in extra-tropical latitudes in the northern hemisphere has fallen by 0.3°C. This in turn can be seen to have increased the low to high latitude air temperature gradient, strengthening the upper air westerly flow in middle latitudes and the influence of this flow into lower lati-

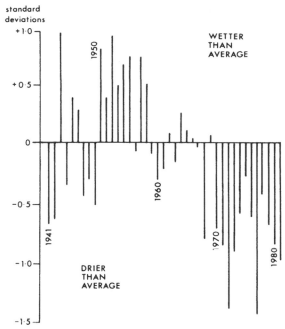

Figure 107 Rainfall trends in the Sahel, 1940–82 (after Musk 1983)

tudes, and in turn displacing southwards the sub-tropical anticyclone and preventing the expansion northwards of equatorial air in the northern hemisphere summer.

Associated with these changes in the general atmospheric circulation is the human misuse and overuse of the land, especially in the sensitive Sahel, which diminishes the level of relief that the occasional less dry year should produce.

Man-aided desertification has increased the dust content of the atmosphere which may help to cause drought by maintaining higher air temperatures at high altitude, thus increasing atmospheric stability. Furthermore, these higher upper-air temperatures will decrease the north-to-south temperature gradient aloft and thus the baroclinicity, and therefore weaken the easterly jet flow. Studies of the 1972–73 drought showed the virtual disappearance of the 850 mb trough near 8°N and the weakening of the easterly jet (Kidson, 1977).

Such a situation could be self-generating, so the loss of the vegetation cover in the Sahel by over-grazing and cash-cropping is particularly serious. Over-grazing will lead to an increase in surface albedo and thus net radiative loss; this in turn will diminish surface temperatures and increase horizontal surface temperature gradients and surface wind outflow (the harmattan in West Africa), but also increase air inflow aloft and subsidence. Such subsidence superimposed on the descending Hadley circulation further diminishes the possibility of rainfall mechanisms operating. Between 1967 and 1973 the mean dry-season albedo in the Sahel increased from 29 to 34 per cent, the wet-season level from 23 to 33 per cent (see also Chapter 11).

Whatever the cause of climatic fluctuations or long-term changes in climate in West Africa, they have major consequences for human activity and well-being. Part Three of this book considers the impact of climate upon man and man upon climate in West Africa.

PART THREE

Application *Applied Climatology in West Africa*

Introduction

Probably the most fundamental need and desire of a human being is good health. Given the absence of physiological malfunction due to congenital or acquired defects or ageing, or of a serious accident, health or illness is usually determined by external factors: by the type, quantity and quality of the other basic requirements of food and drink, clothing and housing; by the presence or absence in the body of debilitating disease-inducing organisms; by the exercise, relaxation and rest that the body needs; by social pressures.

In turn the external factors are very much determined by the climate. Green plants are the basis of human food requirements, being consumed directly or indirectly after ingestion by cattle, sheep, goats or other animals. The type and abundance of vegetation are clearly climate-dependent, whether the vegetation be natural or planted. The availability of water, for drinking and hygiene, also obviously reflects precipitation. Large numbers of people can only survive in desert regions if the precipitation of distant areas is in effect imported in the form of rivers or oases. Seasonal or periodic droughts can inflict hardship even in areas with more than adequate annual precipitation totals.

Most peoples in the world feel the need for clothing and not just for social reasons. Clothing aids independence from the rigours of climate, providing protection from sunlight, from heat or cold, from wind and rain. For similar protection, housing, be it the complex air-conditioned palace or the simple cover of grass or leaves, is built to provide an alternative climatic environment to the one outside.

The serious accident may be the result of weather phenomena, the lightning strike, the flood, the blinding duststorm, the drought or the gale. All diseases are not, of course, climate related, but as Chapter 13 will show, many are. The prevalence of infection in the tropics, of dysentery, malaria, yaws, schistosomiasis and trypanosomiasis, suggests the relevance of climate, as does the general absence of rickets in the sunnier parts of the world. The relationship between weather conditions and tiredness, restlessness, headaches, rheumatism, perspiration, appetite, skin disorders and accident proneness is apparent to all from personal experience.

The recreational activities that may play a significant role in mental and physical well-being can be affected by climate. Most sporting activities decline in the rainy season in West Africa. Those athletic activities requiring prolonged exertion are the least favoured; it is very difficult to train and practise in the exhausting conditions of the dry season.

The United States Center for Environmental Assessment Services (CEAS) of the National Oceanic and Atmospheric Administration (NOAA) makes monthly climate impact assessments in eight categories:

1 industry and commerce, including employment, banking and business;
2 construction;
3 energy;
4 government and taxes (local government spending);
5 food and agriculture, including fisheries;

151

6 recreation and service industries;
7 human resources: health, pollution, education, crime;
8 transportation and communications.

West African nations would do well to approach many of their problems in a similar way, but also to establish climatological and meteorological programmes that are directed to achieving specific objectives. The work of meteorologists and climatologists should be relevant to the people's needs. Working in collaboration with other scientists they should seek to provide the basis of national well-being through a knowledge of climatic conditions, and an understanding of climatic processes, to provide an ability to forecast and foresee, to plan ahead, to minimize the evil consequences of the climatic environment, to maximize the advantages of this vital resource, to encourage its care and protection and vigorously oppose its pollution.

To the basic human needs listed at the start of this section, that of education should be added. To the essentials of reading and writing should be added an awareness of environment and of a need for a careful stewardship of it. It is to help in this direction that this book has been written, and especially this third part.

11 Climate and water resources

Water as a resource

Water is one of the essential requirements for the survival of life on the earth's surface. It comprises about 71 per cent of the earth's surface and occurs in the liquid state in the oceans, seas, lakes, rivers and in underground water; in solid form in the ice packs of the polar regions and mountain tops; and as water vapour in the atmosphere. Except in the rare instances where tidal flow is used to generate electricity, and unless one includes fishing in the seas or recreational pursuits in or on the water, the oceans are not directly utilized. This is essentially due to the high levels of dissolved salts in the water (Tables

Table 24 *Common salt composition of ocean water*

Salt	per cent
Sodium chloride (table salt)	77.7
Magnesium chloride	10.8
Magnesium sulphate (epsom salt)	4.7
Calcium sulphate (gypsum)	3.6
Potassium sulphate	2.4
Calcium carbonate (chalk limestone)	0.3
Magnesium bromide	0.2

23 and 24). Not that the rivers and lakes of this world, providing so-called 'fresh' water are without a dissolved load, as Table 23 also shows, but the preponderance of sodium chloride (NaCl) in the sea water makes all the difference as far as taste and utilization for drinking, industry and agriculture are concerned. However the oceans provide abundant surfaces for the supply of atmospheric water vapour through evaporation. Much of the fresh water, anyway, is frozen in the ice of Antarctica and Greenland.

The study of water provides a logical link between the physical and social environments. Water is a naturally occurring commodity directly used by man and often provides not only the mainspring for extensive economic development but also an essential element in man's aesthetic experience. It is also a major formative factor of the physical and biological environment which provides the stage for human activities. Such activities include agriculture, forestry and fishing, navigation, recreation and the generation of electricity. In the context of West Africa, the water resource will be considered from the point of view of its availability in both spatial and

Table 23 *Mean composition of the solid part of sea and river water*

Solid matter	River (%)	Sea (%)
Carbonate (CO_3)	35.15	0.41
Sulphate (SO_4)	12.14	7.68
Chloride (Cl)	5.68	55.04
Nitrate (NO_3)	0.90	0.01
Calcium (Ca)	20.39	1.15
Magnesium (Mg)	3.41	3.69
Sodium (Na)	5.79	30.62
Potassium (K)	2.12	1.10
Iron and Alumina ($Fe_2O_3 + Al_2O_3$)	2.75	0.01
Silica (S_iO_2)	11.67	0.01
Others	—	0.31
Total	100.00	100.00

seasonal patterns, its utilization and the problems associated with a lack or an excess supply of it. In this regard the problems of drought, flood and soil erosion will be considered and a final section will examine the measures being applied for effective harnessing of available water resources in West Africa.

The hydrological cycle

The hydrological cycle is of great importance not only in the natural world but also in the technological and social systems of mankind (Barry, 1969). The study of the hydrological cycle deals with the transfer of water from one state to another. The transformations involved in the various stages include evaporation and transpiration, moisture transport, condensation, precipitation and runoff. Figure 108a shows a diagrammatic representation of processes while 108b indicates the relative average annual amounts of water involved in each phase of the cycle; it also shows that the atmosphere holds only a very small amount of water although the exchanges with the land and oceans are very considerable.

The greatest sources of moisture input into the atmosphere are the oceans which hold about 97 per cent of all the earth's water. They provide about 84 per cent of the annual total of evaporated water while the continents provide the remaining 16 per cent. The transformation of the water into vapour requires a great deal of solar energy. This water vapour is transferred horizontally and vertically by the planetary winds. As it is transported vertically, as noted in Chapter 5, cooling usually occurs perhaps leading to condensation, cloud formation and eventually, precipitation. Clouds account, however, for only about 4 per cent of the total atmospheric moisture.

Sources of water in West Africa

There are two main sources of water.

Rainfall

The most important source is rainfall. The type of rainfall (its duration and intensity) is of par-

ticular hydrological significance, since it is this which determines the volume of water in the rivers and the river regimes.

It is therefore pertinent to reiterate the following points:

1 Rainfall in West Africa decreases generally from the southern coastal areas northwards to the edge of the desert in the interior; mean annual values ranging from over 2000 mm in the coastal areas to less than 100 mm on the desert margins.
2 The isohyets tend to parallel the parallels, or the south-western coastline.
3 The number of months with rain (more than 10 mm per month) also decreases from the coastal areas to the desert margins ranging from about 10 in the south to about 2–3 on the desert margin.
4 The seasonal rainfall shows a gradual transition from the double maximum type in the south to a short, well defined single maximum type in the interior. Stations south of about latitude 9°N lie under the influence of the weather zone characterized by the 'little dry season' experienced during July and August, while areas north of this latitude experience a peak of precipitation in July and August.

Rainfall intensity and duration

The intensity and duration of rainfall are of great importance to surface drainage particularly in the interior of West Africa where the rainfall occurs in short sharp spells. In most cases the records of daily rainfall may be regarded as the records of individual storms (Ledger, 1964). Rodier (1964), showed that for the areas of West Africa receiving over 200 mm per annum, storm characteristics are very similar throughout the region (Table 25). Hence the overall increase in rainfall from the north to the south is not the result of an increase in storm size but an increase in the number of storms of the same size (plus of course, the greater importance of the monsoonal rains in the south).

Rainfall amount and the water balance

The total amount of rainfall alone is not enough to determine the amount of discharge by rivers; there is a need to take into consideration the losses due to evaporation and evapotranspiration

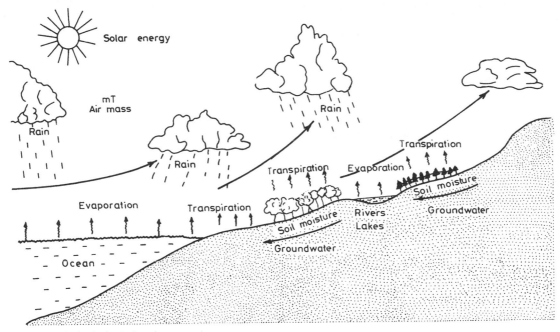

Figure 108a   The hydrological cycle

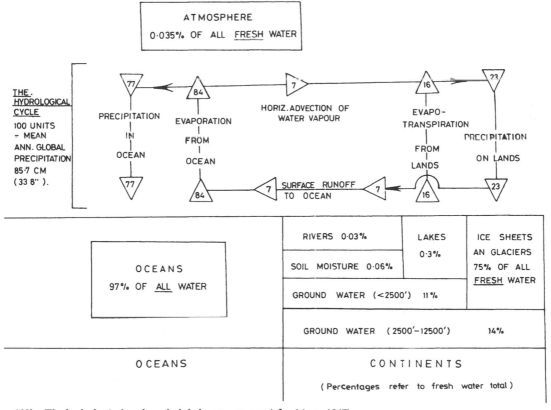

Figure 108b   The hydrological cycle and global water storage (after More, 1967)

and local geology. In other words, the water balance of an area is vital to the hydrological regime. The analysis of the potential evapotranspiration in West Africa given in Chapter 3 shows that the values increase from the southern coast to the interior, the highest values occurring during the dry season. The differences between the rainfall and potential evapotranspiration (PE) are used to delimit areas of moisture surplus and moisture deficit (Ojo 1969).

In January the PE increases from about 100 mm in the coastal areas to about 250–275 mm in the interior. In July the PE is more uniform over West Africa; values are generally lower due to the reduction in the amount of insolation due to cloud cover. Values are generally below 75 mm in areas south of latitude 15°N, while north of this zone there is an increase up to 300 mm in the desert margins (Figure 41, p. 63).

In January, there is an extensive area of moisture deficit in West Africa (Figure 77b, p. 100). The lowest values of moisture deficit are along the coast except in the Togo Gap and the far north. In August the northern limit of moisture surplus is about latitude 12°N. On an annual basis therefore the southern areas north to about latitude 8°N show an annual surplus. This is the area where the mT airmass prevails throughout the year. North of this zone there is a general zonal increase in the moisture deficit; however, the gradient varies from zone to zone.

Rainfall and runoff

Annual runoff is greatest in areas with the highest rainfall and the lowest evapotranspiration, so that both the volume of the runoff and the percentage of rainfall that becomes runoff tend to increase from the north to the south. West African rivers experience marked seasonal alternation of high and low flows (Ledger, 1964). There is, however, a considerable difference between the regimes in the north, where runoff is concentrated into a few floods in August and September, and those of the south where the runoff is spread more evenly. This pattern clearly reflects the seasonal distribution of rainfall to the exclusion of other factors. The little dry season is reflected in a double-peaked hydrograph only in the small coastal rivers whose catchments do not extend

Table 25 *Storm sizes (mm of rainfall) in West Africa*

Mean annual Rainfall (mm)	Percentage Occurrence		
	50	30	10
20–90	3–4.5	5.5–7	8.5–12.5
90–200	3–6	6–10	16–19
200–300	6–7.5	11–15	24–28
300–500	6–7.5	11–15	25–31
500–750	6–7.5	13–19	29–34
750–1000	6–8	13.5–16.5	30–35
1000–2000	6–8.5	13.5–16.5	30–36
1200–1500[a]	4.5–11.5	9.5–21	26–39
1500–1800[b]	6–16	15–27	32–44

Notes: [a] Southern Benin; [b] Ivory Coast
Areas that have a mean annual rainfall of 20–90 mm may expect 50 per cent of storms to produce 3 to 4.5 mm of rain, 30 per cent to produce 5.5 to 7 mm and 10 per cent to produce 8.5 to 12.5 mm.
Source: Rodier 1964.

far inland. The river Niger also shows two peaks in its lower course.

Most of the rivers in West Africa have extremely low dry season discharges. In the heavily wooded areas of the south where the dry season is short, the flow of even small streams is often permanent. With the increasing severity of the dry season northwards, flow conditions deteriorate until at approximately 12°N most small streams are completely dry for at least six months of the year. Large rivers (such as the Black Volta) flowing over extensive areas of permeable rocks often have a small surface discharge; so these too may dry up in years of deficient rainfall.

The hydrological differences of the West African rivers have been used to recognize two major regimes (Rodier, 1964). These are, first, the Equatorial regime of the extreme south, having two separate periods of high water; and second, the tropical regimes of the interior, having a single high water season (Figures 109 and 110). Regimes of the tropical type have been subdivided into a number of smaller groups. The typical tropical regime consisting of a season of

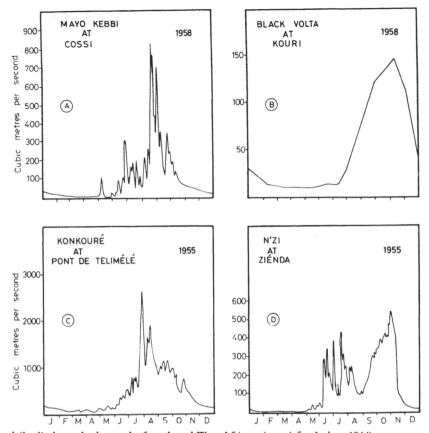

Figure 109   Mean daily discharge hydrographs for selected West African rivers (after Ledger 1964)

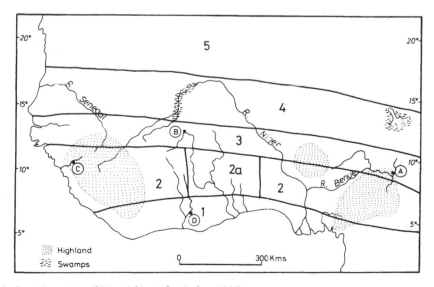

Figure 110   Hydrological regions of West Africa (after Ledger 1964)

Note: 1 – Equatorial type; 2 – Tropical transitional type; 2a – Tropical transitional, Dahomean sub-type; 3 – Classical tropical type; 4 – Sahelian type; 5 – Desert and sub-desert types. A – River Mayo-Kebbi at Cossi; B – Black Volta at Kouri; C – Konkouré at Pont de Télimélé; D – N'Zi at Zienda.

high water from July to October and a low water period from December to June, is restricted to those rivers whose drainage basins lie predominantly in the 750 mm–1250 mm rainfall belt. To the north lies the Sahelian type which, though similar in seasonal pattern to the rivers further south, differs markedly in other respects so as to merit separate consideration. Finally, to the far north lie the desert and the sub-desert regimes; the former characterized by an average of less than one flood per year and the latter by several floods.

The foregoing analysis shows that there is a close correlation between rainfall regime and river regimes in West Africa. This has been used to divide the region into hydrological zones as follows (Ledger, 1964; Figure 110):

Zone 1 This zone is characterized by high annual runoff (5–20 litres sec^{-1} km^{-2}); the seasonal flow pattern shows two periods of high flow from April to August and October to November with a severe dry season from December to March.

Zone 2 In this zone the wet season occurs from May to October and mean annual runoff ranges between 10 and 21 litres sec^{-1} km^{-2}. From 40 to 50 per cent of the total runoff is concentrated between September and October. There is a significant dry season flow.

In Zone 2a, the wet season occurs from July to October; approximately 60 per cent of the total runoff is concentrated between September and October; a long dry season occurs from December to May.

Zone 3 The wet season occurs between July and October; the mean annual runoff ranges between 2 and 8 litres sec^{-1} km^{-2}; over 60 per cent of the total runoff is concentrated in the period from mid-August to mid-October and there is low water from December to June.

Zone 4 This zone is characterized by a very short wet season between July and September and a long dry season with 8–9 months with no flow. Mean annual runoff is very low being between 0.20 and 0.50 litres sec^{-1} km^{-2}.

Zone 5 In this zone flow is limited to one or two flows per year resulting from periodic storms of line squall type, usually in August and September; mean annual

runoff has no meaning at all in view of the sporadic nature of the runoff.

Underground water

Just as rainfall determines the magnitude and intensity of surface runoff so also does the geology control the location of underground water. Basically two main lithologies control the distribution of underground water in West Africa: rocks of the basement complex and the sedimentary rocks. Very extensive regions of West Africa are underlain by the crystalline basement rocks of the continental shield. They are not very productive aquifers. Water can only be obtained in the weathered mantle (regolith) and in regions where faults occur in these basement rocks. Often the water from these sources is of only local significance.

At present groundwater is generally underutilized in West Africa. Recently, however, attention has been directed towards increasing exploitation of this source of water. Results of exploration have confirmed that the basement complex rocks have only restricted aquifers whereas the sedimentary rocks store larger volumes of water underground. The extraction of this water is effected through the use of boreholes or pumps and in some cases these pumps are solar powered (Oguntoyinbo, 1976).

Although the underground water in the basement complex rocks is restricted to favourable areas (Omorinbola, 1964) and of limited quantity wherever such locations are identified, the water extracted is of immense value for domestic use both in the rural areas and in the rapidly expanding urban centres where water is often a scarce commodity. In such cases extraction is from wells or boreholes.

Areas of sedimentary rocks are particularly important where they form artesian basins. Notable among such are the Chad Basin, Sokoto-Rima Basin and extensive areas in Niger. The extraction of water from such sources through wells or boreholes is invaluable in those interior locations where the dry season is long and drought is a common phenomenon. Underground water helps to reduce domestic water shortages and aids the irrigation of crops during the dry season. Other sources of water worthy of mention are the oases which support appreciable populations of both people and livestock and many crops.

Implications for planning

The foregoing analysis shows that rainfall is the most important source of water in West Africa. Its spatial and seasonal distribution is therefore of paramount importance to the main occupation of the region which is agriculture. In areas of high rainfall and long duration it is possible to cultivate tree crops like cocoa, rubber, kolanuts etc. Furthermore, two cropping seasons are possible for such main food crops as maize and vegetables. As the rainy season decreases farther inland it is concentrated into even shorter periods. Evapotranspiration also increases and thus the moisture balance is negative for a greater part of the year. Consequently only one crop can be grown in a year. Such crops include groundnut, millet and guinea corn (sorghum) (see Chapter 12).

As the rainfall decreases inland, its reliability also decreases, consequently drought is a recurrent phenomenon in the interior. The incidence of drought in the recent past has generated a great deal of hardship for the inhabitants of the interior of West Africa particularly in the Sudan–Sahel Zone.

Drought

The word drought is difficult to define; it may be a period of continuous dry weather or the absence of significant precipitation for a period long enough to cause moisture deficits in the soil due to evapotranspiration. For the agriculturist drought may be defined according to the type of crops concerned. For example, certain types of weather conditions may be called dry spells for sorghum but they may be a drought for sweet corn. An agricultural drought may therefore be defined as a period of dry weather sufficiently prolonged for the lack of water to cause a serious hydrologic imbalance (i.e. crop damage, water supply shortage, etc.) in the affected area (Tannehill, 1947).

Drought severity depends upon the kinds of crops, degree of plant wilting, duration of drought and the size of the area affected. Therefore drought designates the period between rainfalls. The longer the interval the greater the severity of the drought. Such has been the case in the Sudan–Sahelian Zone, as noted in Chapter 10, where a 3000 by 700 km zone stretching along the southern edge of the Sahara from Mauritania to Chad suffered a protracted drought from 1969 to 1973.

Drought may be caused by meteorological or human factors or a combination of both. From the meteorological point of view, it has been established that increases in area and persistence of the sub-tropical high pressure cells in the late 1960s and early 1970s led to a southward expansion of the Azores anticyclone (Barry and Chorley, 1982). Such a phenomenon prevented the maritime airmass from penetrating far enough inland to bring rain to the desert margins during successive years between 1969 and 1973 (Oguntoyinbo and Richards, 1977).

The processes of overcultivation and overgrazing have increased the exposure of land on desert margins to high surface reflectivity (albedo). The increase in surface albedo, it has been suggested, may lead to a mechanism which suppresses convective rainfall derived from local evapotranspiration (see Chapters 10 and 15; Charney, et al, 1975). Thus the processes of overcultivation and overgrazing help perpetuate desertification.

Drought in West Africa

It has been established earlier in this volume that the distribution of rainfall in West Africa is closely related to the seasonal migration of the ITD which is linked with global seasonal variations in pressure. The area most affected by the drought is the Sudan–Sahel zone (Figure 111).

Evidence from various sources shows that in the Sudan–Sahel Zone there have been periods of heavy rainfall interspersed by periods of drought for hundreds of years (Schove, 1977). This is clearly demonstrated in Figure 112 which shows fluctuations in the level of Lake Chad for the past 100 years. Even though there had been droughts and famine in the past, the first written record appears to have been from the Arab chronicles of the Western Sudan. These studies showed that there had been droughts in the Timbuktu area in the fifteenth and sixteenth centuries when food prices were phenomenally high. The fluctuations in the levels of Lake Chad have helped scholars to further establish periods of such drought in the region. The general impression is that the frequency of occurrence of droughts has been more rapid since historical records began.

The occurrence of drought is often marked by the shortage of food and lack of pasture, and

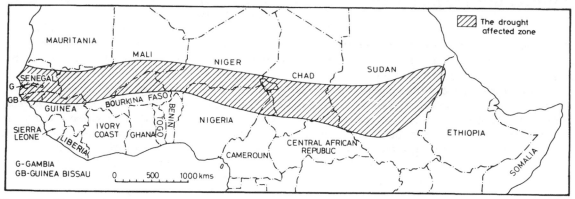

Figure 111   The drought-affected Sudano–Sahelian zone, 1968–73 (after Oguntoyinbo 1981)

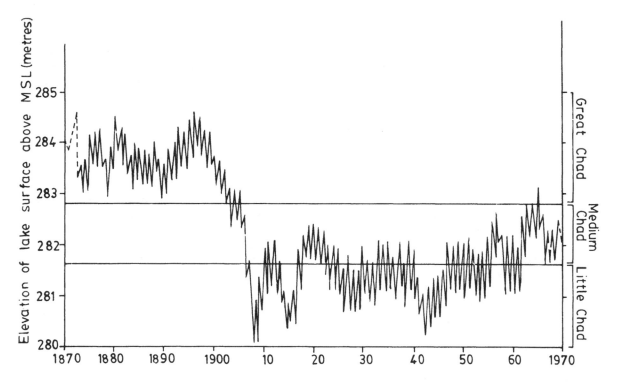

Figure 112   Surface level oscillations of Lake Chad (after Sykes 1972)

consequently overcultivation, overgrazing and famine often occur. The result of all these is that there is a loss of human life and of livestock (Table 26).

In order to alleviate the problems posed by drought the inhabitants resort to a number of emergency measures. These include: a reduction in the number of livestock through sales and slaughter, consumption of grains stored for plan-

ting, and mass migration. In more recent times, international aid has become very important in reducing the hardships caused by drought and famine.

On a long-term basis, efforts are being made to find a permanent solution to the problems of drought through the development of large-scale irrigation schemes. Among notable developments are the schemes at the inland Niger Delta and Upper Niger Valley in Mali, the Sokoto–Rima Basin Scheme, the Lower Niger and Benue Schemes in Nigeria and the Volta River Scheme in Ghana. Authoritative bodies are needed to work on the development of river basins on a comprehensive scale. The example of the Nigerian River Basin Development Authority (RBDA) is described below.

Floods

Floods are water hazards associated with periods of heavy rainfall either in the locality of the flood or sometimes upstream of it. They are most common therefore in river valleys and flood plains. In areas of low terrain, their effects may be both hazardous and beneficial. A flash-flood may sweep away crops, livestock and settlements. In the coastal areas of Sierra Leone and Senegal (Casamance) the seasonal floods provide ideal environments for the cultivation of crops of paddy rice, vegetables, millet and guinea corn. In the upper reaches of the Rivers Niger and Benue the swamp lands (*fadamas*) partially exposed after the seasonal floods provide fertile land for crop cultivation and dry season grazing ground for livestock. In the lower reaches of the same rivers the floodable areas with intricate channels provide fishing grounds for the local inhabitants.

Where settlements are located in river valleys, flash-flooding may create danger to life and property. The situation is made worse by the indiscriminate land use pattern. The constriction of river channels through building works in river beds, narrow culverts across river channels and refuse disposal in river beds, all contribute to aggravate the flood effects. In addition the ac-

Table 26 *Changes in livestock population in Zarmanganda, Niger*

Category of Animals	1968	1972–5	Loss/gain (number)	Loss/gain (per cent)
Cattle	50,648	26,733	−23,915	−47.2
Sheep	16,659	12,895	−3764	−22.5
Goats	64,236	50,615	−12,621	−20.0
Horses	6066	2517	−3549	−58.5
Donkeys	4390	2986	−1403	−31.9
Camels	4228	4672	+444	+10.5

Source: Sidikou 1977.

tivities of town planners by which watersheds are exposed for so-called development purposes, also help to accelerate surface runoff. When large volumes of water reach the constricted channels in the city the flood effects are accentuated. In Ibadan, for instance, the flood of 31 August 1980 claimed about 200 lives, displaced about 5000 people and caused property loss worth millions of Naira (Oguntala and Oguntoyinbo, 1982). The study referred to also showed that as the city size has been increasing since the 1950s so the frequency of flood occurrence has also tended to rise.

Soil erosion

Another hazardous effect of rainfall in West Africa is soil erosion. Soil erosion occurs under natural conditions but the rate of occurrence is often accelerated by such human activities as deforestation, overcultivation and overgrazing among others. Berry et al (1977) showed that the disruption of the traditional social and economic practices which followed the colonialization of the Sahel by the French at the end of the nineteenth century contributed to the severity of the 1971–3 drought. As grazing land was reduced through sedentarization, the available pastures were overgrazed. The exposed lands became more and more vulnerable to wind erosion and flash-flooding.

Areas of heavy rainfall, steep slopes and acid sandy soils are subject to a high degree of erosion. Such a situation is aggravated in areas of high population density where the soil surface is exposed through overcultivation. Typical examples of this type of environment occur in the Ondo, Imo and Anambra States and Abuja in Nigeria. In Imo State, for example, estimates show that of the 12,689 km^2 of the land area in the State, 2961 km^2 is affected by severe sheet erosion, 297 km^2 by gully erosion and 8729 km^2 by slight erosion (*Daily Times*, Nigeria, 24 August 1985). One of the most spectacular gullies is the Amucha gorge which is 2.5 km long and over 130 m deep. As large volumes of soil are washed away each year, the gullies retreat at an estimated rate of 14 m year^{-1}. The cost of rehabilitation has been put at about ₦36 billion (*Daily Times*, Nigeria, 25 August 1985). In the Navrongo–Bawku District of northern Ghana, sheet erosion has been observed to be very significant, leading to the truncation of over 40

161

per cent of the soil profile (Adu, 1972). Severe effects of sheet and gully erosion have also been observed in Senegal (Charreau, 1969), and Ivory Coast (Berger, 1964), and probably every country in West Africa is affected to some degree.

Of great importance is the annual wind deflation of the Saharan dust to the coast of West Africa and as far as the Brazilian coast in South America (Rapp, 1974). It has been estimated that the contribution of the Saharan dust to the total tropospheric aerosols is about 60–200 million tonnes per year (Junge, 1979). The quantity of sand deflated is to some extent due to human activities particularly on the desert margins; overgrazing and overcultivation help to intensify wind erosion particularly in the drier years.

Irrigation schemes in West Africa

The high temperatures recorded all the year round in West Africa mean that crop cultivation can be undertaken all year provided there is adequate moisture supply. As has been established, rainfall is highly seasonal in West Africa and its reliability decreases as one moves from the southern coast to the interior. To increase agricultural production therefore, it is necessary to give attention to the expansion of irrigation schemes as a regular source of moisture for crops.

Large-scale irrigation schemes are new in West Africa. Buchanan and Pugh (1955) noted that there was no such scheme in Nigeria prior to the 1950s; the French had an earlier start through the development of the inland delta of the River Niger from the second decade of this century. In terms of magnitude and experience therefore the practice of irrigation agriculture in West Africa is new when compared with the traditions of the Nile and Ganges valleys, for example, which are thousands of years old. Nevertheless significant steps have been made.

For the development of large-scale irrigation schemes in West Africa, dams have been constructed across major rivers. Such dams have helped not only to ensure adequate water supply during periods of unreliable rainfall, but also to extend the growing season and expand the area under cultivation. Figure 113 shows the location of major irrigation schemes in West Africa. Among the crops grown by these schemes are groundnut, cotton, millet, guinea corn (sorghum) wheat, vegetables and tomatoes.

In addition to providing facilities for irrigation, the major dams provide facilities for generating hydroelectricity. Examples include the Kainji Dam which provides the major source of power not only to Nigeria but also to the Niger Republic and the Volta River Dam which provides power for the aluminium-smelting industry in Ghana and for southern Ghana and Togo. The lakes also provide facilities for transport, fishing, tourism, domestic water supply and navigation.

River Basin Development Authority (RBDA), Nigeria

The increasing awareness of the role of water as a resource which can be harnessed and exploited has led to the development of some of the projects discussed in the previous paragraphs. However, such developments have taken place on a piecemeal basis, the greatest emphasis being on the generation of hydroelectricity and irrigation. In more recent times, attempts have been made to develop a comprehensive river basin development scheme in Nigeria.

The objectives of the RBDA include the socio-economic development of all the resources – water, land, plant, animal etc. – of the major drainage basins of the country. In essence the idea is similar to that of the Tennessee Valley Authority in the USA. Such river basin programmes are therefore multipurpose (Faniran, 1977). Comprehensive river basin development also lends itself to a unified or centralized control. It is for this purpose that Nigeria has set up 11 River Basin Development Authorities to work out comprehensive schemes relevant to each authority. The objectives include a comprehensive assessment and organized development of the potentialities of each river basin.

The RBDAs, in their respective areas, are empowered to acquire land and to exercise the following functions:

1 undertake comprehensive development of groundwater resources for multipurpose use;
2 undertake watershed management schemes for flood and erosion control;
3 construct and maintain dams, dykes, drainage systems, wells/boreholes, irrigation and drainage systems;
4 develop irrigation schemes for the production of crops and livestock;

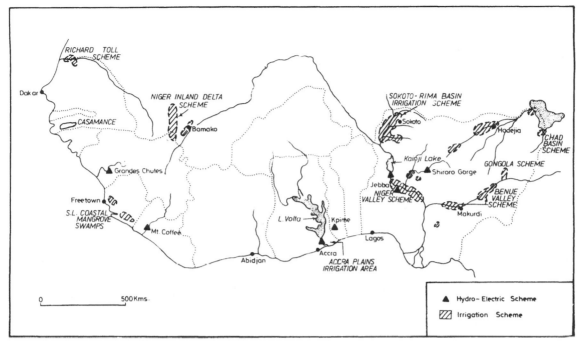

Figure 113 Major irrigation and hydroelectric power projects in West Africa

5 provide water from reservoirs, wells and bore-holes for urban and rural supply schemes;
6 control pollution in rivers and lakes in the authority's area in accordance with nationally laid down standards; and
7 resettle persons affected by the works and schemes in 2 and 4 above (Ayoade and Oyebande, 1983).

The pursuance of such a policy is already yielding good results with respect to irrigation schemes, water supply, and hydroelectric power generation.

Conclusion

A broad review of climate and water resource potentials in West Africa shows that in a predominantly agricultural area, water/moisture supply is the most important factor controlling agricultural activities. In the coastal areas where the rainy season is long, tree crops can be cultivated and double cropping of arable crops is possible. In the interior where the rainy season is short and unreliable, single cropping is possible and livestock rearing becomes predominant. The general lack of moisture has encouraged the harnessing of rivers for irrigation and other purposes.

Such irrigation schemes have helped to increase the length of the growing season and areas under cultivation. Initially the development of river basins was uncoordinated, but now comprehensive river basin development schemes are being undertaken to ensure effective utilization of the water resources. Examples of such schemes include the RBDA in Nigeria and the Chad Basin Commission and Organisation of Senegal River States (OERS). Even though many of these projects are faced with problems (lack of capital, of skilled manpower, etc.) the evolution of the river basin development concept as an agent of water resource development is a step in the right direction, because it provides a basis not only for co-operation among states but also for economic growth and development within such a river basin.

It has been shown that the interior of West Africa is characterized by semi-arid and arid conditions; this same area is covered by extensive layers of sedimentary rocks which are favourable for the storage of underground water. The development of this source of water is now being accelerated especially as drought occurrence is becoming more frequent and its intensity more severe.

12 Climate and agriculture

Introduction

Agriculture is the main occupation of the inhabitants of West Africa, 60–80 per cent of the total population being engaged in it. The location of the region within the northern tropical zone means that it experiences high insolation all the year round so that low temperatures are not a hindrance to crop cultivation throughout the year. The major factor determining the activities of the peasant farmers is rainfall (Figure 114a–c).

Agricultural activities in the zone depend on the seasonal distribution of rainfall. In areas of high and reliable rainfall the cropping season may last for eight to ten months in the year; under such conditions tree crops can be grown and at least two seasonal crops can be raised each year (Figure 114a). But in the areas where the

seasonal rainfall is of short duration, only one crop can be raised in a year and animal rearing becomes predominant (Figure 114c).

The optimal locations of crops in West Africa are determined by such parameters as rainfall and temperature as well as altitude and soil characteristics. Although many of the crops discussed later in this chapter have been introduced from other parts of the world by European explorers and colonial administrators, the ecology of West Africa has favoured their cultivation. Among the few crops whose origins have been traced to West Africa are the oil palm, certain species of cotton, yams, peas and millet (Purseglove, 1968, 1972). Figure 115 shows the locations of the major crops currently cultivated in West Africa. The most favourable conditions for each crop will be discussed later on in this chapter especially as they relate to climate.

However, before discussing the ecological conditions for each crop, the impact of climate

Figure 114a The farming calendar in the Forest Belt

Season	Month	Agricultural Activity	Rainfall in mm 100 200 300 400	Comments
Long dry season	January	Clear and burn farmland. Dig yam holes		Slack period in farming
	February	Prepare compound farms. Plant maize, yams, Okro.		Collect forest products
Long rainy season	March	Plant yams. Plant melons and fluted pumpkin		Peak demand for labour on the farms
	April	Plant cocoyams and cassava. Stake yams Weed Harvest fresh maize & veg.		
	May	Harvest fresh maize. Weed. Plant cassava. Stake and train maize.		
	June	Weed. Harvest dry maize.		
	July	Weed		
Short drier season	August	Top early yam. Train yam. Weed Plant 2nd maize and vegetables		Peak demand
Short rainy season	September	Top early yam. Plant cassava. Weed		
	October	Yam harvest begins		Peak demand for labour
Long dry season	November	Harvest yam. Build yam barns. Store yam		Slack period farming
	December	Clear new farmland Harvest and store yam		

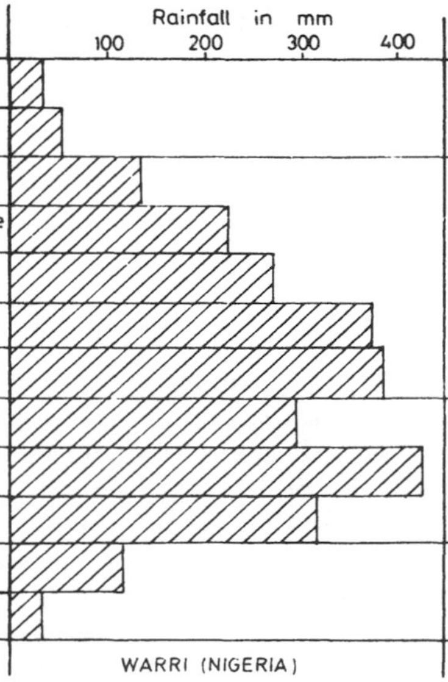

WARRI (NIGERIA)

Season	Month	Agricultural Activity	Rainfall in mm	Comments
Long dry season	January			Livestock migrate southwards to feed on stubble
	February	Plant late yam. Prepare land for early maize		
	March	Plant early maize Plant late yam		
Long rainy season	April	Plant early maize, okro, cowpeas. Plant early tobacco.		
	May	Plant guinea - corn		
	June	Weed. Harvest early maize and early tobacco		
	July	Top early yam Plant late tobacco		
Short dry season	August	Harvest early cowpeas. Prepare land for late crops		
Short rainy season	September	Plant late maize guinea corn and cowpeas. Harvest late tobacco and yam		Second maize crop susceptible to shortage of moisture
	October	Harvest guinea corn and yam		
Long dry season	November	Prepare land for yam Plant early yam		
	December	Plant yam Harvest cotton and cowpeas		

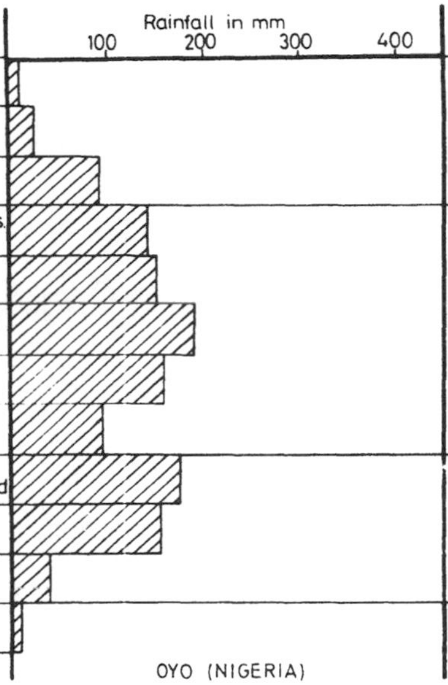

OYO (NIGERIA)

Figure 114b The farming calendar in the Middle Belt

Season	Month	Agricultural Activity	Rainfall in mm	Comments
Dry Season (Harmattan)	January	Clear farms, collect millet, guinea corn stalks. Cassava harvest starts		Livestock migrate southwards to feed on stubble
	February	Bush clearing continues		
	March	Distribution of manure. Collect firewood		
	April	Spread manure, make ridges. Harvest tree crops		
Rainy Season	May	Plant guinea corn and millet Move animals from farmland to stalls		Very busy planting month
	June	Make ridges and plant groundnuts. Weed.		
	July	Weed. Plant cowpeas		Severe floods may result and loss of crops
	August	Plant cowpeas. Weed. Repair ridges		
	September	Millet harvest. Cut grass for fodder. Plant cassava		
Dry Season	October	Groundnut harvest Dry groundnuts		
	November	Guinea corn harvest. Shell groundnut and sell		
	December	Sell groundnuts. Collect fodder. Graze animals on farmland		Slack season begins

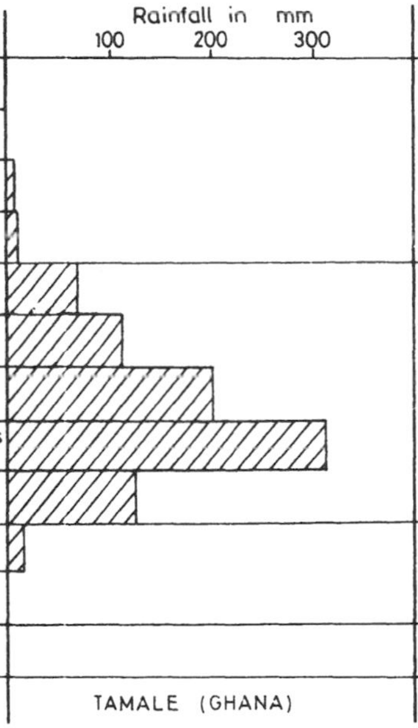

TAMALE (GHANA)

Figure 114c The farming calendar in the Sudan-Savanna zone

on agriculture will be discussed in general. The reader is advised to refer to appropriate sections of this book for additional information on rainfall, evaporation and evapotranspiration and water balance. Finally, in the discussion on climate and agriculture, some attention will be given to pastoral activities in the northern parts of the region. In this respect the role of pests and diseases in crop and livestock production will be treated at some length since these significantly affect productivity, and thus the well-being of even millions of people.

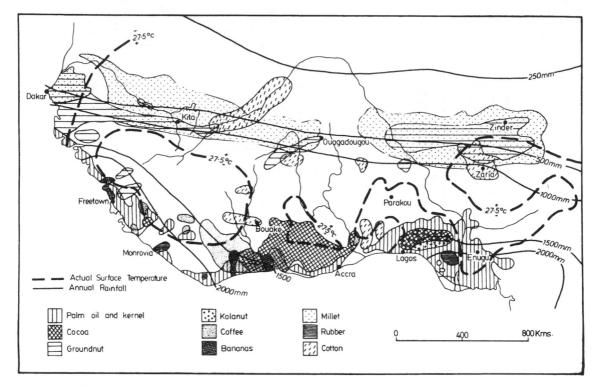

Figure 115 Selected major crops in West Africa with rainfall and temperature characteristics

Environmental factors affecting crop and livestock production in West Africa

The two main ecological factors affecting agricultural activities in West Africa are climate (solar radiation, temperature and rainfall) and soils. Rainfall distribution is most critical especially because most of the inhabitants are peasant farmers and depend upon its availability for their activities.

Apart from rainfall, it must be emphasized that other parameters like solar radiation and wind play a major role in crop and animal growth. Solar radiation provides the energy for photosynthesis, evaporation and evapotranspiration, while strong winds may do considerable damage to crops and affect yield also through evaporative processes and seed distribution of weeds.

Rainfall as a major factor in agricultural activities

Although rainfall in West Africa has been discussed elsewhere in this book, its role in agricultural activities is so vital that a further discussion of the subject is necessary.

It has been established that the temporal and spatial variations in precipitation are well marked (Table 27 and Figure 114). On the macromesoscale, they result in three distinguishable rainfall/cropping regimes: first, the monomodal regime such as occurs in the Sudan Savanna Zone in which Kano is located; second, the bimodal regime such as occurs in the coastal areas around Abidjan, Ivory Coast, and Warri in Nigeria; and third, the pseudo–bimodal regime such as occurs in the middle belt of Nigeria (as at Shaki, Oyo) and Tamale in Ghana (Lawson et al, 1979).

The monomodal regime prevails across northern West Africa. Its southern boundary lies between latitudes 8° and 11°N. With the wet season in the area lasting not more than two to five months, the most significant aspect of the temporal variability of rainfall in this regime is the irregularity of the onset and cessation of the rains. This effectively reduces the length of the cropping season at times or mars it altogether.

Table 27 *Rainfall means, extremes and variabilities at selected stations (mm)*

	Abidjan 1936–60				Ibadan (Moor Plantation) 1941–70			
	Mean	**20%**	**80%**	**Extremes**	**Mean**	**20%**	**80%**	**Extremes**
Jan.	26	0.0	50.0	0–122.0	9.4	0	17.5	0–46.5
Feb.	42	15.5	68.5	0–119.0	21.1	0	44.3	0–93.2
Mar.	120	56.5	150.5	27.0–287.0	79.5	42.7	122.8	8.0–179.3
Apr.	169	87.5	241.5	46.0–283.0	132.5	90.4	164.3	48.5–249.1
May	366	237.0	539.5	131.0–769.0	168.1	129.4	208.7	57.9–272.3
Jun.	608	306.0	764.5	48.0–1150.0	194.1	134.5	251.5	91.7–382.0
Jul.	200	26.0	278.5	0–732.0	156.7	62.5	242.6	2.8–357.4
Aug.	34	7.0	38.0	1.0–192.0	101.1	24.3	151.9	15.5–340.9
Sept.	55	17.0	110.5	1.0–185.0	189.0	95.0	260.4	71.9–508.0
Oct.	225	90.5	349.0	7.0–486.0	176.0	111.4	225.6	92.7–318.8
Nov.	188	106.5	375.0	93.0–358.0	40.5	28.1	66.2	0–135.1
Dec.	111	54.0	163.0	6.0–253.0	7.1	0	16.0	0–39.6
Total	2144				1275			

	Kano 1931–60				Daru 1941–60			
	Mean	**20%**	**80%**	**Extremes**	**Mean**	**20%**	**80%**	**Extremes**
Jan.	0	0	0	0–0.8	12	0	27.1	0–40.1
Feb.	0.3	0	0	0–6.3	32	1.5	73.7	0–116.1
Mar.	2	0	0.5	0–34.5	106	61.9	140.1	35.1–242.6
Apr.	8	0	16.7	0–61.0	152	111.7	186.9	82.3–247.7
May	71	25.9	117.5	0.5–224.3	251	174.3	330.2	143.8–424.4
Jun.	119	79.3	149.9	40.6–248.4	290	208.8	371.9	161.3–482.1
Jul.	209	108.9	260.5	39.1–369.0	285	226.0	326.7	161.5–424.9
Aug.	311	252.0	395.6	158.0–499.1	375	297.2	477.5	122.9–544.3
Sept.	137	81.1	205.3	30.5–276.1	408	321.3	498.9	257.9–665.7
Oct.	14	0	26.4	0–114.5	334	266.1	440.2	204.2–510.0
Nov.	0.3	0	0	0–4.3	206	151.1	268.6	61.0–437.9
Dec.	0	0	0	0–0.3	63	12.5	103.3	0–185.9
Total	872				2514			

Notes: 20 per cent of the years have rainfall below and 20 per cent rainfall above the 20% and 80% values respectively.
Source: Lawson et al. 1979.

The data for Kano City in Table 27 illustrate this regime. Here animal rearing has been well integrated into the agricultural activity in which only one grain crop can be grown in a year.

The bimodal regime lies south of the preceding zone between 8°N and the Gulf of Guinea and is restricted in longitudinal extent to between 5°E and 7°W. It is characterized by an appreciable drop in rainfall in July/August which effectively divides the period March/April–October into two cropping seasons (Figure 114b). The first and major season is plagued not only by uncertainties in the start of the rains but also by intraseasonal dry spells particularly in the month of May, well into the cropping season (Lawson, 1977). The data for Ibadan and Abidjan in Table 27 clearly depict this pattern.

The second season on the other hand is too limited in duration but can easily accommodate short season varieties of maize and a good crop of cowpea. Premature cessation of the rains during this season is, however, a possibility. Thus, from the moisture point of view primarily, neither of the two seasons can be considered optimal for maize or rice. No adverse effects have been demonstrated on root crops which grow through the period of break in the rains.

The third rainfall regime, the pseudo-bimodal regime, depicted by Warri and Daru in Table 27, covers the area east and west of the bimodal regime but differs from it in that the decline in the July/August rainfall is not sufficiently pronounced to result in a period of moisture deficit. The rainy season here is therefore much longer than in the other regimes and the number of raindays greater.

All these regimes are marked by high rainfall rates, often reaching upward of 50 mm hr^{-1} with short interval intensities in excess of 100 mm hr^{-1}, (Charreau, 1974; Kowal and Kassam, 1976; Lawson et al, 1979). The adverse impact of these high intensity rains in terms of structural deterioration of the soils, excessive runoff, and erosion has been well documented, as has their ultimate effect on yields (Charreau and Nicou, 1971; Lal, 1973).

Solar energy

This section assumes that the reader is familiar with the characteristics of solar radiation already discussed in this volume so that emphasis is now placed on the relation of solar energy to crops.

Solar radiation is of fundamental importance in the growth and development of crops. In agriculture, information on the amount and distribution of solar radiation incident on the crop surfaces is required for determining the available energy. The available energy is also of fundamental importance for the processes of evaporation, evapotranspiration and photosynthesis. Oguntoyinbo (1971b) and Stanhill et al (1966) have shown that of all the parameters of the radiation balance equation, the albedo (reflection coefficient) of the surface is the most important discriminant for the estimation of the radiation balance of natural vegetation and agricultural crop surfaces.

Surface reflectivity also has an important effect on the long wave radiation balance because absorbed solar radiation directly relates to the surface heating. Multiple reflection within the plant stand contributes to the shaping of the radiation profile; thus the reflectivity of individual plant elements merits consideration as does that of the canopy as a whole.

In short the reflection of whole plant communities and their individual elements is a factor of major importance in determining the plants' radiation balance, directly in the case of solar radiation and indirectly in the case of long-wave radiation. To a lesser extent reflectivity determines the distribution of solar radiation within the plant community and within the atmosphere, as well as the spectral composition of these fields (Oguntoyinbo, 1974).

The importance of albedo has been well recognized and attempts have been made to study it not only at the microscale (Rijks, 1965, 1967) but also at the macroscale (Davies, 1973; Kung et al, 1964; Oguntoyinbo, 1970a). Seginer (1969) has shown that evaporation and evapotranspiration can be reduced appreciably by increasing the surface albedo of plant surfaces. This technique is of great value to moisture conservation, especially in arid and semi-arid areas.

Photo-period

Because of the effects of the cloud cover, global radiation is generally low during the cropping season. This factor, which reduces photosynthetic activities in plants, has been shown to reduce yield in some crops by about 25 per cent (Lawson, 1977, 1978). In the area of monomodal rainfall regime in the interior, the high light intensity is conducive to considerably high photosynthesis but at the same time the higher amount of global radiation increases evapotranspiration in the area, which has been shown to be a moisture deficient zone.

Variation in daylength hours over West Africa is not pronounced as a result of the latitudinal extent of the region. A maximum difference of about 30 minutes in the south and 2 hours in the northern reaches of the area does exist over the normal cropping season (Table 7a, p. 26). This is, however, sufficient to induce important photo-periodic response in crops, a mechanism that has been used effectively by farmers in traditional farming systems.

Local short-day photosensitive varieties of sorghum planted in the savanna zone of northern Nigeria flower at the end of the rains in September/October thus avoiding conditions conducive to moulds and insect pests (Curtis, 1968; Bunting and Curtis, 1968). Similarly photosensitive cowpeas planted into millet and sorghum are triggered by short daylength to flower at the time the leaf area in the main standing crop starts to decrease, allowing increased light interception by the intercrop. The crop also sets pods and matures in the drier, sunnier environment free of many diseases (Steele, 1972; Kowal and Kassam, 1973).

Air temperature

Maximum near-surface air temperatures are high over the whole of West Africa at the onset of the rainy cropping season. The optimum range of temperature for most crops, 25–30°C, is often exceeded particularly in the northern areas. Concurrent photosynthesis may consequently be presumed to be at the sub-optimal level, even if allowance is made for species and varietal differences. In cowpea, for example, photosynthetic decline has been shown at temperatures above 30°C (IITA Annual Report, 1976).

The prevailing high temperatures also induce faster rates of growth and development of crops given adequate moisture, causing reduction in total dry matter accumulation and subsequently in yield. Such effects have been reported for certain varieties of rice (Yoshida, 1977) in which yield reduction was found to be accentuated by the shortening of the grain-fill period and also as a result of the higher ambient temperatures. Similarly daytime temperatures of 33°C in the post-flowering period were found to be detrimental to cowpea yields, causing a 45 per cent reduction over plants maintained at 27°C (Summerfield et al, 1978).

High temperatures during the day are also conducive to high rates of evapotranspiration. At the initial phase of growth much of the excessive moisture loss, in consequence, is from the soil with little surface cover. This rapid loss of soil water hastens the burning of young plants should the rains prove irregular. Crop yield is therefore adversely affected.

Relatively high night-time temperatures contribute to the reduced productivity of crops in the West African region by enchancement of dark respiratory losses, and reduction in net photosynthesis (De Wit, 1965; Monteith, 1977; Kassam and Kowal, 1973). Enhancement of vegetative growth with high night temperatures (24°C as compared to 19°C) has however been reported in cowpea (Huxley and Summerfield, 1974), but a strong negative effect in the reproductive phase was again noted. The overall effect therefore remains consistent with that on other crops.

Soil temperatures

The effects of soil temperatures on crop physiological and morphological processes have been widely reported. For example, a decrease in photosynthesis can occur in certain varieties of rice at root temperatures between 30°C and 40°C, with leaf temperature maintained at 30°C. Nishiyama (1976) gave temperatures of 18°C–40°C, 25–30°C and 25–28°C as optimum for seed germination, seedling emergence, and rooting respectively. Lal (1974) reported nutrient imbalance in young maize seedlings attributable to high soil temperature, and a decline in cowpea and rice growth under controlled conditions at constant root temperatures above 34°C and 32°C respectively. High soil temperatures have also been found to have a negative effect on nodulation in cowpea, nodule formation decreasing linearly between 31°C and 42°C (Philpotts, 1967).

Observed soil temperatures at different locations in West Africa exceed the values reported above, again particularly at the onset of the cropping season. Mid afternoon (15.00 LST) soil temperatures at 5 cm at the University of Ibadan over a 15-year period averaged 32.0°C in February

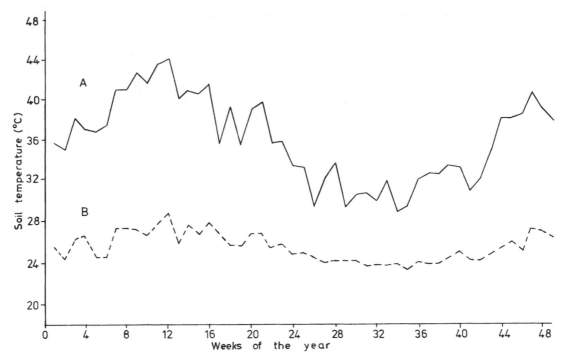

Figure 116 Soil temperature at 5 cm (IITA, 1977) (after Lawson et al. 1979)

Note: A – Max. temps.; B – Min. temps.

and 32.5°C in March. The extreme value recorded during the period was 39.8°C in February.

A typical annual trend in temperature at 5 cm in bare soil at the International Institute for Tropical Agriculture (IITA), Ibadan, is shown in Figure 116. In all cases, values on specific days certainly reach well above the averages for the periods. This would be particularly true after successive days without rain as soil moisture becomes depleted. It is thus reasonable to deduce that these high temperatures also possibly contribute to poorer crop growth and yield although their effect cannot be separated from the normally concurrent and well-established effects of moisture deficits. The rapid decrease in the temperature values as the rains are established is noteworthy in this respect.

Winds

Following the discussion in Chapter 2 on the seasonal shifts of the two major airmasses which constitute the large-scale circulation over West Africa, it is pertinent to consider here the medium-

and small-scale surface flows that more directly affect the meso/microclimate of crops.

High wind velocities of hurricane or gale force are uncommon in West Africa. Surface winds are generally light, frequently averaging less than 2 ms⁻¹. Wind speeds however increase considerably at the approach and passage of rainstorms and mesoscale disturbances or squall lines, reaching peaks of over 30 knots (56 km hr⁻¹). The direct effects of the surface winds on crops may be viewed as twofold. Turbulent transfer in crop canopies is sufficiently restricted during periods of calm to impede the exchange of heat, moisture and carbon dioxide between the crop and the surrounding atmosphere. Leaf and canopy temperatures may therefore increase beyond the optimum for high CO_2 concentration in the canopies of certain crops like maize (IITA Annual Report, 1976).

At the other end of the scale, winds of high speed passing over a field may blow the crops over from their vertical position and, in some cases, the crops may be broken; this is known as lodging. Lodging usually tends to be more drastic in cereals and the adverse effect on yield may be considerable (Table 28).

Surface wind may also affect crops on the mesoscale through their advective influence, especially as regards evaporative losses. This is much less pronounced in the south than in the north partly because of the lower mean speeds (4.2 km hr^{-1} to 5.8 km hr^{-1}) but also because of the smaller vapour pressure deficits in the south (Lawson, 1977). Although latitudinal difference in wind speed increases in July (3.3 km hr^{-1} in the south, 8.3 km hr^{-1} over the interior uplands in Nigeria (Ayoade, 1980)) it is in the dry season as represented by the month of January that the differential impact on evapotranspiration is well marked because of the low humidities, even though the mean speeds are more comparable at 3.3 km hr^{-1} in the south and 5.8 km hr^{-1} further north.

Humidity and dew

The higher humidities that persist through most of the year in the south are a determinant factor in the prevalence of many diseases in this zone as compared to the drier areas of the north. This high moisture content of the air even during the dry season is very much evidenced by the incidence of dew and precipitating fogs which have been observed to occur at maximum frequency in December and January (IITA, 1976). The resident time of the resulting moisture on leaves may reach 14 to 16 hours per day (IITA, 1975); during this time the rate of transpiration from the moist leaf (crop) surfaces may be reduced thus playing an important role in the water economy of the atmosphere. On the other hand the moisture available on the leaf (crop) surfaces may provide

Table 28 *Mean grain yield per plant with respect to time of lodging of maize (variety TZB), 1976*

Time of lodging	No. of samples	Mean grain yield per plant	
		Actual (g)	Relative (%)
4 wks after planting	77	99.5	93
6 wks after planting	74	87.2	82
8 wks after planting	80	66.6	62
10 wks after planting	78	62.3	58
12 wks after planting	80	102.3	96
Not lodged	311	106.7	100

Source: Adapted from Lawson 1977.

favourable sites for the development of pathogens like *Cladosporium fulvum* (grey mould) on tomato leaves, *Cercospora cruenta* (leaf spot) on cowpea and *Xanthomonas campestris Pv Vignicola* (bacterial blight) also on cowpea.

The ecology of selected crops in West Africa

This section highlights the basic environmental conditions which are optimal for the cultivation of the major crops in West Africa. The crops have been grouped into two main categories: the dicotyledons and the monocotyledons.

Dicotyledons

Rubber (*Hevea braziliensis*)

Rubber grows wild in the tropical rain forest in the Amazon basin. It was introduced into West Africa early this century (Purseglove, 1968). It grows successfully in areas between 15°N and 10°S where the climax vegetation is lowland tropical rain forest. Here, the climate is hot with temperatures ranging from 23° to 35°C and a well–distributed rainfall of 2500 to 5000 mm or more per annum; the lowest limit is 1270 mm. It can tolerate a wide range of soils with a pH of 3.8–8.0. It does best on well–drained loams with a pH between 4.0 and 6.5. Figure 115 shows that rubber is grown in areas having over 2000 mm of rain per annum evenly distributed with no less than 50–75 mm in any month and a uniformly high temperature of 27°C or more. Areas of southern Nigeria and Liberia satisfy these conditions.

Cassava (*Manihot escutenca*)

Cassava is believed to have originated in Central America (Purseglove, 1968). It appears to have been taken by the Portuguese to São Tomé and Fernando Po (now Bioko) and thence to Warri in southern Nigeria in the sixteenth century. From there it spread to the rest of West Africa where it has now become the most important staple food.

It is a lowland tropical crop but it can grow at elevations up to 1700 m on the equator. It thrives

171

in areas with rainfall ranging from 500 to 5000 mm. Except at planting it can withstand prolonged periods of drought, when it sheds its leaves, but 'comes alive' again quickly with the rain. It is therefore valuable in areas with low and uncertain rainfall.

It grows best on sandy or sandy loam soils and will produce an economic crop on exhausted soils unsuitable for other crops; hence it is often the last crop taken in the rotation in the shifting cultivation. Cassava is grown almost everywhere in West Africa where the mean annual rainfall is above 500 mm.

Groundnuts (*Arachis hypogaea*)

Groundnuts originated in South America and were brought from Brazil to West Africa by the Portuguese in the sixteenth century. They grow between latitudes 40°N and 40°S. They are a warm season crop and are killed by frost. Most of the crop is produced in areas with 1000 mm or more annual rainfall and there should be at least 500 mm of rainfall during the growing season. Dry weather is required for ripening and harvesting. The most suitable soils are well-drained, loose, friable, sandy loams well supplied with calcium and with moderate amounts of organic matter. This explains why the crop is grown in the Sudan-Savanna Zone across West Africa where the rainfall is adequate and the loose sedimentary soils of riverine and aeolian origins provide favourable conditions. In West Africa groundnuts grow best in the light sandy soils of the Sudan Zone with a rainfall of 640–1000 mm and a marked dry season for harvest. They form the main export crop of Senegal and Gambia. They used to be a major export crop in Nigeria but the 1968–73 drought and domestic consumption by local industries have reduced its importance as an export product.

Cotton (*Gossypium* spp.)

There is controversy as to the exact origin of cultivated cotton: whether it comes from the arid regions of tropical and sub-tropical Africa or from South East Asia (Saunders, 1961; Smith, 1967). Cotton is a low altitude crop. The optimum temperature for germination is 34°C, for growth of the seedlings between 24° and 29°C, and for later continuous growth 32°C. Low temperatures encourage vegetative growth while high temperatures increase the number of fruit-

ing branches and reduce the cropping period.

Cotton is a heliophyte which cannot tolerate shade or heavy rainfall, and where it is grown as a rainfed crop the average annual rainfall is usually about 1000–1500 mm per annum. In arid areas it is watered by irrigation.

It can be grown on a variety of soils from light sandy soils to heavy alluvium. Soil aeration, moisture and temperature are more important in germination and early growth than the seeds, which are starchy with a high oil and protein content. Figure 115 shows that the best cotton-growing areas in West Africa are those with a mean annual rainfall of 760–1400 mm, where most of it occurs during the growing season. There is abundant sunshine and a well-marked dry season when the bolls dry and are harvested. These conditions are generally well satisfied in the Zaria–Katsina regions of Nigeria, in the inland Niger Delta of Mali and other small pockets between these two major growing areas.

Coffee (*Coffee* spp.)

Arabica coffee originated from the Ethiopian highlands (Wellman, 1961). This is an upland species found between 900 and 2000 m above sea level. It occurs as an under-tree storey at temperatures approximately 16°C to 24°C with a well distributed annual rainfall of 1900 mm but with a drier period of 2–3 months during ripening and harvests. It grows well on slightly acidic, well-drained, fertile loams of lateritic or volcanic origin. The most important coffee-growing area in West Africa is in Ivory Coast where the rainfall is between 1100 and 1520 mm annually. Other smaller growing areas occur in Nigeria, Guinea and Sierra Leone.

Kola (*Cola* spp.)

Kola is a native of tropical Africa found in Cameroon, Ashanti (Ghana), Ivory Coast, Sierra Leone and Nigeria. It occurs naturally in the rainforest belt of West Africa between latitudes 6° and 7°N where the annual rainfall is between 1500 and 1800 mm. It requires well-marked wet and dry seasons. It grows best on well-drained fertile soils rich in humus.

Cocoa (*Theobroma cacao*)

Cocoa is a tropical crop which originated from the lower slopes of the Andes. The limits of cultivation are 20°N and S of the equator but the bulk of the crop is grown between 10°N and S. It was introduced by the Spanish and Portuguese to the islands in the Gulf of Guinea in the seventeenth century and was taken to Ghana in 1879 from Fernando Po.

It is grown mainly at low altitudes, usually less than 300 m. Areas of evergreen and semi-evergreen rainforest are the more suitable ecological zone for cocoa. The optimum temperature ranges between 21° and 32°C with small seasonal and diurnal range. Rainfall in the cocoa belt varies from 1000 to 2500 mm but most cocoa without irrigation is grown in areas with rainfall above 1250 mm. The rainfall should be well distributed, preferably 100 mm or over per month.

Cocoa can survive with dense shade but can also survive considerable exposure. Seedlings grow best under shade with approximately 25 per cent of full light. Self-shading occurs in mature trees, thus modifying the light relations. The amount of light may gradually be increased to 50 per cent as growth of the young trees occurs. Later, provided that the crop has optimal conditions of rainfall, drainage, soil aeration and nutrition, overhead shade may be gradually removed and the crop may be grown in full sunlight when considerable self-shading has been established. High winds may cause damage to young trees, therefore windbreaks may be desirable.

Cocoa requires a well-drained, well-aerated soil with good crumb structure and adequate supplies of water and nutrients. Soil depth should be at least 1.5 m to allow for ample root penetration. The lower limit of pH is 4.6 while the optimum range is 5.1–6.0.

The major cocoa-producing areas in West Africa are found in south-western Nigeria, Ivory Coast and Ghana, where the soil is rich and well drained and the rainfall is about 1270 mm annually. Many of these areas have suffered a decline in production due to old age of the cocoa trees and attacks by pests and diseases.

Monocotyledons

Banana (*Musa* spp.)

Banana is one of the most important tropical fruits. Its origin has been traced to Malaysia, and Pemba and Zanzibar. It was introduced into West Africa via the Zambesi valley, the East African Lakes and then Zaire (Simmonds, 1962). It is a tropical, humid, lowland crop found mostly between 30°N and 30°S. A mean monthly temperature of 27°C is optimal; at very high temperatures banana suffers from sun scorch but temperatures of 21°C or less result in a check in growth. The time required from planting to shooting is 7–9 months in the tropical lowlands, but this may be extended to 18 months at 1000 m or in the subtropics. Banana requires high light intensity.

It also requires a high amount of water and approximately 25 mm per week is minimal for satisfactory growth. An average of 2000–2500 mm that is well-distributed is considered most satisfactory. Light winds may cause characteristic tearing of leaves but this is probably not deleterious. Severe winds may damage or destroy the crop; windbreaks are therefore desirable.

Banana can be grown on a variety of soils as long as drainage is good and fertility and moisture are adequate. In West Africa, banana grows best on deep well-drained, retentive loams with high humus content originating from volcanic or alluvial sources with rainfall between 1900 and 2540 mm. On peasant farms in Nigeria and Ghana it is grown along with cocoa where it offers shade for the young cocoa. In the Cameroons and Ivory Coast, it is grown commonly on plantations.

Yam (*Dioscorea* spp.)

Yams originated from Asia and were brought into Africa in the sixteenth century by the Portuguese and Spanish travellers through the Malagasy Republic (Purseglove, 1972). Yams require a temperature of 25–30°C, a dry season of not more than 2–4 months in the year and a rainfall of at least 1500 mm evenly distributed throughout the year. They require deep, friable, fertile soil. In West Africa the northern limit of yam cultivation is 10°N beyond which the dry season is too long. The southern limit is set by the dry coastal areas of eastern Ghana, Togo and the swampy lagoons in the Niger Delta.

Yams require reasonable soil fertility and they are usually the first crops taken in the cropping cycle during shifting cultivation. They grow best in loose deep free-draining soils and they cannot tolerate waterlogging. Yam grows over a wide area of West Africa (Waitt, 1961) from the coastal areas where the rainfall is over 3000 mm to the middle belt where it is about 1000 mm per annum. Different species have become adapted to different rainfall conditions.

Rice (*Oryza sativa*)

Rice is the staple food of about half of the human race. It is believed to have been cultivated in India and south-east Asia for a long time but the date or place of its domestication is uncertain (Zukovskij, 1962). It shows a great tolerance of wet habitats.

It thrives over a wide range of climatic conditions from sea level to about 3000 m. The chief limiting factor to its growth is water supply. Hill or dry-land rice requires a high rainfall for successful cultivation. Swamp rice tends to be concentrated in flat lowlands, river basins and deltas with high temperature, abundant sunshine and adequate water. Rice is usually grown in monsoon regions particularly during the wet season when there is ample supply of water. Nevertheless, higher yields are usually obtained during the dry season when there is a greater amount of solar energy than during the wet cloudy monsoon months.

Cultivars vary in their resistance to drought, flooding and salinity and in their reaction to day-length and soil fertility. In the tropics two or even three crops can be gathered in a year, provided there is adequate water supply. The average temperature during the growing season varies from 20 to 38°C. Long periods of sunshine are essential for high yields. Grain yields are correlated with the amount of solar energy received, particularly during the last 45 days in the field.

Successful swamp paddy cultivation is dependent upon the supply and control of water to provide adequate inundation during the greater part of the growing period. The rice is usually transplanted in a well-soaked field with little standing water after which the depth of water is increased 15–30 cm as the plants grow. At plant maturity water is gradually reduced until the field is almost dry at harvest. Both rainfall and irrigation are required to provide a good crop.

Rice may be grown on a variety of soils ranging from sandy loams and shallow lateritic soils to heavy clays. Mangrove swamps have been successfully used for rice cultivation.

Both upland and swamp rice are grown in West Africa. In the Casamance and also in the swamp lands of Sierra Leone and the riverine basins of Nigeria, swamp rice is predominant, whilst in south-western Nigeria, upland rice predominates.

Bulrush millet (*Pennisetum typhoides*)

Bulrush millet is the staple food of the drier parts of tropical Africa, particularly in the northern territories of West Africa. Its origin has been traced to tropical Africa (Sampson, 1936). It is a rainfed crop in semi-arid regions particularly in the Sahel Zone of West Africa; in the Sudan Zone to the south it is of equal status with sorghum. The northern limit of sorghum in West Africa is marked by the 375 mm isohyet; that of bulrush millet is further north at the 250 mm isohyet. An even distribution of rainfall during the growing season is more important than the total precipitation. Bulrush millet is drought-resistant and suited to sandier and drier areas than sorghum. Too much rain at flowering can cause a crop failure and high temperatures are required to mature the crop. It is said to have one of the lowest water requirements of any cereal. It matures in 60 to 90 days and can be grown when the climate is too hot, the rainy season too short and the soil too poor for most other cereals. It is therefore not surprising that it is the most important crop through the length and breadth of the Sudan–Sahel Zone where the rainfall ranges between 500 and 1000 mm per annum.

Sugar cane (*Saccharum* cvs)

Sugar cane is believed to have its origins in eastern northern Africa. It requires high temperatures, plenty of sunlight, large quantities of water with at least 1500 mm per annum unless grown with irrigation. Optimum temperatures for the sprouting (germinating) of stem cuttings or settings used in planting are 32–38°C. Growth does not progress at a regular rate. Its growing season is bimodal, with a reduction in growth during the adverse conditions of the dry season. The ideal climate is a long, warm growing season, fairly dry and sunny but frost-free during ripening and harvesting. Duration of the crop ranges from 10 months to two years for the plant crop and 12 months for the ratoons. High winds are detrimental to the crop as uprooting and lodging may affect yield. It grows on a wide variety of soils but very heavy soils are usually preferred. It is grown mainly in the riverine swamps in West Africa or with the aid of irrigation.

Sorghum (*Sorghum bicolor*)

Sorghum originated from north-east Africa and was domesticated in Ethiopia (Doggett, 1970). It adapts to a wide range of ecological conditions. It can tolerate hot and dry conditions but can also be grown in areas of high rainfall. It may be grown under irrigation. The optimum temperature for growth is about 30°C. Its great merit is its drought-resistance. It is most extensively cultivated in the drier savannas and grasslands of Africa. The drought-resistance of sorghum is due to the following morphological and physiological properties:

1 the plant above the ground forms slowly until the root system has become well established;
2 sorghum produces twice as many secondary roots as maize;
3 silica deposits in the endodermis of the roots prevent collapse during drought stresses;
4 the leaf area is about half that of maize;
5 the leaves have a waxy coating;
6 the leaves roll in times of drought;
7 evapotranspiration for sorghum is about 20 per cent less than that of maize to produce an equivalent amount of dry matter;
8 sorghum can compete well with weeds once it has become well established; and
9 the plant can remain dormant during periods of drought and resumes growth when conditions become favourable.

Sorghum can tolerate a wide range of conditions; it will grow well on heavy soils and can endure temporary waterlogging. For this reason sorghum is found in West Africa mainly in the region where the rainfall ranges between 1500 mm in the south and 500 mm in the north.

Maize (*Zea mays*)

Maize is believed to have originated from central America (Mangelsdorf, 1965). It has a wide range of environmental conditions. It is grown in greatest amounts in regions with temperatures ranging between 21 and 30°C at tasselling. The optimum temperature for germination is 18–21°C. The most critical period is the 30 days of maximum growth before pollination when warm wet weather is required with 100–125 mm of rain. High temperatures and deficient moisture at this time may result in the pollen being shed before the silks are receptive. Hail can do great damage to the crop while strong winds can cause lodging.

Maize can be grown on a variety of soils but performs well on well-drained, well-aerated, deep, warm loams and silt loams containing adequate organic matter and well supplied with available nutrients. Maize grows successfully on soils with a pH of 5.0–8.0, but 6.0–7.0 is the optimum.

Because of the shortness of its life cycle it is possible to raise two crops of maize in the southern parts of West Africa where bimodal rainfall characteristics prevail. However, the late maize may suffer from water shortage and reduction in yield during those years when the late rains end abruptly. To the north where the rainfall characteristic changes to the monomodal type, maize is replaced by sorghum. Maize cultivation takes place in regions with rainfall ranging from 1000 mm in the middle belt to over 3000 mm in the coastal areas. The Niger Delta is excluded owing both to the heavy rainfall and to the permanently waterlogged conditions in the freshwater and mangrove swamps near the coast.

Livestock production

Large-scale livestock production in West Africa is predominantly extensive rather than intensive (Mortimore, 1983). It involves nomadic cattle breeding which has developed to utilize the environment to the utmost (Figure 117). This is limited by two environmental constraints: the

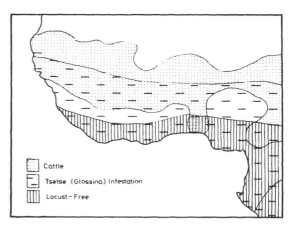

Figure 117 Distribution of cattle and major insect pests in West Africa

175

long dry season and the tsetse fly, the vector of *trypanosomiasis*. The first restricts the use made of the northern savanna during the period of low humidity and the second restricts the use that can be made of the southern savanna during the period of high humidity.

The adaptation to the system has resulted in the process known as pastoral nomadism involving a cycle of periodic movements; travelling in the dry season from the north to the south or from upland to lowland and back again in the wet season. Although the herds normally return to the same area every wet season, there is no permanent settlement.

Rangeland in the northern tsetse-free savannas provides abundant grass in the wet season but, except in the riverine swamps (*fadamas*) where grass roots can reach the subsurface water, the grass rapidly dries out with the onset of the dry season. At the same time the surface water gradually disappears and the herds may have to be watered from wells, a process which requires a great deal of labour. The nomadic herds move south for distances which range from 100 to 350 km although movements of up to 480 km are not unknown. An alternative course is to move to the large *fadamas* in the major river valleys. When the rains return many of the southern pastures and *fadamas* have to be evacuated because the tsetse multiplies rapidly and extends its area of distribution as soon as humidity increases.

One of the major nomadic tribes, the Bororo, recognizes five seasons of the year which are intimately bound up with these movements (Hoppen, 1958). These are:

1 *dungu*: This is the wet season period occurring between July and September. Grazing is generally good and the period is spent in the home grazing area.
2 *yawal* or *yamnde*: This is the hot season period between October and December after the rains. Surface water begins to dry up.
3 *dabbunde*: This is the cool harmattan season between December and February when crop residues are grazed after harvest; grass becomes scarce and trekking takes place to the south or to the *fadamas*.
4 *cheedu*: This is the hot dry season occurring between February and April when grazing and water are most scarce and the condition of both the animals and their herders weaken.
5 *seeto*: This is the stormy season occurring between May and July when the rains make an

uncertain start and herds return to their wet season grazing grounds; balancing the risk of the tsetse fly against the need to allow home grazing to recuperate.

Thus the seasonal routes of the Bororo are determined by the distribution of pasture, water and the tsetse fly.

Pests and diseases

The hot and humid climate of West Africa provides a favourable environment for a wide variety of pests and diseases which are deleterious to human beings, livestock and crops. In general pests and diseases proliferate when the temperature is high and water supply at a maximum. It was with this realization that Gourou (1966) identified the humid tropical climate as the fundamental cause of the low state of health in the area. He contended with ample justification that:

> The steady high temperatures, the high humidity of the air, the many water surfaces fed by rains are necessary for the maintenance of pathogenic complexes in which man, insect and a microbe are closely associated.

The effects of climate on the diseases and pests of crops are shown not only by the localization of the diseases and the occurrence of the pests, but also by the geographical limits of the range of cultivation of the crops, by the variations in yields due to differences in the seasons or date of planting and by the annual variations in yield corresponding to changes in weather (Adejuwon, 1983). Even though the limits of certain crops have been shown to correspond to the patterns of rainfall distribution, crop performance and range of distribution are not always directly related to climate but may be determined by disease. For instance, the incidence of *rosette* and other diseases of groundnuts increases from the north to the south in conjunction with rainfall. It is as a result of this, rather than as a result of climatic influence, that areas further south do not grow this crop extensively (Booker, 1963). It has also been noticed by Booker (1965) that the beetle *Ootheca mutabilis*, which eats the leaves of cowpeas and carries the *yellow mosaic* disease of the same crop, is not important in the drier northern areas. The effort to increase production in areas further south concentrates mainly on the use of insecticides.

Types of pests and diseases

Diseases of crops can be broadly grouped into three types; virus, fungal and bacterial. Bacterial diseases are the least significant economically but almost every crop grown in West Africa has its own virus and fungal disease. While fungal diseases affect a much larger number of crops than virus diseases, they seem to be more amenable to treatment.

The best known virus disease of cash crops in West Africa is *swollen shoot*. It is predominant in areas of high temperature and high humidity. The cash crop most susceptible to this disease is cocoa. The first recognizable symptoms of the disease are swellings on the branches, and twigs and leaves become asymmetrical. Later chlorosis takes the form of blotching or spotting. All these lead to poor tree development and reduced yield. *Swollen shoot* was first reported in Ghana in the 1930s and it later spread to Nigeria. Initially many cocoa trees had to be destroyed before suitable insecticides were developed for spraying the affected trees.

Most of the food crops are affected by one or other of a variety of virus diseases. In the main they affect the leaf parts, destroying the green matter and thereby reducing the capability of the plants to photosynthesize. Among the best known of these virus infections are *lanceolate mottled leaf*, *leaf spot*, *brown* and *red leaf spot* and *leaf mosaics*. *Lanceolate mottled leaf* and *mosaic* affect yams; *leaf spot* affects cassava and yams, while *leaf mosaics* are well known diseases of tobacco, kola nuts, groundnuts, cassava and yams.

The best known fungal disease also affects cocoa. This is *black pod* whose severity has been known to be closely related to the intensity of rainfall, particularly in the months of July and August. Cotton is also susceptible to fungal diseases especially in the wetter areas. *Rusts* are peculiar diseases of grain crops, most commonly in wet weather.

Among the most notorious pests are stem borers which do a great deal of damage to such crops as maize, rice, millet, kola, coffee and oil palm, and worms which destroy large fields of maize, yams and tobacco. Boll worms enter

cotton bolls and reduce the quality of the lint. Grasshoppers are perhaps the most dreaded insect pests for two reasons: first, they tend to affect large areas at the same time; second, they damage a large variety of crop types. For this reason an outbreak of grasshopper infestation leaves the population with few alternative food sources and the result has been widespread famine. Perhaps the best known grasshopper is the locust. Under certain conditions locust numbers can build to plague proportions and they move long distances consuming green vegetable matter *en route*. For about 13 years, from 1928 to 1941, tropical Africa suffered from a plague of migratory locusts (Batten, 1967). It was established (Lean, 1931) that the swarms originated in the flood plains of the Middle Niger where the mean annual rainfall is between 800 mm and 400 mm. From there the locusts migrated to all parts of West Africa and reached as far as eastern and southern Africa. As a result of certain control measures sponsored by international organizations, locust plagues have now become much less serious in West Africa (Figure 117).

However, other species, for example the variegated grasshopper (*Zonocerous variegatus*) have become a menace to most perennial food crops (Barker et al, 1977). Their numbers, which usually build up in the dry season, tend to be reduced at the onset of the rains. Consequently, the rainy season crops which include the grain crops are less susceptible to attack, while root crops like cassava, which remain in the field through the dry season, are most susceptible.

Among the animal diseases *trypanosomiasis* is the most widespread in West Africa. The disease, carried by the tsetse fly, is endemic in the forest areas as well as in the sub-humid areas with guinea savanna vegetation. It is because of this disease that extensive herds of cattle are limited to the northern areas of West Africa. Only the N'dama and Muturu breeds of cattle, which are resistant to the disease, are more commonly found in the south. Other animal diseases associated with climatic influence include *tuberculosis, rinderpest* and *bovine pleuro pneumonia*.

It is evident from the foregoing that pests and diseases play a major role in reducing the quality and quantity of crop and livestock yields in West Africa. Much research has been done to develop varieties of insecticides and pesticides suitable for each pest and disease. Some of the new varieties of insecticides and pesticides are *Perenox, Dieldrex 5, Alder T., Gamaline 20*, to mention a few.

Furthermore, research is being undertaken to identify the weather conditions under which crop spraying can be most effective (Johnstone and Huntington, 1977). In addition, aerial spraying of the variegated grasshopper and desert locust is usually undertaken during major outbreaks.

Conclusion

This chapter has demonstrated the role of the physical environment in agricultural activities in West Africa. To a very large extent, climate, especially rainfall, plays a major role in the location of the crops predominant in any particular zone.

The ecological characteristics of the major crops have also been noted. It is evident from this analysis that although most of the crops now being cultivated in West Africa are exotic, they have been introduced into zones in West Africa which are ecologically very suitable for their cultivation. The adaptation of the pastoral system to both the climatic conditions and the occurrence of pests and diseases has led to a system known as pastoral nomadism.

The high temperatures and abundant humidity over large areas of the region are conducive to the breeding of various pests and diseases which cause a great deal of damage to crops and livestock. The distribution of these pests and diseases and the gravity of their deleterious effects are, to a very large extent, controlled by climate.

Finally, the impact of drought and the role of irrigation, flood and forest fires, though not discussed in this chapter, cannot be underestimated in the agricultural experience of the inhabitants.

Agriculture is one of the most risky professions in the world. Natural hazards affecting farm output in West Africa include uncertainties of weather, such as moisture deficiency or drought, excessive moisture, in the form of flooding and waterlogging; strong winds, natural fires and possibly lightning. These may be called meteorological hazards. In the previous chapter the drought problem has been considered, as have its devastating effects in the interior of West Africa where the rainy season is short and rainfall duration is unreliable. In order to reduce the severity of such droughts large-scale irrigation is becoming widely practised in the drought-prone areas.

Additionally, the role of forest or bush fires deserves a brief mention. Even though the economic costs have not been quantified, there is no doubt that bush fires cause great damage to crops in particular, not only in the Sudan Savanna where forest fires are deliberately lit for various purposes during the dry season but also in the forest regions where accidental fires during the dry season may destroy large areas of tree crops.

13   Climate and health

The body and meteorological stimuli

Great care must be taken in associating ill-health with climate. So many other factors, including social and work pressures, life-styles, diet and especially levels of hygiene, must obviously be taken into consideration. In this chapter, however, and in the context of this book, it is the relationships between the West African climate and human physical well-being that are to be noted primarily, particularly under the headings of heat stress, disease and discomfort; radiation inputs and humidity are the crucial meteorological factors.

That these are the main factors may be appreciated when one considers the main parts of the human body which register meteorological stimuli. The *eyes,* exposed to over-bright light may become inflamed, even irreparably damaged; glare can induce headache. Prolonged periods in darkness, on the other hand, can produce changes in body metabolism, glandular activity, and blood sugar levels. The *respiratory tract and organs* react swiftly to changes in atmospheric humidity and temperature (as all readers in West Africa will know if they have experienced the sudden arrival or departure of the harmattan) or to air pollution levels, the pollutants ranging from dry season dust to pollen. The *skin* rapidly responds to thermal changes, especially to direct solar radiation, although the human body also receives energy inputs from reflected, refracted and scattered radiation. Skin surface and tissue temperatures are important. High temperatures will encourage some viral, bacterial and fungal infections; however it is the cooler parts of the body, the nose, ears, feet and hands, that suffer most from leprosy. Skin albedo is significant here. The increased number of cell layers in the epidermis and the high density of the pigment melanin in dark coloured skin, as opposed to that in pale skin, means a lower albedo (15 to 20 per cent in the visible light range for the dark skin compared with 30 to 40 per cent in the pale) and reduced penetration of short-wave radiation (0.4 mm compared with 2 mm). Five times less ultra-violet radiation reaches the upper dermis in the dark skin than in the pale. This lack of penetration affects the synthesis of vitamin D in the dermis. Only one quarter as much vitamin D can be formed in so-called black skin as in white at the same levels of ultra-violet radiation (Weihe, 1982). Rickets would be much more prevalent therefore in West Africa if the people kept indoors or were more heavily clothed.

Erytherma (inaccurately termed sunburn) or blistering of pale skins, especially of middle-aged males, will readily occur following exposure to ultra-violet radiation. This involves actual damage to the skin cells with possible serious consequences to health. A slow exposure, however, will bring the limited melanin pigments to the surface, affording greater protection from the radiation and leading to the darkening of the skin so sought after by the European tourist to West Africa. Excessive exposure can cause cancer of the skin however.

At low levels of humidity dry lips and skin may crack, lessening resistance of the body to invasion by dust and microflora. At high levels pores narrow, the skin swells and becomes soggy, providing ideal growth areas for bacteria and fungi. Slight cuts and abrasions that would probably be ignored in cooler climatic environments, may readily, therefore, become infected in tropical conditions. Exposed sites of infection may be slow to heal and may also attract disease-bearing insects.

179

Heat stress

Regulation of body heat is essential for health and the body not only receives heat from external sources but produces heat for itself, up to 73,000 gcal (gram-calories) hr^{-1} in an adult male and up to eight times this amount during short spells of vigorous activity. The body *gains* heat by:

1 absorption of radiant energy, as seen, from sunlight, from radiating and reflecting nearby objects and surfaces;
2 conduction, when the surrounding air temperature exceeds the skin surface temperature, or by contact with objects hotter than the skin;
3 condensation; and
4 bodily heat production due to bodily functions or physical activity (Table 29).

Table 29 *Metabolic rates of healthy adults*

Function	Metabolic rate	
	In Kcal m^{-2} hr^{-1}	In watts min^{-1}
Sleeping	35	70
Resting	40	
Office work	45–60	130–160
Driving a car	50–100	200
Walking (5 km hr^{-1})	160	290–400
Lorry driving	160	
Farm labouring	250	440–580–700

Sources: Fanger 1970; Koenigsberger et al. 1974.

The body *loses* heat, however, by:

1 outward radiation (at a still air temperature of about 20°C, 60 per cent of body heat is lost by radiation);
2 conduction, when the air or object in contact with the body is cooler than the skin; and
3 evaporation, from lungs and skin.

Clearly air temperature, humidity and movement and radiation inputs and outflows are all important in producing a bodily heat balance, controlling a balanced heat flow which may be summed up simply:

$$R + Cv + Cd + E + M = S = 0$$

where R is the level of radiation, Cv is convection, Cd is conduction, E is evaporation, M is metabolic heat production and S is heat storage.

It is less easy to protect the body against heat than it is against cold, but it is crucial to counter heat stress. If gains of heat exceed losses the rare condition of heat stroke can be achieved in which irreversible pathological changes in vital organs of the body will occur (Figure 118). Automatic bodily responses take place in a healthy person to seek to hold the body temperature at the optimum 37°C. Blood vessels in the skin dilate to allow more blood to pass through, but this can lead to a loss of blood to some organs including the brain, so early symptoms of heat stroke include lassitude, headaches, dizziness and fainting. The surface area of the body increases to expose more surfaces for cooling, muscle tone diminishes, lethargy increases, urine output falls. A feeling of thirst is an encouragement to take in moisture to reduce dehydration, but appetite for food, the digestion of which creates heat, is reduced. The need for water intake is especially marked because of the most obvious bodily response to overheating, that of sweating. At a still-air temperature of 32°C, 90 per cent of bodily heat production is lost by sweating. The rapid loss of moisture from the skin will lead to a loss also of sodium (Na) from the blood, which is replaced by potassium (K). This readily leads to cramps and nausea.

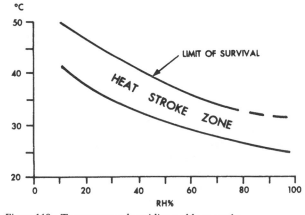

Figure 118 Temperature, humidity and heat stroke

The climate and human water needs

The normal body water requirement (adult male) is about 3000 cc per day, coming mainly from hydrated foods and drink. The body normally excretes, perspires and respires 10 per cent more moisture than it takes in, the excess coming from the oxidation of foodstuffs, hence the danger of increasing the outflow. A rise of rectal temperature only 1°C from 37.5° to 38.5°C will produce an increase in the heartbeat rate of even 40 per cent. A rapid regulatory reaction by the hypothalmus starts or increases the sweating mechanism, reaching an output up to 1.7 litres per hour, particularly dangerous to children and the elderly.

In a hot desert environment, such as the Sahara, the mean sweat output would be about 5 litres per day, twice the level of the losses from people in the forest zones. Table 30 (from Gates, 1972) shows likely losses of body water by evaporation at two different air temperatures and for different activities. The surface temperature of the skin is normally between 33 and 35°C. A naked man, unable to find shade and exposed to air temperatures above these skin temperatures, would lose some 19 litres of water in a 12 hour day. Hard physical work could lift this loss to 48 litres, impossible to replace, for after rapid consumption of 1 litre the desire to drink more disappears. A loss of 19 litres per day would mean a loss of

Table 30 *Body water losses (g hr^{-1}) by evaporation*

Function	Losses (g hr^{-1}) at dry-bulb air temperatures	
	35°C	40°C
Sitting in the shade naked	220	360
Sitting in the shade clothed	245	305
Sitting in the sun naked	385	615
Sitting in the sun clothed	280	460
Marching in the sun clothed	640	910

Source: Gates 1972.

Table 31 *Days of expected survival of resting man without water at given temperatures*

Daily maximum temp. °C	21	27	32	38	43	49
Daily minimum temp. °C	16	21	27	29	35	39
Days of survival		18	14	7	4	2.5	2

23.7 per cent of body weight in one day from a 'standard' man of 80 kg. With a 4 per cent loss apathy is marked; with an 8 per cent loss speech becomes very difficult; at 12 per cent it is not possible to swallow; at 18 per cent death will occur. Dehydration of 10 per cent raises the body temperature of a resting man by 2°C. The same rise will occur in an active man with dehydration of 6 to 7 per cent. When the body temperature reaches 41°C death is inevitable (Adolph, 1947).

Note that humidity does not add to *heat* stress, although this will be greater in the windless forests than out of them in areas of similar humidity levels. Illnesses in desert areas are commonly related to heat stress: heat exhaustion, dehydration, cramp and related psychological disorders. To offset the problems it is necessary to clothe one's body, despite the high temperatures. Clothing actually diminishes losses of moisture by sweating and reduces heat gain. Pale coloured loose garments might seem the obvious choice, yet many peoples in Saharan countries wear black clothing. This disadvantageously intercepts radiation more thoroughly than white clothing, and converts it more readily to heat, but, if the clothing is loose, this is then lost by convection into the atmosphere, not into the skin.

Because of the low atmospheric humidity levels in desert areas sweating may seem not to occur significantly, the evaporation of the moisture off the skin being so rapid. It does take place of course, so all efforts should be made to reduce heat gain and maintain fluid intakes. To rest in shade is beneficial, but to lie down on hot ground is harmful. 5 to 6 litres of water per day are required by the inactive man, and up to 10 litres by the active, in daily mean shade temperatures of 32°C. Without water therefore, even a resting man would be unlikely to survive more than two days in temperatures exceeding 39°C (Table 31) (Gates, 1972).

Acclimatization to heat includes a quicker onset of sweating, a decrease in salts, especially NaCl in the sweat, and an increase in blood flow

181

to bodily extremities to facilitate heat loss. Pulse rates and lung ventilation tend also to decrease. However acclimatization to dehydration is not possible (see, *inter alia*, Landsberg, 1969, 1972 and Tromp, 1980).

The climate and food value

The levels of body water loss noted for the hot, dry tropics could result in over 20 g day^{-1} of salt (NaCl) loss. In more humid areas 10 to 15 g day^{-1} of NaCl is needed. This is best obtained from food, not tablets, and foods rich in Na such as fish, cheese, bread and eggs are valuable here. However, more commonly consumed in West Africa are beans, groundnuts and bananas, foods rich in potassium (K) which is less needed, as noted above, as it replaces the Na lost in sweating. Iron (Fe) is another element essential for good health but lacking in some people if suffering blood losses due to such ailments as hookworm or haemolytic anaemia of malaria. Also the concentration on a vegetable diet in West Africa lessens iron absorption, for from vegetable sources only about 5 per cent of iron is absorbed, compared with a higher proportion from meat sources (Waterlow, 1982).

As already stated, it is not unusual for appetite to diminish as air temperatures rise. Calorific intake may decrease by 5 per cent for every 10°C rise in temperature above a reference base of 10°C.

Vitamins are essential to health and, sadly, these too are often deficient in the diets of many people in West Africa, despite the consumption of fruit, because of the over-dependency upon rice, millet, maize, cassava and plantain. These same staples also provide less than 4 g protein per 100 calories and are also deficient in one or more essential amino-acids. Yet a growing child needs even more protein than an adult. Groundnuts, beans, pulses and milk are valuable in the absence of fish, meat and eggs. But a child is too often weaned from mother's milk to the deficient grains or tuber crops. A child needs 4 g protein per kg of body weight. Most West African countries could do with increasing the protein of the national diets by 8 to 10 per cent. Although none of the crops mentioned is indigenous to West Africa, they are nevertheless the product of tropical climates.

In a study in Gambia (Waterlow, 1982) children suffered from infections for 14 per cent of the rainy season and there was a deficit of weight gain of 3 g kg^{-1} day^{-1} in the period of infection. In the remaining 86 per cent of the time an extra weight gain, on top of normal growth, is desirable if the child is not to be stunted, but even the relatively low requirements of 2.5 kcal (kilogram-calories) and 0.13 g protein kg^{-1} day^{-1} were not being achieved.

Lassitude and irritability

An inadequate energy intake, or the presence of disease organisms or worms in the body consuming the energy absorbed, will lead to diminished ability to fight disease, to increased lethargy and even to risk of retardation in social and mental development.

It has been suggested (Sulman, 1976) that lethargy in tropical climatic conditions may result from diminished adrenaline. This product of the adrenal glands normally raises blood pressure, stimulates the heart muscles and accelerates the heart rate. Initially warm weather induces the output of adrenaline, but in prolonged hot climate conditions the production is exhausted. Apparent lassitude may be purely psychological, but here again weather conditions may be significant. For some people the grey, hazy atmosphere associated with the harmattan or the absence of sunshine in the rainy season may be psychologically depressing. Conversely the drier or cooler periods may encourage physical and mental activity. The relevance of this to academic, sporting or industrial/ commercial performance should not be ignored.

On the other hand irritability and aggressiveness can increase in prolonged spells of hot weather, especially if the humidity is low. In humid West Africa positive and negative ions in the atmosphere may perhaps range from 500 to 1000 per cubic centimetre. With the arrival of the harmattan numbers may rise to 1500 or higher. This rise triggers off an increase in serotonin neurohormone in the intestines, raising the blood pressure, producing allergy symptoms and what has been identified as serotonin irritation syndrome, a marked increase in irritability. It is suggested (Sulman et al, 1974) that a quarter of the population subjected to these climatic conditions may suffer from this ailment. Similar conditions, however, can prevail during thunderstorms when the air is rich in positive ions.

Climate and disease

The impact of the harmattan does not stop with the results noted above. Other reactions to the hot dry wind include: thyroid complaints, hypertension, ataxia, adynamia, hypoglycemia, presumed to be due to catecholamine deficiency; and also, due to serotonin release, migraine, vomiting, scotoma, edema, heart pain, asthma, rheumatism, hyperperistalis, hay fever, conjunctivitis, laryngitis, pharyngitis, tracheitis and sinusitis! The average human adult circulates about 12 m³ of air through the lungs daily. In dusty, polluted urban areas 10^{12} particles of dust per day may pass into the respiratory system, many not to be exhaled again.

Irritation in the throat, cold symptoms and catarrh are common. Figure 119 shows the analysis of respiratory diseases for the period 1960–75 at two cities in different climatic zones in Nigeria (after Adefolalu, 1984b). These figures show that in the north of Nigeria the number of respiratory problems for which patients were admitted to hospitals tended to approach maximum during the dry season. At Kano, the minimum number of patients treated correlated with the rainy season, between March and August. The peculiar maximum of September coincided with an anomalous dry spell, a prelude to the drought of 1973. In the south (the case of Ibadan), although the number of hospital patients being treated for respiratory disease was at a maximum during the dry months, the actual number was lower than that for the northern example.

In more recent times (1976–80) there has been a rapid increase in the number of cases of pneumonia and bronchitis reported in the southern parts of Nigeria (Table 32). Generally the pre-monsoon maximum number of infections suggests that the transition months (April to May) between the harmattan and the monsoon are critical periods in the south.

Table 33 shows the situation at Zaria from 1976 to 1980. In this table respiratory infections are categorized under three main headings: ordinary or general respiratory infections, pneumonia and bronchitis, in accordance with the International Code for Diseases (ICD). With reference to general respiratory infections, the percentage frequency of infection during the harmattan period (HP in the table) as a function of the yearly total (YR) is 58.8 per cent in the five-year period. Pneumonia showed an average

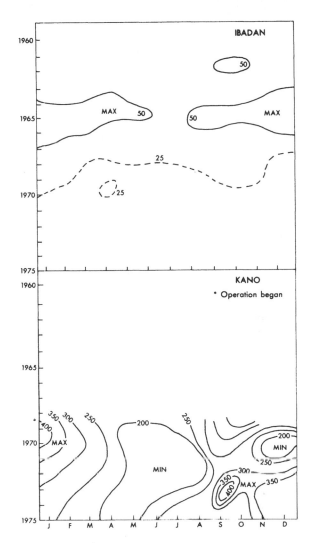

Figure 119 Respiratory diseases at Ibadan and Kano, Nigeria (after Adefolalu 1984b)

Table 32 *Total number of patients (in-and out-patients) treated for respiratory diseases at Ogoja, south-eastern Nigeria, 1977–9*

Year	J	F	M	A	M	J	J	A	S	O	N	D
1977	43	57	54	44	58	668	50	66	29	53	32	30
1978	17	40	82	486	224	177	168	298	253	314	367	187
1979	258	320	258	329	1366	757	1021	562	416	388	396	252

Table 33 *Respiratory infections during the harmattan period, October–March (Hp) and yearly totals (Yr) in Zaria, northern Nigeria, 1976–80*

	1976			1977			1978			1979			1980		
	Yr	Hp	%	Yr	Hp	%	Yr	Hp	%	Yr	Hp	%	Yr	Hp	%
General respiratory infections	372	224	60.2	181	113	62.4	216	126	58.8	167	96	57.4	91	45	57.2
Pneumonia	1653	1143	69.1	1724	816	47.3	1865	700	37.5	1807	859	53.0	1809	711	39.3
Tonsilitis, bronchitis, sinusitis, asthma, etc.	198	122	61.6	125	74	59.2	155	72	46.4	1558	115	9.1	245	116	47.3

Source: Ahmadu Bello University Teaching Hospital, Zaria.

of 48.8 per cent occurrence during the harmattan. The third category of ailments had an average harmattan occurrence of 53.1 per cent.

So much for the wind sometimes described as 'the doctor' because with its arrival there is usually a decrease in those illnesses associated with humidity and the presence of water!

Worst of all, however, in West Africa are the marked outbreaks of cerebrospinal (meningococcal) meningitis that are associated annually with the onset of harmattan conditions. It is suggested (Waddy, 1952) that the main climatic factor responsible is low absolute humidity, the critical value being 10 g m^{-3}. The meningococcus bacterium is transmitted through the populations that tend to huddle together in ill-ventilated houses during the cold nights experienced when the surface air flow is from the Sahara. The drying of the nasal passages reduces resistance to infection.

The most infamous diseases of West Africa, however, are those most prevalent in the humid south and south-west, where the main scourges are probably malaria, tuberculosis, malnutritional problems, leprosy, helminthiasis (diseased condition due to worms in the body, including filariasis, loa loa, onchocerciasis (river blindness), elephantiasis and ankylostomiasis (hookworm)), schistosomiasis (bilharziasis), diarrhoeal diseases, trypanosomiasis (sleeping sickness), venereal disease, yaws and meningitis. To these might be added dengue fever, yellow fever, typhoid, cholera, hepatitis and measles, making a depressingly long yet not exhaustive list.

The incidence of some of the diseases is indicated in Table 34 and Table 35 provides some interesting but saddening statistics concerning life expectancy and death rates in West African states. The data in Table 34 are hardly recent,

but it is highly probable that the situation has changed little since 1957. It is particularly difficult to obtain or estimate levels of incidence of disease for most African countries, as World Health Organization publications indicate. Records of cases in hospitals hardly equate with actual numbers of cases, too few sick people having ready access to modern medical care. The wide ranges of some of the figures in Table 34 illustrate the uncertainties concerning a knowledge of the incidence of disease in West Africa. They do, however, indicate that too many people suffer from the diseases noted.

Table 34 *Incidence of certain diseases in West Africa, 1957*

	Schistosoma (%)	Malaria	Filariasis (%)	Hook-worm (%)	Yaws
French West Africa	10–20	20–80%	10–90	10–90	heavy
Nigeria	47	100,000+	27	15–30	30,000
Gold Coast	10–20	60,000	21	40–50	144,000
Ivory Coast	11	20–80%	10–20+	30–90	63%
Liberia	under 10	21–34%	14–20	59–80	107,000
Sierra Leone	under 10	10–30,000	10–20	over 20	7400
Guinea-Bissau	10–20	20–100%	49	90–95	heavy

Source: Lee 1957.

Table 35 *Life expectancy, birth and death rates in West Africa*

	Life expectancy		Birth rate (per 1000)	Death rate (per 1000)	Infant mortality (per 1000 live births)
	Male	Female			
Benin	39	42	51.1	24.6	110
Bourkina Faso	32	31	48.1	24.0	182
Cameroon	44	48	43.2	19.2	127
Gambia	32	35	48.3	30.4	204
Ghana	48	52	47.1	15.9	156
Guinea	38	40	46.9	25.3	216
Guinea-Bissau	39	43	40.9	21.9	154
Ivory Coast	43	47	45.9	19.5	138
Liberia	46	44	49.8	20.9	159
Mali	47	50	43.2	18.1	121
Mauritania	40	44	50.0	22.5	187
Niger	39	42	50.9	25.0	200
Nigeria	37	37	50.6	18.5	124
Senegal	40	43	47.9	22.5	93
Sierra Leone	31	33	47.8	31.9	215
Togo	32	38	45.5	18.6	127
For comparison					
Japan	74	80	12.8	6.0	7
USA	71	78	15.5	8.6	11
England and Wales	70	77	12.6	11.7	11
USSR	64	74	20.1	10.3	28

Source: UN Demographic Yearbook 1983 (published 1985).

Of the diseases listed above venereal disease is universal and unrelated to climatic conditions, and others, for example tuberculosis, measles, meningitis, dysentery and hepatitis are by no means purely tropical problems. However, directly or indirectly, climate does play a part in producing these illnesses in West Africa. Some, for example cholera, typhoid, diarrhoeal diseases, tapeworms, hookworm, schistosomiasis and hepatitis, are the result of infection due mainly to poor hygiene, but there is a close correlation between the presence of the cause of cholera, the bacteria *Vibrio cholerae* and *Vibrio eltor,* and high air temperatures and still waters rich in organic matter, sheltered from sunlight: conditions found to perfection in the rainy season in West Africa, to which area it has spread in recent years from Asia.

However, this disease is not confined to the wet season. There was a significant outbreak of cholera in the period October to March 1971-2, in Ibadan, Nigeria. Adesina (1981) attributed the occurrence to the fact that during the dry season many of the city's inhabitants resort to drinking and using highly contaminated water from streams (which at best have turned into unconnected pools of stagnant water or become brackish, muddy and polluted), since the flow of potable water through public pipes is very irregular during these dry periods. To aggravate the situation, the hot, dry season is the period when people have a greater propensity to drink water or consume raw drinks such as palm wine (usually diluted with water from dubious sources). It is not surprising that the diffusion of cholera may increase in the dry season.

Gastroenteritis due to bacterial infection may also be closely associated with warm air conditions and is usually transmitted through infected water. *Schistosomiasis haematobium* is the commonest form of bilharzia in West Africa and is on the increase. The main host, the snail *Bulinus*, prefers calm, warm water conditions typical of many quiet streams and pools and irrigation channels in West Africa.

Other diseases listed, for example, measles, tuberculosis, yaws, meningitis, trachoma, are contact infections and so, again, may seem to have little to do with climatic conditions. The indirect connection between meningitis and the dry season in West Africa has already been mentioned, however, and measles and trachoma are also prominent at the same time. In many lands the drier part of the year is the main season for festivals, when large numbers of people gather together, providing the appropriate conditions for the spread of infections such as measles, a very serious illness for the young in Africa.

Most of the remaining listed diseases are insect-borne infections and here the climate plays a more obvious role. Mosquitoes are the transmitters of malaria, yellow fever and dengue fever, and require water in which to lay their eggs for development through larvae and pupae to adult insects. Female anopheline mosquitoes, the carriers of malaria which still kills one million people per year in the world, require temperatures in excess of 16°C and relative humidity about 60 per cent for optimum breeding conditions, whereas temperatures in excess of 35°C or relative humidity levels below 25 per cent are lethal. Culicine mosquitoes of the genus *Aedes* transmit yellow fever and dengue fever, but prefer temperatures over 24°C.

Gamble (1952) established that in Gambia infant mortality reaches its peak during the latter part of the rainy season, particularly during August, September and early October. He ascribed this increase in infant mortality to a number of reasons, the most important of which is the increase in malaria towards the end of the rains when conditions in the low-lying lands bordering the river beds and creeks become swamp and are therefore ideal for mosquito breeding. Consequently, the population of mosquitoes in the villages builds up to enormous numbers by the end of the rains and decreases when the swamps begin to dry up and the colder weather comes in December. In order to reduce the population of the mosquito in the homes, straw (*Digitaria* spp) is burnt in the compounds; this creates a foul atmosphere in the houses that might well aggravate any lung trouble developed by a small child. The result is a very high rate of infant mortality in the rainy season.

Loa loa ('Calabar swellings') is a filarial infection, transmitted by the female *Chrysops silacea* and *dimidiata* which is essentially a forest dwelling insect of eastern West Africa, preferring the lowland forest climate. Acanthacheilonema infection, so common in eastern Nigeria and Cameroon, is transmitted by the midge *Culicoides milnei,* whose larval and pupal stages occur in rain water in cut banana stems. Bancroftian filariasis, causing elephantiasis, is also conveyed by the bite of some species of mosquito. Onchocerciasis, or river blindness, is transmitted by the small black fly *Simulium*. Its need for fast-flowing, oxygenated water for breeding purposes means that this disease is mainly seasonal in the savannas, though perennial in the West African coastlands. The prevalence of leprosy in West Africa (45 per 1000) is perhaps the world's highest, and the correspondence between high incidence and high rainfall is marked (Hunter and Thomas, 1984).

Climate and discomfort

Despite all the above the climate of West Africa has unjustifiably earned for itself the reputation for being 'unhealthy'. It is also commonly regarded as uncomfortable. Although heat stress and discomfort are not the same, persistently high temperatures can be debilitating and such conditions are typical of coastal West Africa except in the far north-west. Inland by day, temperatures near the limit of human endurance can be experienced.

Thermal comfort is a complicated and highly subjective matter however. It is desirable for maximum intellectual, perceptual and manual performance, but it varies widely for different people. Everyone experiences discomfort at some time, but age, health, sex, stature, skin colour, clothing, physical activity and housing all influence the sensation. Older people, having a slower metabolism, tend to prefer higher temperatures. Women have lower metabolic rates than men and therefore generally prefer temperatures slightly higher than those appreciated by men. Thin people, having a greater ratio of body surface area to volume, are able to tolerate higher temperatures more than fat people who, with their greater insulation need cooler air to dissipate body heat. A dark skin may absorb three times more solar radiation than a light skin, but is able to withstand damage from ultra-violet radiation. But the dark skin also emits heat in the same proportion, so the colour has little effect on thermal stress. At very high temperatures, and especially if it is windy, clothing is necessary for protection, as a counter against dehydration. Loose clothing with insulation for legs and arms is desirable. As evaporation rates fall so clothing will probably need to be discarded, for when skin surface and air temperatures are at the same level some 750 grams of sweat per hour may need to be evaporated from the skin surface. If the clothing is damp this evaporation level may have to double. Thermal comfort requires the evaporation of all sweat.

So the thickness and porosity of clothing is obviously significant. The total thermal resistance from skin to outer surface of a clothed body (I_{cl}) can be measured. I_{cl} is called a clo–unit.

$$I_{cl} = \frac{R_{cl}}{0.18}$$

where R_{cl} is the total heat transfer resistance to outer surface of a clothed body in kcal m^{-2} hr^{-1} at a given temperature. The higher the clo–unit the worse the discomfort. If the nude body is given a clo–unit value of zero then the values in the same climatic conditions for various clothing types are as follows:

Clothing	I_{cl}
Shorts	0.1
Shorts + open-neck, short-sleeved shirt + sandals + light socks	0.3 to 0.4
Long lightweight trousers, open-neck short-sleeved shirt + socks + sandals	0.5
Cotton long-sleeved shirt + trousers + shoes and socks + jacket	0.9
Business suit	1.0
Suit + underwear + woollen socks + shoes	1.5
Heavy wool suit	3.0 to 4.0

(after Fanger, 1970)

If a sedentary, naked person is comfortable at 30°C, the same person in a business suit would prefer a temperature some 9°C lower. It is interesting that in Europe and North America to greet people 'warmly' is to greet them in a friendly manner, but in parts of Africa, the writer understands, a 'cool' greeting is a friendly one!

For many people in West Africa discomfort is correlated subjectively with conditions of sultriness, in which high temperatures are associated with high humidity levels. Knoch and Schulze (1956) published a map of mean sultriness for Africa. For each value of relative humidity a critical temperature can be calculated at which vapour pressure equals 14.08 mm of mercury in the hygrometer. The difference between the critical temperature and the actual dry-bulb air temperature is the 'thermic sultriness value'. If this is positive the air temperature is still below the critical limit and the climatic conditions are said to be 'comfortable'. Figure 120 shows the pattern of sultriness for West Africa and suggests that the least comfortable climate is to be found along the southernmost Guinea coast, with conditions steadily improving towards the Ahaggar mountains, with the Fouta Djalon and Jos Plateau also ameliorating conditions, as do northerly

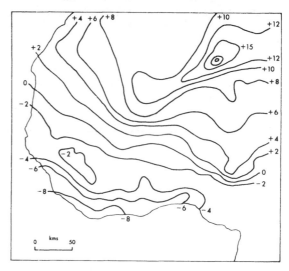

Figure 120 Sultriness in West Africa (after Knoch and Schulze 1956)

Note: The lower the value, the greater the sultriness.

Table 36 *Brazol's (1954) scale of climatic sensation*

Class	Sensation	Wet-bulb temp. °C	Classification
12	lethal heat	above 31	
11	intolerable heat	26–31	
10	suffocating heat	19–26	hyperthermal climates (a)
9	uncomfortable heat	12–19	
8	very warm	11–12	
7	summer warm	10–11	
6	autumn, spring comfort	8.5–10	homeothermal climates (b)
5	winter acceptably cool	7.5–8.5	
4	cool	6–7.5	
3	moderately cold	3.5–6	hypothermal climates (c)
2	cold	2.5–3.5	
1	icy	below 2.5	

winds along the coast of Western Sahara.

Interesting as the map is it does not, of course, really indicate those areas of West Africa most suitable, climatically, to live in. The wind factor is not built into the picture and discomfort from cold will be a marked feature of the northern and highest areas, especially in the northern hemisphere winter and at night. Purely on the grounds of temperature, (wet-bulb), Brazol (1954) suggested a scale of climatic sensation as in Table

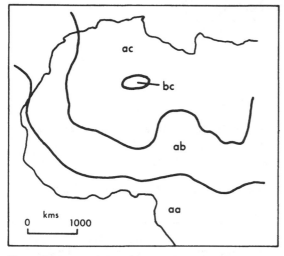

Figure 121 Air enthalpy climates of West Africa (after Gregorczuk 1967)

Table 37 *Gregorczuk's (1967) scale of climatic sensation*

		Enthalpy kcal kg⁻¹	
Class	Sensation	Warmest month	Coldest month
aa	permanently hyperthermal	above 11	above 11
ab	hyperthermal/ comfortable	above 11	7.5–11
ac	hyperthermal/ hypothermal	above 11	below 7.5
bb	permanently comfortable	7.5–11	7.5–11
bc	comfortable/ hypothermal	7.5–11	below 7.5
cc	permanently hypothermal	below 7.5	below 7.5

36. Gregorczuk (1967, quoted in Landsberg, 1972) expanded Brazol's three major categories of climate into six as in Table 37 and, applied to West Africa, produced Figure 121.

Airflow over the body is of particular importance. If the relative humidity reaches 90 per cent the water evaporating from the skin will saturate the layer of air that is in contact with the skin, preventing further evaporation. If this thin layer of air can be removed however, by air flow over the body substituting in its place unsaturated air, then evaporation from the skin can recommence. In most parts of West Africa the highest humidity levels tend to occur in the cloudy periods when air temperatures are generally below skin temperature.

Discomfort indices

In the early 1920s Houghton and Yaglou (1923) sought to devise a discomfort index by exposing clothed and half-naked adults to varying degrees of humidity, temperature and air flow and asking them to assess their feelings of warmth or cold. From the results of this experiment conducted in the USA, in restricted conditions, an Effective Temperature index was devised:

$$ET = 0.4 (Td + Tw) + 15$$

where Td is the dry-bulb air temperature and Tw the wet-bulb in °F.

Modifications to the methodology were made by Houghton and others by 1926 to permit wider applicability of the system and in the 1950s further developments were carried out by the American Society of Heating and Ventilating Engineers (Thom, 1959). Thom reused the early 1923 formula substituting DI (Discomfort Index) for ET (the Effective Temperature). Further modifications, depending on input data available, produced alternative indices as follows:

$$THI \text{ (Temperature-Humidity Index)} = 0.55 \, Td + 0.2 \, Tdp + 17.5$$

$$\text{or } THI = Td - (0.55 - 0.0055rh)(Td - 58)$$

where Td is, as before, the dry-bulb temperature in °F, Tdp is the dewpoint temperature and rh is the relative humidity.

Gates (1972) expressed the US THI as in Figure 122. For instance, at a relative humidity of

100 per cent and a temperature of 70°F (21.1°C), or at a relative humidity of 20 per cent and a temperature of 80°F (26.7°C), or at any combinations of relative humidity and temperature on the straight line joining these two points on the graph, the THI will be 70. It is suggested that 10 per cent of normal healthy adults will feel uncomfortable with a THI of 70, 50 per cent will feel discomfort with a THI of 75, and all people will be uncomfortable if the THI reaches 80.

Terjung (1967), using daily mean, maximum and minimum temperature data, mean daily highest and lowest relative humidity, mean hourly highest and lowest wind speeds where known, and percentage of possible sunshine or hours of actual sunshine from which radiation data could be computed, applied the Effective Temperature Index (after Thom) to Africa. The outcome for West Africa is shown in Figure 123 a and b. An activity level of 175 kcal m^{-2} hr^{-1} was used as a base (cf Table 118). The maximum ET then was calculated to be 82.6°F (28.1°C), above which all people would feel uncomfortable.

In the map for July (Figure 123a) it will be seen that nearly all parts of West Africa north of 14°N have values above this suggested critical value of 28.1°C. (At 85°F (29.4°C) the work rate should be reduced to 125 kcal m^{-2} hr^{-1}.) The ET is reduced in the Ahaggar mountains and along the north-west coast and the southern Guinea coast, because of cooling sea breezes. In January (Figure 123b) pockets of high ET value occur in eastern Gambia and Senegal, in southern Mali and northern Benin. Values in southern West Africa do not vary significantly from July, indicating that uniformity of climate which some people find enervating.

In 1947 McArdle et al published a further index of heat stress, the P_4SR, the predicted four-hour sweat rate. Climatic factors, metabolic heat production and clothing were all considered. It was assumed that stress inflicted by the environment is limitless, but that there is a limit to the amount of sweat the body can produce. The rate of sweating is taken to be a sound measure of heat stress. The P_4SR was calculated to be 4.5 litres, although losses of 2.5 litres would be seriously damaging to health.

Terjung (1967) also applied the P_4SR index to Africa and Figures 123c and d illustrate the conditions for January and July again. In January the daytime P_4SR nowhere exceeds the critical 2.5 litres level, but the same three areas with the highest ET noted above also have P_4SR values of 2 litres. In July once again areas north of 14°N experience highly stressful conditions, the area around Taoudenni in northern Mali being particularly bad. The ameliorating effects of sea breezes are again prominent in the north-west. Of course the maps show daytime conditions only; night relief (even cold stress) is possible in desert regions.

Two further indices were adopted by Terjung and applied to Africa: Relative Strain (RS) (after Lee, 1965) and Still-Air Temperature (SAT) (after Burton and Edholm, 1955), but only the former is now briefly mentioned. The index is stated:

$$RS = \frac{E_R}{E_m}$$

where E_R is the heat stress registered as the evaporative cooling required to compensate for the conditions which include metabolic heat rate, air-temperature, radiation heat load, insulation of the air in clo-units, insulation of clothing when saturated, vapour pressure, air resistance to passage of water vapour, maximum respiratory volume and surface area of the body! E_m is the maximum stress possible as reflected by the maximum cooling possible, that is the maximum compensation that can be tolerated. Clearly the higher the value of RS the greater the strain and

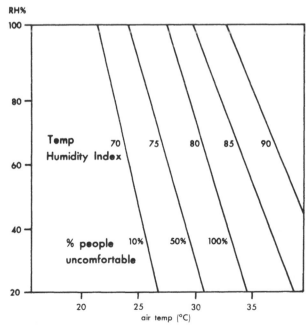

Figure 122 A temperature–humidity index (after Gates 1972)

Note: For explanation see text

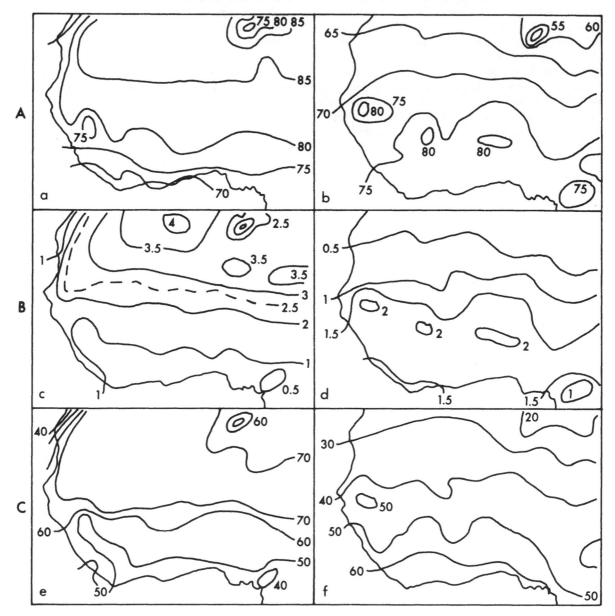

Figure 123 A Effective temperature (°F) in *a* July; *b* January;
B Predicted four-hour sweat rate (litres) in *c* July; *d* January;
C Relative strain in *e* July; *f* January (after Terjung 1967)

Figures 123 e and f again illustrate the situation for January and July. In January conditions are more stressful along the southern Guinea coast than in the cloudier, cooler month of July. Once again the area around eastern Gambia has an anomalously high value. In July again West Africa north of 14°N has extremely stressful values above 70, excluding as

usual the north-west coast and the Ahaggar mountains.

Although all the indices devised have their limitations and their critics, the comparable patterns illustrated in Figures 123 a, c and e, or b, d and f suggest that the methods applied may have real value in attempting to measure and map levels of heat stress and discomfort. Although Terjung used data from 162 stations in West Africa, even this number enables only a very general view of the pattern of stress or discomfort for the region to be seen. Ayoade (1978)

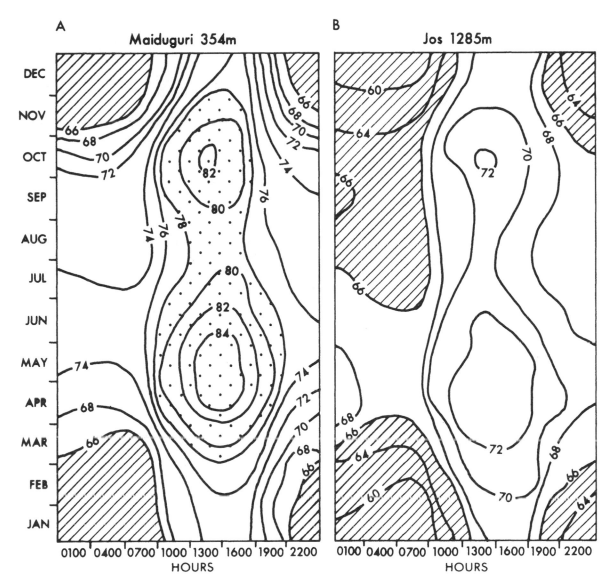

A
Maiduguri 354m

B
Jos 1285m

HOURS

HOURS

Figure 124 Physiological comfort at *a* Maiduguri; *b* Jos,
Nigeria (after Ayoade 1978)

applied the Thom ET index to Nigeria alone,
using data for 13.00 hours when climatic conditions are least comfortable. He noted ET values
greater than 78°F (25.6°C) for the southern half
of the country in January. In July the northern
areas and the Niger–Benue valleys in central
Nigeria had an ET greater than 78°F (Figure 124a).

ET index values rise through the morning, as
one would expect, attaining highest levels between 13.00 and 16.00 hours. Minimum values

tend to be between 22.00 and 10.00 hours.
Values are lower in the dry season than the wet,
especially between 10.00 and 19.00 hours. The
highest values everywhere are just before the
rains. For long periods Jos can attain ETs below
66°C (18.9°C), regarded as the minimum value
for comfort; below this value cold stress is experienced, particularly in the evenings (Figure 124b).
Ayoade defined four comfort zones for Nigeria:
I, the south; II, the centre; III, the Kaduna–Jos
area; IV, the north. Zone III is the least stressful.
The Niger–Benue valleys are the most uncomfortable. Kaduna is the most favoured town in

191

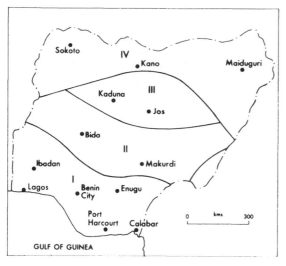

Figure 125 Comfort regions for Nigeria (after Ayoade 1978)

Nigeria, using these criteria, and Lokoja and Maiduguri are the most stressful places. Jos is at times too cold for comfort (Figure 125).

By 1968 Terjung had extended his studies and by superimposing a system of subjectively derived categories of comfort on a psychrometric chart produced the nomograph of comfort index reproduced as Figure 126. The divisions defined, with bounds shown as on the nomograph, were: +4 extremely hot; +3 sultry; +2 hot; +1 warm; 0 mild; −1 cool; −2 keen; −3 cold; −4 very cold; −5 extremely cold; −6 ultra cold. The categories −3 to −6 do not apply to West Africa. The index for a place is derived by plotting mean dry-bulb temperatures against relative humidity (or vapour pressure or dew-point temperature or wet-bulb temperature). Terjung subdivided his main categories however on the basis of daily mean maximum (daytime) and mean minimum (night) temperatures. Thus, if daytime temper-

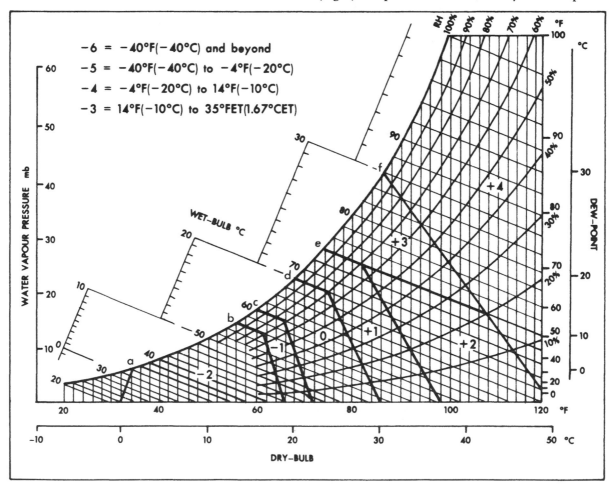

Figure 126 The Terjung Nomograph of Comfort (after Terjung 1968)

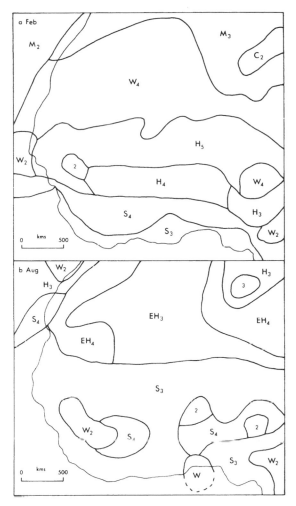

Figure 127 Comfort indices for West Africa in *a* February, *b* August (after Terjung 1968)

Note: See text and Table 38.

atures fall within the extremely hot range (+4), but nights are sultry (+3) (extremely hot nights are considered as absent), the conditions are classified as EH$_1$ (Table 38). Applied to West Africa the comfort index patterns for February and August are as shown in Figure 127. The reader may wish to compare results with others shown above. For further extensions of Terjung's method and for his map of annual climates reference should be made to his 1968 paper. Attempts to define comfort have not stopped with Terjung, however, and in more recent years interesting work has been published by Koenigsberger et al (1974) and Pagney and Besancenot (1982).

Koenigsberger defined comfort ranges in the tropics as lying between 22°C minimum and 27°C maximum, with 25°C the optimum. But again, air movement must be taken into consideration and when added to the nomogram (Figure 128) provides the Corrected Effective Temperature (CET). The comfort zone delimited includes air movement up to 1.5 ms^{-1}. At this level dust will be raised and papers blown about, but these non-climatic irritations are regarded as acceptable for the sake of thermal comfort. Wind speeds greater than 1.5 ms^{-1} would be unacceptable, however. In his nomogram example Koenigsberger shows a dry-bulb temperature of 30°C and a wet-bulb temperature of 26°C (which values just about equate with the mean February values for Cotonou in Benin). These temperature points are joined by the line A–B. With little or no air movement the CET is 27.5°C, which is outside the comfort zone. Airflow of 0.5 ms^{-1} produces just tolerable conditions and an airflow of 1.5 ms^{-1} produces an upper limit of comfort CET of 26°C. At night the dry-bulb and wet-bulb temperatures may be the same (they commonly

Table 38 *Terjung's day/night Comfort Index*

Extremely hot	Sultry	Hot	Warm	Mild	Cool	Keen
+4/+3 = EH$_1$	+3/+3 = S$_1$	+2/+2 = H$_1$	+1/+1 = W$_1$	0/0 = M$_1$	–1/–1 = C$_1$	–2/–2 = K$_1$
+4/+2 = EH$_2$	+3/+2 = S$_2$	+2/+1 = H$_2$	+1/+0 = W$_2$	0/–1 = M$_2$	–1/–2 = C$_2$	–2/–3 = K$_2$
+4/+1 = EH$_3$	+3/+1 = S$_3$	+2/+0 = H$_3$	+1/+1 = W$_3$	0/–2 = M$_3$	–1/–3 = C$_3$	–2/–4 = K$_3$
etc.	etc.	etc.	etc.	etc.	etc.	etc.

Note: The categories: Very keen (VK); Cold (CD); Very cold (VC); Extremely cold (EC); and Ultra cold (UC) do not apply to West Africa.

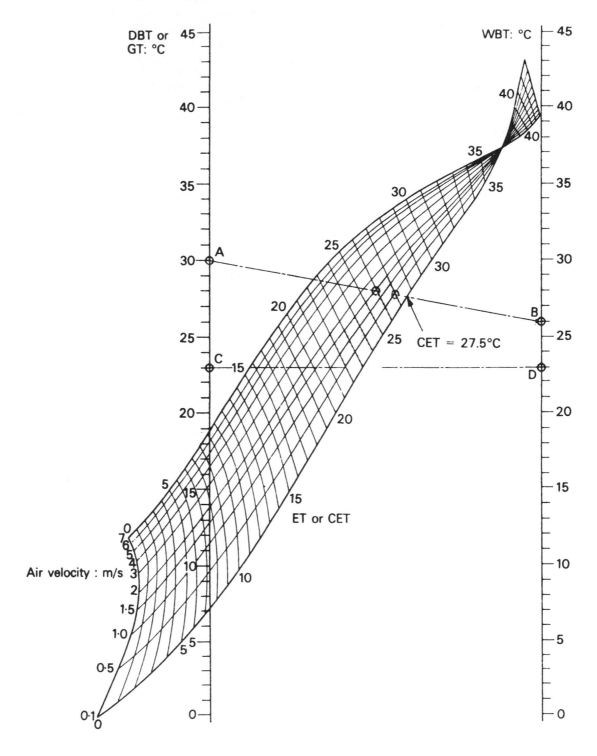

Figure 128   Koenigsberger Comfort Nomogram (after Koenigsberger et al. 1974)

Note: See text.

are in coastal West Africa). Conditions are comfortable then with little or no airflow, but if this exceeds 0.5 ms⁻¹ conditions become too cool. As noted in Chapter 3, if the dry-bulb temperature and relative humidity of a place are known, the wet-bulb temperature can be readily obtained from psychrometric tables or charts. An example of a chart is given as Appendix 8 and Appendix 9 provides a set of meteorological data for 38 stations in 16 countries of West Africa as defined in this book. Readers may wish to apply some of these data to the equations and nomograms reproduced above.

Pagney and Besancenot (1982) used three bioclimatic indices. (1) The K index.

$$K = (10.45 + 10\sqrt{v} - v) (33 - t°)$$

where v is the airflow velocity in ms⁻¹ representing the chill factor, 33 is the skin temperature in °C and t° is the dry-bulb temperature in °C. Measurement is in kcal m⁻² hr⁻¹ of body surface. Values of 899 to 600 kcal m⁻² hr⁻¹ are termed Tonic conditions, 599 to 300 are Relaxing, 299 to 150 are Hypotonic, 149 to 0 are Atonic. If K is less than zero, conditions are called Endothermic. In this case the air temperature is higher than the skin temperature so that there is danger of overheating. The suggestion is that when K is less than 150 the average adult will feel discomfort. (2) The THI index. Here Thom's equation is used:

$$THI = t° − [(0.55 − 0.0055 \, U\%) (t° − 14.5°C)]$$

noted before, where °t is the dry-bulb air temperature in °C and U% is the relative humidity. The greater the THI the greater the discomfort. 26.5°C is taken to be the critical value. (3) The Ū index. This relates to the gas exchanges to and from the lungs. A theoretical balance in water vapour terms is achieved at 31.3 mb water vapour pressure. However actual balance is achieved at a much lower level and it is assumed that body discomfort occurs in the lungs when Ū exceeds 26.5 mb (when atmospheric water would enter and dilute the blood plasma).

Using these three indices Pagney and Besancenot produced an accumulative index of discomfort for the middle of the day for tropical Africa, and mapped the results as in Figure 129 for West Africa, which might be used to sum up this discussion of discomfort. The accumulative (annual) index was produced as follows: if no

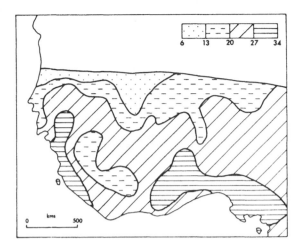

Figure 129 An accumulative index of discomfort for the middle of the day (after Pagney and Besancenot 1982)

Note: See text.

place has K, THI or Ū beyond the discomfort level in any month between noon and 15.00 hours ST, it is given an accumulative index value of zero. On the other hand if a place has discomfort values for all three indices for all 12 months of the year it would have an accumulative index of 36 (three indices over the critical values multiplied by 12). Values range then from zero to 36 and for Figure 129 have been grouped according to the legend. Again, even in this crude map, the influence of altitude and of cooling sea breezes in the north-west are apparent.

Calculating the index values for each month of the year on the basis of monthly means, permits seasonal patterns of discomfort to be demonstrated as well as the annual picture. Pagney and Besancenot illustrated these by means of piegraphs and in West Africa for four stations (Figure 130). Lastly the data used permit daily rhythms of discomfort also to be assessed, illustrated by graphs for two stations in West Africa: St Louis in Senegal and Gao in Mali (Figure 131).

Despite the arbitrariness of the indices and thresholds of discomfort chosen, Pagney and Besancenot indicate the practical benefits that might be derived from their approach, benefits perhaps to be obtained from many of the systems discussed in this chapter:

> Even on a small scale, our methods can show trends which may benefit man. Furthermore they enable us to gain insight into diurnal and

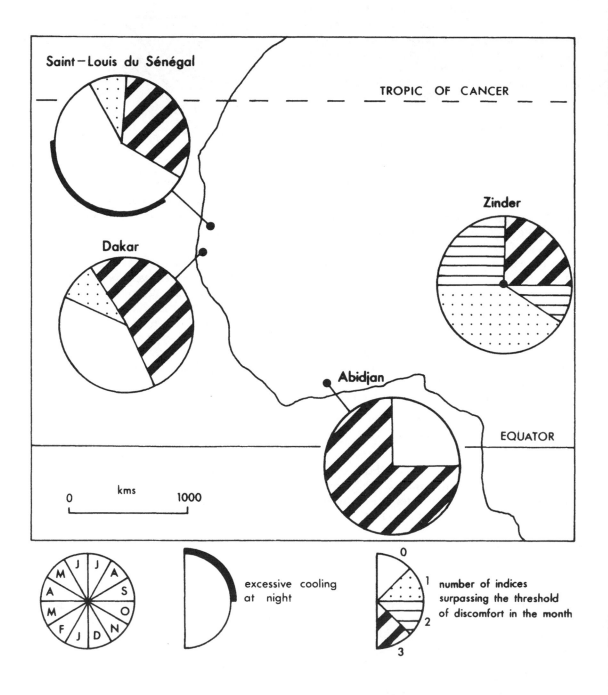

Figure 130   Seasonal bioclimatic rhythms for four stations in West Africa (after Pagney and Besancenot 1982)

Note: See text.

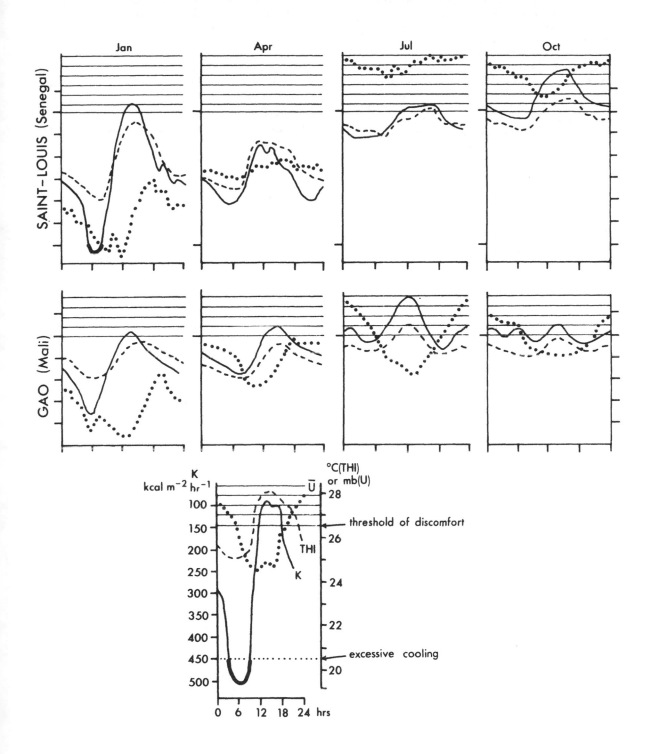

Figure 131   Hourly bioclimatic rhythms for St Louis and Gao in four months (after Pagney and Besancenot 1982)

seasonal patterns, and to draw conclusions about life rhythms (work and rest periods). In the domain of urban geography we can likewise benefit from the precision of bioclimatic responses to different altitudes, exposures and even to the location and interior structure of buildings. If we possess a well adapted network of measurements ... we should be able to examine and help solve some problems of natural and artificial air conditioning in urban centres: by seeking optimum air movement conditions in the location of housing and the most favourable bioclimatic elements in the choice of sites for schools and hospitals. This should result in a rationalised plan for the location of public works and also in the optimum use of energy (artificial air conditioning). Bioclimatic maps of urban centres would provide a useful starting point.

Here they are noting the application of their method to indoor climates as well as outdoor. People in urban areas, working in offices, shops, factories, may spend two-thirds of their lives indoors. Even in village communities, and especially in the rains, people may pass half their lives indoors. The indoor climate merits some attention, therefore, and this it receives briefly in the next chapter.

14   Climate and building design, industry and transportation in West Africa

Buildings

Introduction

Buildings for people in West Africa range from the simplest grass hut to the office tower block or modern presidential palace. They are essentially constructed for shelter, comfort, storage or cultural requirements. It follows, therefore, that in designing buildings there are three main considerations: first, the people and their needs; second, the climate in which the people live; and third, the available, affordable, building materials. In addition the site on which the building is to be erected should be considered and, possibly, the cost and level of upkeep or maintenance required.

For most people the crucial building is the house and in West Africa the rural house, mainly used for shelter from the rain and, at night, for privacy and storage. The availability of construction materials usually reflects the climate, from the skin or cloth tent of the desert, to clay and mud, wood and grass in the seasonally or more permanently humid areas, as does the way in which the materials are used. For, as seen in the previous chapter, amongst the needs of people is some level of physiological comfort and this, given the environmental constraints, has been achieved to a fair degree by vernacular architecture. In recent years pressures to abandon the old methods of building and to substitute modern alternatives has had some regrettable consequences, as many school children and office workers will know.

Indoor comfort

The discussion of comfort in Chapter 13 was primarily concerned with outdoor conditions.

For southern Nigeria, Ambler (1955) calculated the Effective Temperature (cf. Koenigsberger's work outlined in Chapter 13) as:

$$WB + \left(\frac{DB - WB}{3} \right) - V$$

where WB is the wet-bulb temperature (°F in Ambler's work), DB is the dry-bulb temperature and V is the wind speed (miles per hour). He claimed that for mean indoor conditions and for a sedentary male of 102 kg the Comfort Zone lay between 71°F (21.7°C) and 74.6°F (23.7°C).

For the permanently humid tropics Webb (1960) stated the Comfort Index as:

$$I = tw + 0.447 (t - tw) - 0.231 \, v^{\mathsf{i}}$$

where t is the dry-bulb temperature (°F), tw the wet-bulb temperature, v the wind speed in feet per minute. Significant discomfort, he claimed, commences indoors at 78°F (25.6°C) wet-bulb temperature. Stephenson (1963) added a ventilation factor to his calculations determining the most suitable temperature to be 20.6°C and the highest acceptable temperature for sedentary persons to be between 24.4 and 25.6°C. The reader's attention is also drawn to the work of Brooks (1946) and especially Olgyay (1952, 1963).

Thermal comfort depends upon the ambient and radiant temperatures, air movement and humidity, all highly variable factors. Protection is required from radiation, glare and dust. To the climatic elements in confined spaces will be added the energy and humidity emitted by the occupants and resulting from cooking and bathing. The internal environment is influenced also by the absorption of heat by the external surfaces of the building, in part determined by the colour of those surfaces and the thermal capacity of the materials from which the walls and roofs are made, and by the ventilation. Thus the number, size, shape and location of windows and doors and the level of shading of openings, walls and roofs are all important.

Building needs in humid West Africa

In southern West Africa with its almost permanent high levels of humidity and cloudiness, with prolific vegetation growth and abundant animal and insect life, it is desirable to insulate buildings against the incoming radiation and to facilitate air movement, to counter thermal stress and to protect against the rainfall and insects.

Roofs are more important than walls. They should be of low mass to reduce heat gain and re-radiation into the living space below. They should preferably be of light colour to maximize reflection and of double construction to provide insulation. They should be inclined for rapid removal of the rainfall directly into ground drains, not roof gutters which might provide mosquito breeding sites, to reduce the risk of soil erosion around the building. They should overhang the walls to protect walls from the rain and from direct insolation.

Full advantage should be taken of cooling breezes, so villages should be sited appropriately; buildings should not be close together; individual houses, if possible, should face the breeze. (Compound houses providing pockets of still air are often undesirable for climatic and medical reasons.) Buildings should be spaced at distances six times the height of the buildings in order to provide adequate air movement, especially as eddy effects behind structures can produce 'wind shadows'. It is also desirable then to stagger the layouts of buildings and not to put them in long, straight rows aligned with the wind. However, building sites should not be exposed to storms or driving rain, or placed downwind of air-polluting industrial complexes or odiferous rubbish dumps or by polluted waters.

Windows should be large and on opposite walls, 0.5 to 1.5 m above the floor, facing north and south. In the northern hemisphere summer the north-facing windows can be closed with blinds or shutters, in winter the south-facing windows. Louvres can direct the airflow to bed or floor level. Internal louvres are not recommended, however, unless they are used to control energy flow into the room before internal heating has occurred from the penetration of direct insolation. External, horizontal shades, set off from the walls to reduce conduction, are best for a high-angle sun; vertical shades for the low-angle rays. They may reduce radiation penetration by up to 90 per cent (the internal blinds by only 30 to 60 per cent). Glass in windows should be avoided as it increases the 'greenhouse effect' in the building, and internal curtains can also absorb radiation and reduce ventilation. Mosquito netting, although desirable to counter insect invasions, may reduce air velocity by 35 to 70 per cent.

Verandahs are useful as cool extra living space and for added protection against driving rain.

External shading from bushes and trees is also desirable, although the vegetation will reduce ventilation. Tall trees are best, providing the shade required yet permitting airflow beneath the high foliage. A further benefit provided by vegetation is as a counter to the problem of noise which can be considerable in open buildings. Shading must not be excessive, however, as interior darkness will increase the risk of accident. Windows are designed to let in light as well as breeze, although how much enters depends upon cloudiness, time of day and year, reflection of light from the surroundings as well as the shade cast by foliage.

Elevation The rural building particularly in humid West Africa may beneficially be elevated on stilts, so to catch the airflow, but also to provide some protection against the incursions of insects and animals. The aim is to keep out the heat and the glare but not the light, to keep out the rain but not the breeze.

Buildings in arid West Africa

In arid, northern West Africa the requirements for building design are very different from those in the south. The basic environmental conditions include low humidity, a lack of cloud and vegetation, much glare and dust. Breezes are here not beneficial as the flowing air is too warm and dust-laden. It is now advantageous to have enclosed, compact, inward-facing buildings, close together to decrease radiation on walls and to maximize shade, aided by verandahs and trees. Inner courtyards will also reduce air turbulence and dust. Temperatures under shade may be 20°C lower than in direct sunlight. Twisting lanes between houses are helpful to provide shade. Markets (souks) are advantageously roofed, if only with palm fronds.

Roofs are now less important than walls and can be flat as rainfall is limited, providing additional spaces for sleeping and storage, for drying clothes and food. Like the *walls* they should be of thick earth, stone or brick, so to absorb heat over a prolonged period before passing it to the interior of the building. This provides cooler daytime conditions therefore, but hot nights as the stored energy is re-radiated inwards at this time (Figure 132).

Windows should be kept to a minimum, especially on outer walls, and placed high up to reduce glare and dust. However this arrangement is poor for ventilation and, as noted in Chapter 13, not good for health in crowded family units. Buildings should be light coloured to increase reflection. White adobe walls absorb about 10 per cent of incident radiation, light grey walls 40 to 50 per cent, dark colours 60 to 70 per cent and black walls 80 to 90 per cent.

In the Sahel with its seasonal alternation of intensely hot, arid and warm, wet conditions, a mixture of the building designs of both the humid south and arid north is desirable, involving courtyards, day and night rooms, overhanging roofs, verandahs and surrounding foliage. Dwellings in the coastal north-west will advantageously be open to ameliorating sea breezes from the west by day, but closed to chilling night winds from the east.

Traditional rural house types in West Africa

These commonly have the characteristics described above, but especially in the south the round or rectangular hut of lattice pole-work infilled with mud, with or without a clay or chalk plaster, or the hut of moulded mud brick, with grass or raffia palm thatch, is giving way to the rectangular dwelling made from concrete blocks with corrugated iron roof.

Something of the variation in interior climate that can occur in these various buildings can be deduced from observations made by Peel (1954). He studied the temperature conditions over a 24-hour period in four dwellings in Bo in Sierra Leone. In each the windows were closed at night, primarily for fear of thieves, snakes and insects. This lack of nocturnal ventilation, plus the inward radiation from the thick mud walls, caused temperatures between 20.00 and 08.00 to be up to 3°C higher on the inside of the walls than on the outside, and room temperatures to be higher than air temperatures outside. It was a metal-roofed, *non-insulated* house that showed the greatest discrepancy and produced the highest internal temperatures. In the roof space beneath the corrugated iron pitched at 30°, the midday dry-bulb temperature reached 37°C (14/15 February). The roof in this study, however, was painted black, hence its absorption of solar radiation would have been greater than that of a white or new, shiny roof. Fortunately the metal in question has a low temperature emissivity, that is energy absorbed is not readily re-radiated, and a ceiling, if only of straw mats, beneath the roof would significantly lower still further the levels of radiation into the room. The downward radiation from a new metal roof, unpainted, can be 70 kcal m^{-2} hr^{-1}. If the roof is painted white the value falls to 10 kcal m^{-2} hr^{-1}. Under clay tiles the figures would be more than twice as great, so the relatively cheap metal roof, more durable than thatch and less of a fire risk and harbouring fewer snakes, is probably the most suitable roofing material, especially if treated or ventilated, and provided that the rooms beneath are fitted with ceilings (Holmes, 1951).

On the Bauchi Plateau in Nigeria at $10\frac{1}{2}$°N and at elevations over 600 m, where uncomfortably cool nights can be experienced in the dry season, some traditional huts are cleverly constructed with two roofs, an inner one of clay to conserve the heat absorbed by day and to re-radiate it downwards at night, and an outer one of thatch to shed the rain and to protect the clay. Between the two is an insulating air space to further reduce daytime temperatures in the rooms beneath. Small doors and windows also help in this purpose.

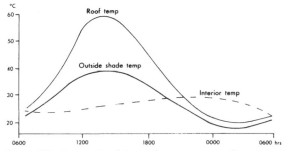

Figure 132 External and internal temperatures of an adobe desert dwelling (after Fitch and Branch 1960)

Urban climates

The comments above relate essentially to rural buildings and especially houses. In urban areas, and particularly in the major cities of West Africa, the climatic conditions can be significantly different from those in the countryside. Mean *wind speeds* tend to be lower because of the friction on the air flow of many buildings of various sizes, but turbulence is normally increased. Wind velocity can increase to the point of discomfort around the base of the slab blocks so characteristic of many urban centres. *Temperatures* can be uncomfortably higher in the towns and cities with the extra heat generated by thousands of people, kitchens, motor cars, by factories, refrigerators, dehumidifiers and air conditioners, and by the 'greenhouse effect' due to re-radiation from the concrete and asphalt into a polluted atmosphere. Thus even in the tropics cities may be heat islands and along the coast may exacerbate land and sea breezes (see also Chapter 15).

Many buildings in cities are made from concrete and glass. Concrete walls store heat readily by day, but lose it quickly by night. With the help of glass windows, if unprotected by shading devices such as those noted above, unbearably high temperatures will soon be attained indoors. With an external air temperature of 27°C the outer surface of a 50 mm thick concrete slab will readily reach 50°C and within an hour and a half the interior surface will reach 47°C. The worst thermal conditions for buildings are probably experienced in the Sahel in December, when shade temperatures by day can still attain 30°C but night temperatures may fall to 10°C. In March or April a concrete roof may attain surface temperatures in excess of 60°C, which, within an hour in a rain storm, may fall to 15°C. Thus the thermal load on buildings can be considerable, and the rapid heating and cooling noted leads to further stress. Walls facing different directions, too, have different times of peak heating and differential heating and cooling rates.

Wind pressure on tower blocks can be excessive, especially during the passage of line squalls. It is proportional to the square of the wind speed multiplied by a factor depending upon the shape of the building. For a wind of 13.5 ms^{-1} the pressure can be 28 kg m^{-2}.

Given the exposure of tall buildings in treeless urban areas and the air turbulence, *rain penetration* can also be a problem. It can be expressed:

$$p = rv^2$$

where r is the maximum rainfall in mm in five minutes; v is the wind speed in ms^{-1} at the time the rainfall occurs. Penetration spots are likely to occur if p exceeds 100. Then, further, in urban areas, on the flat roofs of offices, over the concrete or asphalt surfaces, rainfall absorption is minimized, runoff maximized. 5 millimetres of rain on a roof some 46 m^2 equates with 0.22 m^3 or 227 kg of water. Hence the need for storm drains in most of West Africa.

It is for these reasons that apartment blocks are generally unsuitable in the tropics, unless very expensively built and air-conditioned. And the problems noted will pertain to most offices, schools, universities, libraries and hospitals. If air conditioning cannot be afforded then walls in humid West Africa have to be limited to increase ventilation. This is particularly disadvantageous in schools where penetrating noise, dust and wind can be so detrimental to study, and the lack of wall space limits the display of notices, pictures, charts, students' work, etc.

Temperatures in motor cars

Many people spend many hours by day in motor vehicles that are not air-conditioned. The metal surfaces of dark-coloured vehicles may reach 30 to 40°C above the ambient temperature, and inside, with windows shut against the rain, temperatures may easily exceed external air temperatures by

Table 39 *Temperatures in parked cars at Ilorin, Nigeria*

| Time | Without shade | | | | Under tree shade | |
| | Air temp. °C | | Inside car °C | | Air temp. °C | Inside car °C |
	Max.	Mean	Max.	Mean	Mean	Mean
07.00	27	24	29	27	27	30
09.00	33	31	47	42	29	31
11.00	38	36	59	54	31	34
13.00	40	37	58	54	33	36
14.00	39	37	54	51	34	37

Source: Oyegun 1982b.

10 to 30°C. In southern Nigeria, with windows closed but the car in motion, face-high temperatures of 35°C were experienced in one measured test and temperatures of 42°C registered at feet level (Ambler, 1955). Conditions in parked cars have been recorded by Oyegun (1982b) in south-west Nigeria (Table 39).

Building design

In urban or rural areas modern architects would do well to study traditional building designs and materials, and to take heed of the details of local climate and weather, appreciating that available meteorological data are usually pertinent to screened conditions unaffected for the most part by nearby buildings or vegetation, and will differ from those that would be recorded in alleyways and streets, under skyscrapers and in sprawling suburbs.

Attention to detail is essential. Oyegun (1982a) noted discomfort in the use of domestic tap water in Ilorin, Nigeria, because the water pipes were buried no deeper than 30 cm below the surface. Soil temperatures at that depth commonly exceed 30°C and water temperatures by 17.00 hours reached 39°C, more than 3°C above the indoor air temperature. Water stored in pots, however, was nearly 10°C cooler. It is recommended, therefore, that domestic water pipes be buried to a depth of at least 60 cm.

Whatever decisions town planners or designers of individual buildings make, they affect the lives of many people, for better or worse. Clearly the aim should be to avoid changes for the worse and all effort should be made to prevent financial limitations, incompetence or greed from damaging the environment. As a preliminary step towards providing the basis of rational building or settlement design, Carl Mahoney developed a set of tables for recording and analysing climatic information (UN Department of Economic and Social Affairs, 1971). Full discussion of the Mahoney approach cannot be offered here, but the data required and their use are indicated in the application of the system to the university site at Freetown (Table 40) and at Timbuktu, Mali (Table 41).

Borderline situations will arise and compromises will then be required by the designers, but the basic needs for buildings in different climates are probably well indicated by the method outlined. Readers may wish to apply their local meteorological data to the system and see to what extent their own homes, or colleges, or offices meet the recommendations that result!

Industry and transportation

Introduction

The impact of climate upon economic life anywhere in the world is considerable and, in many cases, very obvious. Climate functions as an economic 'timing mechanism', as Chapter 12 has already indicated, the seasonal pattern of climate

Table 40 *Mahoney tables for Kortright, Freetown, Sierra Leone (8°29'N, 13°13'W altitude 335 m a.s.l.)*

a *Temperature (°C)*

	J	F	M	A	M	J	J	A	S	O	N	D		
													Highest	Annual Mean
Mean monthly max.	29	29	29	29	29	27	26	25	26	28	28	28	29[a]	25.5
Mean monthly min.	22	22	22	22	22	21	21	21	21	21	22	22	21[b]	8
Mean monthly range	7	7	7	7	7	6	5	4	5	7	6	6	Lowest	Mean Annual Range

Notes: The *annual mean* = (a + b)/2; the *mean annual range* = a − b.

b Relative humidity, precipitation and wind

	J	F	M	A	M	J	J	A	S	O	N	D	
Mean monthly max. RH(%)	82	81	85	84	86	89	93	95	95	90	88	86	
Mean monthly min. RH(%)	64	65	67	71	80	84	87	88	88	83	81	70	
Mean	73	73	76	77.5	83	86.5	90	91.5	91.5	86.5	84.5	78	
Humidity group	4	4	4	4	4	4	4	4	4	4	4	4	
Rainfall (mm)	6	1	10	6	178	384	763	836	607	269	109	28	Total 3197
Prevailing wind	NW	NW	NW	NW	NW	SW	SW	SW	SW	SW	SW	NW	
Secondary wind	NE	SW	W	W	SW	NW	W	W	W	NE	NW	NE	

Notes: Humidity group 1 – mean RH below 30 per cent; 2 – mean RH 30–50 per cent; 3 – mean RH 50–70 per cent; 4 – mean RH over 70 per cent. The prevailing and secondary winds are obtained from first and second peaks of frequency figures using simple compass points.

c Day and night comfort (°C) and thermal stress

		J	F	M	A	M	J	J	A	S	O	N	D
Day comfort	max.	27	27	27	27	27	27	27	27	27	27	27	27
	min.	22	22	22	22	22	22	22	22	22	22	22	22
Night comfort	max.	21	21	21	21	21	21	21	21	21	21	21	21
	min.	17	17	17	17	17	17	17	17	17	17	17	17
Thermal stress	day	H	H	H	H	H	H	—	—	—	H	H	H
	night	H	H	H	H	H	—	—	—	—	—	H	H

Notes: The day and night comfort limits are obtained from the following table using the appropriate humidity group and relevant annual mean temperature:

Humidity Group	Annual mean over 20°C		Annual mean 15–20°C		Annual mean under 15°C	
	Day	Night	Day	Night	Day	Night
1	26–34	17–25	23–32	14–23	21–30	12–21
2	25–31	17–24	22–30	14–22	20–27	12–20
3	23–29	17–23	21–28	14–21	19–26	12–19
4	22–27	17–21	20–25	14–20	18–24	12–18

If the mean monthly maximum temperatures exceed the day comfort maximum enter H in the thermal stress day column. If the mean monthly maximum temperatures fall within the day comfort limits enter a dash (= comfortable) in the thermal stress day column. If the mean monthly maximum temperatures fall below the day comfort minimum enter C in the thermal stress day column. Repeat for the mean monthly minimum temperatures and the night comfort maxima and minima.

d *Humid indicators*

	J	F	M	A	M	J	J	A	S	O	N	D	Totals
Humid													
H1 air movement essential	√	√	√	√	√	√				√	√	√	9
H2 air movement desirable							√	√	√				3
H3 rain protection						√	√	√	√	√			5
Arid													
A1 thermal storage needed													0
A2 outdoor sleeping space needed													0
A3 Cold-season problems													0

Notes: H1 When day thermal stress = H, is combined with high humidity (= group 4) or when day thermal stress = H, is combined with humidity group 2 or 3 and the mean monthly temperature range is less than 10°C.

H2 When day thermal stress = – and is combined with humidity group 4.

H3 Precautions against rain penetration essential when monthly rainfall exceeds 200 mm;

A1 When mean monthly temperature range is 10°C or more and coincides with moderate or low humidity (humidity groups 1, 2 or 3).

A2 Outdoor sleeping space desirable when night thermal stress = H and humidity is low (group 1 or 2), or when nights are comfortable outdoors but uncomfortable indoors as a result of heavy thermal storage (day = H, night = –, humidity group 1 or 2 and temperature range over 10°C).

A3 When day temperatures fall below the comfort limits (day thermal stress = C).

In Table 40**d** the months when these indicators apply are ticked and the number of ticks totalled.

e *Building design recommendations*

Indicator totals from Table 40 d						Recommendations
Humid			Arid			
H1	H2	H3	A1	A2	A3	
						Layout
			0–10			1 Buildings orientated on east–west axis to reduce exposure to sun
			11 or 12		5–12	
					0–4	2 Compact courtyard planning
						Spacing
11 or 12						3 Open spacing for breeze penetration
2–10						4 As 3, but protect from cold/hot wind
0 or 1						5 Compact planning
						Air movement
3–12						6 Rooms single banked. Permanent provision for air movement
			0–5			
1 or 2			6–12			7 Double-banked rooms with temporary provision for air movement
	2–12					
0	0 or 1					8 No air movement requirement
						Openings
			0 or 1		0	9 Large openings, 40–80 per cent of N and S walls
			11 or 12		0 or 1	10 Very small openings, 10–20 per cent
		Any other conditions				11 Medium openings, 20–40 per cent
						Walls
			0–2			12 Light walls; short time lag
			3–12			13 Heavy external and internal walls
						Roofs
			0–5			14 Light insulated roofs
			6–12			15 Heavy roofs; over 8 hours' time lag
						Outdoor sleeping
				2–12		16 Space for outdoor sleeping required
						Rain protection
		3–12				17 Protection from heavy rain needed

Notes: The indicator totals from **d** are applied to **e** to obtain design recommendations relating to layout, spacing, air movement, openings, walls, roofs, outdoor sleeping and rain protection.

In the case of Kortright, Freetown, nine ticks for H1 provide recommendations 4 and 6; three ticks for H2 lead to recommendation 7; five ticks for H3 show protection from heavy rain needed; no ticks for A1 suggest recommendations 1, 6 again, 9, 12 and 14. A2 does not apply, and no ticks for A3 indicate recommendations 2 and 9 (not 10 as this recommendation would contradict that resulting from column A1).

206

Table 41 *Mahoney tables for Timbuktu, Mali*
(16°46′N, 3°01′E; 301 m)

a

Temperature (°C)	J	F	M	A	M	J	J	A	S	O	N	D	Highest	AMT
Monthly mean max.	31	34	38	42	43	43	39	36	39	40	38	32	43	28
Monthly mean min.	13	14	19	22	26	27	25	24	24	23	18	13	13	30
Monthly mean range	18	20	19	20	17	16	14	12	15	17	20	19	Lowest	AMR

b

Humidity (percentage):		J	F	M	A	M	J	J	A	S	O	N	D	
	Monthly mean max.	39	33	34	27	34	55	74	83	76	50	36	35	0600 HRS
	Monthly mean min.	22	19	18	15	18	31	45	57	45	23	17	19	1200 HRS
	Average	30.5	26	26	21	26	43	59.5	70	60.5	36.5	26.5	27	
Humidity group		2	1	1	1	1	2	3	3	3	2	1	1	
Rainfall (mm)		<1	<1	2.5	2	5	23	79	81	38	5	<2	<.1	237 (Total)
Wind:	Prevailing	NE	NE	NE	NE	NE	SW	SW	SW	NE	NE	NE	NE	
	Secondary	E	E	E	E	E	S	S	S	E	E	E	E	

c

Temperature (°C)		J	F	M	A	M	J	J	A	S	O	N	D
Monthly mean max.		31	34	38	42	43	43	39	36	39	40	38	32
Day comfort:	Max.	31	34	34	34	34	31	29	29	29	31	34	34
	Min.	25	26	26	26	26	25	23	23	23	25	26	26
Monthly mean min.		13	14	19	22	26	27	25	24	24	23	18	13
Night comfort:	Max.	24	25	25	25	25	24	23	23	23	24	25	25
	Min.	17	17	17	17	17	17	17	17	17	17	17	17
Thermal stress													
	Day	–	–	H	H	H	H	H	H	H	H	H	C
	Night	C	C	–	–	H	H	H	H	H	–	–	C

d

	J	F	M	A	M	J	J	A	S	O	N	D	Total
H1 Air movement (essential)													0
H2 Air movement (desirable)													0
H3 Rain protection													0
A1 Thermal storage	√	√					√	√	√				5
A2 Outdoor sleeping			√	√	√	√				√	√		6
A3 Cold-season problems												√	1

Notes: Applying the indicator totals to Table 40e the design recommendations for Timbuctou buildings include: buildings orientated on an east to west axis, compact courtyards and planning (wide open streets in desert towns may be good for the motor car but not for people), little provision for air movement, small openings only, heavy walls but light, insulated roofs, provision for outdoor sleeping.

determining the farming calendar. It is the 'great controller' of agricultural activity and productivity, of international trade in agricultural produce. Its relevance to economic life is implicit in Chapters 11 and 13 also, as it is in the above discussion of the design of buildings, and the use and misuse of building materials. But in West Africa the combination of high temperatures, high humidity, sunlight, insects and moulds affects not only agriculture, water supply and the physical ability to work, but nearly every other facet of economic life, from the products used or created to the food processed and stored and to the transportation of all of these.

Fungi and insects

In this part of the chapter, attention is to be concentrated on the direct effects of atmospheric conditions on materials, industry and transport, but in the tropical situation particularly it is the indirect effects operating through fungi and insects that may often be most apparent. Fungi are difficult to suppress. Their minute spores, 1μ to 100μ in size, are in the soil and the air and may drift on to all exposed surfaces, clothes, glass and wood. Many may remain alive with relative humidity levels under 70 per cent but should the level exceed 90 per cent and the temperature be between 25 and 35°C, then growth can be extremely rapid and damaging. The problems may be summarized as:

1 deterioration, e.g. to electrical equipment, leatherwork, lenses;
2 decomposition, brought about by chemical changes created by enzymes produced by the fungi;
3 corrosion, as many fungi also produce acids.

To counter these problems with materials such as some plastics, treated wood and steel that are resistant to fungal attack or do not support fungi, or with dehumidifiers, desiccating agents or fungicides, can be expensive, adding to national economic burdens.

The main destructive insect pests are termites and beetles. The former feed on wood and other materials containing cellulose, such as canvas; species of the latter also attack wood, leather and some textiles, mainly by their larvae. Again, to use resistant timbers, such as *Mansonia altissima*, *Lophira alata*, *Terminalia ivoirensis* or preserva-

tives and insecticides is costly. Not to use them, however, in some circumstances can be even more economically disastrous.

The deterioration of materials due to climate

Metals

The main problems here are corrosion due to atmospheric moisture, pollutant gases and salt, and warping due to heating and cooling. At 99 per cent relative humidity, air with 0.01 per cent SO_2 is 35 times more corrosive to iron than is 'clean' air.

A study in southern Nigeria of the corrosion of iron and zinc ingots, exposed for one year (Hudson and Stanners, 1953), produced the results shown in Table 42, with corrosion rates in an arid environment (Khartoum) also noted for comparison. The significance of salt as a corrosive agent is clearly demonstrated. Within 250 metres of the surf at Lagos corrosion rates were of the same order as in a highly corrosive industrial atmosphere. The rate of rusting of unprotected machinery, cars, bicycles, corrugated iron roofs etc. can be up to 10 times faster in the humid tropics than in cool, dry environments.

Although corrosion may become less of a problem with increasing distance northwards from the humid Guinea coastlands, the dangers of warping are exacerbated by the higher levels of solar radiation and greater diurnal temperature ranges that are experienced in the Sahel and the Sahara. The high surface temperatures that can be achieved by sheet metal (unspecified) in West Africa, and the importance of the albedo of the metals, are indicated in Table 43 (based on data in Höller and Kerner, undated).

Brick, stone, concrete

Some indication of temperatures that can be attained by concrete under solar radiation has been given in the first part of this chapter. Corrosion within cement can also occur, however, as reinforcing metal rods rust. Acidic waters produced by decomposing vegetation can also attack some cements, as can the spread of black and red algal growths on lower walls. Cracking and spalling under salt attack is an all too familiar experience in many structures along coastal West Africa. Brick and stonework is also prone to

Table 42 *Corrosion of ingots of iron and zinc, exposed for one year*

| Location | Rate of corrosion, mils yr^{-1} | | | | | | | |
| | Iron | | | | Zinc | | | |
	No. of tests	Max.	Min.	Mean	No. of tests	Max.	Min.	Mean
Near Port Harcourt	2	0.2	0.2	0.2	3	0.03	0.01	0.02
Apapa, Lagos	16	1.6	0.7	1.1	17	0.11	0.02	0.05
Beach at Lagos	2	26.6	22.3	24.4	3	0.60	0.49	0.56
Khartoum	9	0.1	0.0	0.1	9	0.03	0.01	0.02

Source: Hudson and Stanners 1953.

crypto–efflorescence due to salt crystallization below the surface or the penetration of sulphates. Bricks properly made from local clays and laterite, however, can be very durable.

Wood

Timber and wooden furniture and structures are also prone to shrinkage and expansion due to temperature and humidity changes. Many a resident in West Africa will have experienced the damage to doors and furniture inflicted by the sudden arrival of the harmattan after a prolonged spell of humid weather. Warping readily occurs, glues are vulnerable. The drying and cracking of timbers can occur under direct sunlight.

Paint

The need to protect walls, roofs and woodwork from the ravages of the humid tropical climate is particularly important. But here again paints are also prone to deterioration under strong, direct sunlight, to photochemical change, to expansion and contraction. Holmes (1951) noted that a paint which offered protection in wet and windy Scotland for nine years lasted only five or six years at Apapa, Lagos. Modern paints are highly durable and special paints for tropical conditions and for special purposes, such as protecting oil storage tanks, are available, but at a cost.

Roofing materials

The rapid rusting of metal roofs, especially at the points where nails have been driven through, has

Table 43 *Range of highest temperatures (°C) of exposed surfaces of black and white sheet metal*

Location	Black	White
Timbuctou (central Mali)	74 (May)–60 (Jan)	59 (May)–44 (Jan)
Zinder (southern Niger)	72 (Apr)–63 (Aug)	57 (Apr)–49 (Aug)
Garoua (northern Cameroon)	71 (Mar)–61 (Aug)	54 (Mar)–45 (Aug)
Ouagadougou (Bourkina Faso)	71 (Apr)–59 (Aug)	55 (Apr)–45 (Aug)
Bauchi (northern Nigeria)	68 (Mar)–59 (Aug)	52 (Mar)–44 (Aug)
Bamako (southern Mali)	68 (Apr)–56 (Aug)	54 (Apr)–43 (Aug)
Odienné (northern Ivory Coast)	66 (Feb)–57 (Aug)	48 (Feb)–41 (Aug)
Monrovia (Liberia)	64 (Apr)–59 (Sept)	47 (Apr)–44 (Sept)
Yaoundé (southern Cameroon)	63 (Mar)–59 (Aug)	47 (Mar)–43 (Aug)
Banjul (Gambia)	63 (Apr)–58 (Aug)	49 (Apr)–44 (Aug)
Lagos (coastal Nigeria)	62 (Mar)–58 (Aug)	45 (Mar)–41 (Aug)
Abidjan (coastal Ivory Coast)	62 (Apr)–57 (Aug)	46 (Apr)–41 (Aug)
Nouadhibou (coastal Mauritania)	62 (Oct)–56 (Jan)	49 (Oct)–42 (Jan)
Lomé (coastal Togo)	60 (Feb)–54 (Aug)	45 (Feb)–40 (Aug)
Conakry (coastal Guinea)	58 (Apr)–53 (July)	44 (Apr)–40 (July)

Source: Höller and Kerner.

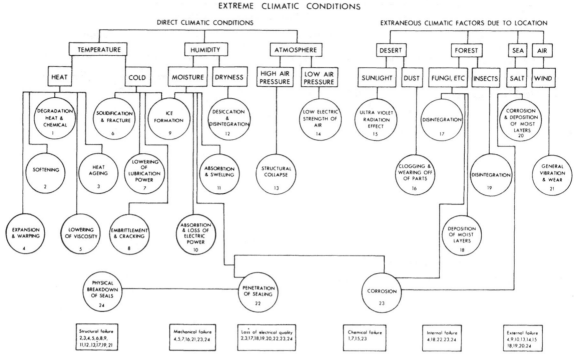

Figure 133 The deterioration of materials under extreme climatic conditions (after Greathouse and Wessel, 1954)

been noted above. Nevertheless corrugated iron roofs may serve for eight to ten years, especially if painted. A thick rust cover also affords some protection for the surfaces beneath. Thatch roofs, however, will significantly deteriorate within 18 months to two years. Bituminous products and lead, much used on modern office buildings, are seriously prone to blistering, cracking and flowing. If Atkinson (1950a) is correct in noting mud roofs on Ghanaian houses to be 75 years old before needing to be replaced, then perhaps more use should be made of this local material.

Textiles

The main problems for textiles are moulds, creasing and moisture absorption. Direct sunshine can also bleach coloureds or turn whites yellow. The drying of clothes after washing in the rainy periods can sometimes be difficult.

Paper

Due to heat and aridity paper can readily become brittle. In humid conditions discolouration and damage from insects and fungi are to be expected. Newspapers particularly deteriorate quickly: libraries, therefore, and offices, colleges and schools using and storing much paper, should ideally have air conditioners and/or tropicalized papers. In the absence of these expensive items regular cleaning, dusting and treatment by fungicides and insecticides are required.

Summary

Figure 133 (after Greathouse and Wessel, 1954, adapted from Coursey, 1944) summarizes the deteriorating effects of extreme climatic conditions on materials, and can still be relevant despite modern advances in technology. In West Africa the main effects are due to:

1 high temperatures, producing mechanical changes, speeding up chemical reactions, decreasing fluid viscosity and raising the level of bacterial activity;
2 high temperature combined with humidity, leading to the corrosion of exposed surfaces, organic changes (moulds, fungi), and the 'breathing' of partly sealed containers (see below);
3 impurities in the air, especially the presence of salt.

Figure 134 seeks to summarize the climatic conditions that are particularly damaging to metals, furniture, textiles and flour.

Brooks (1946) sought to produce a general expression for both chemical and organic effects involving temperature, water vapour, wind speed and concentrations of impurities, and in simpler form claimed that experiment showed that the sum of the values of 1.054t (H − 65)/10 (t being the mean monthly dry-bulb air temperature in °C, H the mean monthly relative humidity) giving a value of about 100 for the worst tropical conditions for that month. Applied to West Africa the result is as shown in Figure 135. According to this map the deterioration of materials, such as those discussed above, is likely to be four times greater in southern Cameroon than elsewhere in West Africa north of about 12°N, and is significant all along the Guinea coast. Applying Brooks' Index of Deterioration to Freetown, the monthly values (rounded off) are: Jan. 23, Feb. 20, March 30, April 34, May 47, June 54, July 61, August 61, Sept. 64, Oct. 54, Nov. 49, Dec. 34. Values in excess of 40 probably suggest that problems with the deterioration of materials are likely.

Climate and the deterioration of food

It is again those areas of West Africa that experience both high temperatures and humidities that suffer most obviously from the deterioration of food. With high temperatures comes an increase in enzymes and micro-organisms. The rapidity with which moulds appear on cheese, fruit and bread, and milk turns sour, is astonishing. Some fats readily liquefy creating problems for the pastrycook. High humidity levels lead to the absorption of moisture into flour, cereals, sugar and salt. Very low humidity, on the other hand, causes rapid staleness of bread and cakes.

Diurnal changes in temperature can be problematical. The interior of a food container, if not airtight, may experience a diurnal range of temperature even over 40°C. During the hot day 10 per cent of the air in the container may be expelled by expansion, to be replaced by fresher, cooler but moister air at night, which might be damaging to the contents of the container.

Direct sunlight can have deleterious effects, destroying vitamin C, especially in liquids stored in clear glass jars or bottles. Green vegetables will rapidly wilt and turn yellow. Problems of

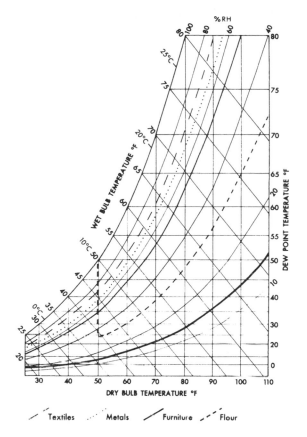

/ Textiles   ⋯ Metals   / Furniture   ⁄ Flour

Figure 134 Climatic conditions damaging to textiles, metals, furniture and flour (after Urdahl 1952)

Note: Dangerous conditions are those to the left of all lines, and, in the case of furniture, also to the right of the lower heavy line.

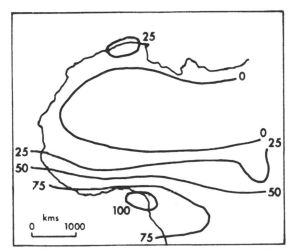

Figure 135 Proneness to the deterioration of materials (after Brooks 1946)

Note: See text.

211

storing drinks and foodstuffs in cool conditions are serious in the parts of West Africa that lack refrigeration. Danger of food poisoning is ever present given the speed with which foodstuffs can spoil.

Climate and manufacturing industry

It follows from the above discussion, brief as it is, that the West African climate poses problems for manufacturing industry. Optimal indoor conditions for selected industries are given in Table 44 (after Grundke, 1955/56). Without expensive cooling and dehumidification few, if any, of the industries can be satisfactorily operated in the more populous parts of West Africa. Many industries, such as canning, paper making and textiles, need copious water supplies. Where are they to come from in arid West Africa, without prohibitively expensive tapping of ground water resources?

Mining

The mining industry is not seriously affected by the climate, although in the rains the working of open-cast mines and quarries and transportation of minerals can be hampered. Much of the equipment used by the industry is permanently exposed to the weather, with all the related risk of rapid decay. In arid areas dust may clog machinery and thicken lubricants.

Construction industries

The same problems of dust in dry conditions, of flood, mud and erosion in wet, apply to most aspects of the construction industry, although the building trade in West Africa does not suffer from the major problems of frost and freezing experienced in cold climates. Rainfall is probably the main problem, hampering access to sites and movement, spoiling newly finished surfaces and hindering drying. It can damage excavations or turn them into ponds; it creates hazards for workers. If high wind accompanies the rain the damage and dangers are increased.

The cost of constructing high quality roads is generally too great for West African states. Those built, with thin asphalt surfaces and unsealed edges, lacking adequate culverts to carry off the storm rains, are soon subject to deterioration. The persistence of high temperature con-

Table 44 *Optimal indoor climatic conditions for selected industries*

Industry	Temp °C	Relative humidity %
Cotton manufacturing	20–25	60
Milling	18–20	70
Flour storage	15.6	50–60
Bakery	25–27	60–75
Processed cheese production	15.6	90
Paper manufacturing	20–24	65
Paper storage	15.6–21	40–50
Printing	20	50
Drug manufacturing	20–24	60–70
Rubber production	22–24.4	50–70
Photographic film manufacturing	20	60
Electrical equipment manufacturing	21	60–65

Source: Grundke 1955/56.

ditions leads to disturbance of the tarmacadam. Temperatures up to 50°C may be experienced 4 cm beneath an asphalt surface. The elasticity of that surface, and viscosity of the tar, are seriously affected and ridges and waves readily form under pressure from vehicles. Earth roads, by far the most common type still in West Africa (90 per cent of all roads in Sahelian states and 82 per cent of all roads in southern countries), are dangerously dusty and sometimes slippery in dry conditions. Shrinkage of the subsoil too may damage the road, creating longitudinal cracks upon the surface. Bridge design and construction must allow for the expansion of steel structures by day, perhaps contraction by night. Wooden constructions are subject to warping and termite attack.

Railway builders must allow for flooding, landslides and mudflows, and seek to minimize dangers from these climatically induced phenomena not only to the tracks but also to signalling, embankments, gorges and tunnels. Many of the railway lines in West Africa are of old design in which the highest temperatures can cause buckling of the rail. For each rise in temperature of 1°C a rail may expand 0.004 mm per 0.30 m length of track. In drought conditions old locomotive stock may be deprived of water. Sand and dust hampers visibility for drivers, enters working machinery and reduces traction on rails. Humid conditions cause the usual corrosion and may lead to lubrication problems and cooling difficulties for some diesel engines. For the transportation of perishable produce, specially designed wagons are required.

Climate and transportation in West Africa

References to transport difficulties on road and rail have been made above. Earth road surfaces in particular, in dry or wet situations, can be hazardous. But probably more accidents occur because of carelessness and sleepiness, the consequence of driving in hot, enervating conditions or due to poor visibility in dust clouds produced by preceding or passing vehicles, or in heavy rain. These same weather conditions, restricting visibility, present hazards for *air transportation*. In the USA the weather is responsible for about 33 per cent of all aviation accidents involving fatalities, and thunderstorms are the commonest cause of flight delays (26 per cent). Unfavourable winds and wind shear are responsible for 19 per cent of flight hold-ups.

High-flying and fast modern jet planes are less susceptible however to danger in storms or from high-level wind shear (clear-air turbulence), or hailstones, than subsonic aircraft. Acceleration, produced by wind gusts on supersonic aircraft, is only 60 per cent of that produced by the same gust on a slower 'plane, owing to the initial high speed of the jet and the lower air density at higher altitudes. Briggs (1972) has noted at Freetown that the higher the intensity of rainfall in a storm, the shorter the duration of the storm, but also, of significance to aviation, the smaller the rain cell diameter. The mean cell diameter of a storm producing an intensity of rainfall of 25 mm hr^{-1} is 4 km, but for intensities of 50 and 100 mm hr^{-1} the diameters are three and two kilometres respectively. The faster aeroplanes are thus able to avoid these storms without too much difficulty or delay in the flight. Dangers from lightning strikes on aircraft are very small.

The main peril for aviation in West Africa, as was implied at the end of Chapter 4, is poor visibility. Morning mists are not uncommon in the humid southern areas, harmattan dust can spread extensively, dust storms are the menace in the north. Almost anywhere the line squall or local thunderstorm with driving rain and low cloud will present a hazard despite modern technology, and this is not yet available at all airports in West Africa. One advantage experienced throughout the region is that the prevailing wind, mainly south-westerly or north-easterly, permits most airports to function with only one runway, aligned with these dominant winds.

Shipping is, of course, also subject to interference by the weather and climate, be it at sea or on rivers, lakes, lagoons or swamps. Sea fog in the north-west is a hazard, as are storms and strong winds especially on rivers. Even the prevailing winds, by generating long-shore drift of sediment across the mouths of navigable rivers or lagoons, can create expensive problems. The surf barrier off the Guinea coast, generated by winds sometimes far to the south, has always been a hindrance to shore access for shipping and is a perpetual danger for local fishermen.

The seasonality of West African climate has important consequences for shipping on many rivers. Figure 136 shows mean monthly discharges (m^{3}s^{-1}) for the periods indicated at Koulikoro on the Upper Niger river near Bamako, and at Timbuktu on the Middle Niger, downstream of the inland Delta. In the low discharge periods shipping is often seriously restricted. These periods do not coincide at the two places either. The recent Sahelian drought has significantly reduced the use of the river by the larger vessels that are so important for trade and public transport in Mali and Niger (see also Chapter 11).

High temperatures and fluctuations in temperature and humidity can also be relevant. Metal decks on ships may experience surface temperatures above 50°C and radiate heat into holds to the detriment of cargoes. Corrosion, rotting and even spontaneous combustion are risks in inadequately protected vessels. Goods packed in cold conditions in Europe without refrigeration, then transported into West African waters may suffer damage. Foodstuffs, cement and chemicals are the most common casualties. Conversely, cargoes packed in warm, humid holds in West Africa are prone to condensation damage as the ship enters cooler climates. Höller (1959) records the loss of one-third of a cargo of cocoa from the Gold Coast en route to Europe.

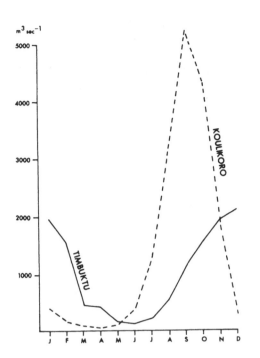

Figure 136 Discharges on the River Niger at Timbuktu and Koulikoro

When packed by the farmer the cocoa beans registered only 7.5 per cent water content, a very good level. Storage in the open at Takoradi raised the content. In the ship the beans were stored at an air temperature of 29°C. North of Cape Verde the air temperature dropped 9°C in four hours, 12°C in 22 hours. Ventilation of the hold brought in this cooler air to the cargo leading to heavy condensation on the sacks and beans. Losses of freight on ships held up in harbours in West Africa or on docksides have been and are regrettably still too common.

Communications

Transport by air or sea is very dependent upon radio communications. The old problems that fluctuations in atmospheric pressure used to create for such contact have been largely overcome by modern technology (though all with access to television may be aware of the interference to picture reception that can occur in high pressure conditions), but in many parts of West Africa old equipment is still in use, and telephone systems especially are prone to all the problems summarized in Figure 133. The inconvenience, the losses to business and even the losses of life that may result are incalculable.

Trade, tourism and sport

From comments made above the importance of climate in trade, tourism and sport is obvious and little elaboration is required. There is marked seasonality of trade in the transport industry and in sales of umbrellas, insect sprays and certain drugs.

Tourism particularly reflects the climate. Conditions in coastal West Africa in the northern hemisphere winter are ideal for the European visitor, and the economies of some nations such as Gambia and Sierra Leone have benefited considerably from this fact. However, the up-keep of empty, expensive hotels in the rainy season is a problem of no small magnitude.

Industrial power and climate

The main sources of industrial power in West Africa at the present time, as elsewhere in the world, are expensively produced or imported fuel oils, supported by hydro-power in limited areas. The cost of petroleum and diesel fuel is a crippling burden for West African states. When, as happened in 1984 in Ghana due to drought and the excessive fall in the level of Lake Volta, supplies of electricity from hydro sources are diminished, the consequences to national well-being can be calamitous.

Given the abundance of sunshine, especially in northern West Africa, and the persistence of pre-vailing winds at certain seasons, especially on the north-west coast, it is a pity that research into the production of power from solar energy and wind flow has not been more extensive in the region and that the nations that could most benefit have lacked the financial resources and technological experience to put the sun and the wind to greater use. Aerogenerators are perhaps the most economic to install at present, but the costs of power distribution and maintenance are not inconsiderable. A sound knowledge of local wind flow is also required, allowance being made for diurnal and seasonal variation. Areas with high mean wind speeds, say in excess of 4.5 ms^{-1}, and for 50 per cent of the time, are most suitable, although it is uneconomic to design aerogenerators for windflow over 16 ms^{-1}. Small, 3.5 m diameter, propellers can supply 1.5 to 6 kW with wind speeds up to the maximum of 13.5 ms^{-1}. Aerogenerators with diameters of 15 m can produce 100 kW, of 60 m up to 3000 kW (Golding, 1955).

Very brief consideration has been given to the many aspects of economic life noted above. The relevance of weather and climate to national well-being is apparent, however. There is clearly much to be gained from a sound understanding or regional climate. There is a meteorology of resources, of transport, trade and development. The money spent by nations on their meteorological services should be seen as part of the cost of economic advancement, as finance well spent. It should be appreciated that maintaining an efficient economy depends not insignificantly on a knowledge of applied meteorology and climatology. There is a need to improve and safeguard agriculture, water resources, the people's health and the industrial base. This can only be done through maximum efficiency, improved weather measurement, recording and forecasting, and with an increased awareness of the value of weather and the need to protect the atmosphere from pollution and adverse modification. Some of the sad consequences of failing to cherish our habitat are noted in the final chapter.

15　Human impact upon climate in West Africa

Introduction

Man affects weather and climate in West Africa at three levels of scale, in three main ways and by three main processes. Changes can be effected at:

1 the microscale, by altering the local vegetation cover, by the growing and harvesting of crops, by draining wet lands or irrigating dry;
2 the mesoscale, primarily by building extensive cities (Accra, for example, and not the largest urban area in West Africa, extends over 150 km^2 and may influence the atmosphere for 200 to 300 m depth);
3 the macroscale, primarily by polluting the air over large areas.

Thus the three main ways suggested are:

1 changing the composition of the atmosphere;
2 releasing heat into the atmosphere; and
3 changing the physical and vegetational conditions on the ground.

And thus the primary processes are:

1 emitting gases and particulate matter into the air from industry and urban areas;
2 destruction of the forest and savanna by burning; and/or
3 by overgrazing the land leading to a rise in atmospheric dust content, affecting the radiation balance and altering the 'natural' levels of condensation nuclei.

Atmospheric pollution

There are many causes, some suggested above, for atmospheric pollution, but globally the main ones are burning oil and coal, the addition of aerosols, and moisture, dust and chemical output (in the latter case particularly hydrocarbons, carbon monoxide, nitrogen oxides, lead compounds and carbon from smoke) from industry and from farming: spraying, burning and overgrazing. Perhaps around the world 700 million tonnes per year of particulate matter are added by human activity to the 1200 million tonnes that are naturally in the atmosphere, mostly sea salt and the output of volcanoes.

In West Africa field burning is probably the most common cause of atmospheric pollution, but unfortunately the region does not escape from the madness of people even thousands of kilometres distant. In the period 1958 to 1974 a rise of 14 ppm of atmospheric CO_2 was noted from observations in even less 'polluted' areas than West Africa, i.e. in Hawaii and Antarctica (Keeling et al, 1976). An annual rise now of 1 ppm will mean levels of CO_2 exceeding 350 ppm by AD 2000, an increase of some 20 per cent in the twentieth century with half of this rise in the last quarter (Landsberg, 1979; Silverberg, 1969). It is thought that the outcome might be a general rise in temperature of the earth's atmosphere by 1 to 2°C. As populations continue to grow rapidly in the tropical world, and as development moves apace often without, for economic reasons, the safeguards against pollution that are commonly enforced in the industrialized western world, so the CO_2 concentration in the atmosphere might even exceed the estimates noted.

Global responses, it has been suggested (WMO, 1981b), will include a greater rate of warming in higher latitudes, especially in the Arctic, and in winter and spring, thus increasing water circulation within the hydrologic cycle, drastically affecting the general circulation of the atmosphere, and increasing the rate of sea level rise. Growing populations will be forced to live on a reduced land area, although the marine incursions into continents and especially North Africa, may reduce the size of deserts.

A benefit of higher CO_2 levels might be enhanced photosynthesis and crop production; however, it is predicted by the WMO that the yields of crops such as sorghum, maize, cotton and sunflowers might decline. Higher temperatures will also mean more problems from insects, fungi and bacteria. An increase in soil bacterial activity will lead to over-rapid decomposition of vegetation litter and thus less humus and poorer soils. Aerosols might lower the albedos of some surfaces leading to increased absorption of radiation and thus further warming up. Additional risks might accrue to fish and other marine life as the acidity of waters increases. Atmospheric aerosols heating by day, cooling by night, may lead to increased daytime stability thus reducing convective activity and precipitation. Aerosols in the clouds reduce their reflectivity, so also increasing the warming of the atmosphere.

The destruction of the forests, so appalling in West Africa, can also raise atmospheric CO_2 levels, not only as the consequence of burning and pollution but also because forests are absorbers of CO_2 and important producers of O_2. They are being replaced by urban or agricultural ecosystems which store less CO_2. It might be thought that the increase in numbers of condensation nuclei that burning of the bush or forest will create should lead to more precipitation. It has been suggested, however (Warner, 1968) that the nuclei result in a reduction in the size of cloud droplets, thus hindering coalescence and rainfall. Much burning takes place in dry periods when cloud cover is minimal anyway.

Having noted the perhaps worrying statements made above, it is only fair to admit that man's knowledge of tropospheric chemistry is very limited and the impact of the pollutants mentioned on the content of the atmosphere, on albedos, on the 'greenhous effect' and on atmospheric heating, is uncertain. There are human influences which will tend to bring about cooling rather than heating. The coalescence of jetplane contrails on some busy air routes, forming high level cirrus cloud, may reduce radiation inputs at the earth's surface and lead to cooling.

The foreseen risk of adding chlorofluorocarbons from aerosol cans to the atmosphere, perhaps to the amount of 0.5 million tonnes per year in the early 1980s, threatening to break down the stratospheric ozone layer so important as a global screen against excessive inputs of ultra-violet radiation $(Cl + O_3 \rightarrow ClO + O_2)$, has been averted by substituting hydrocarbons for the chlorofluorocarbons as aerosol propellants. But then hydrocarbons enhance CO_2 levels! However there is no sign yet of significant changes occurring to reduce the NO_2 that is part of the effluent of aircraft. This dissociates photochemically and the nitric oxide formed also reduces the high level ozone to O_2. A 10 per cent reduction in ozone means a 20 per cent increase in ultra-violet B radiation to earth. A 5 per cent increase, it is thought, could lead to a 10 per cent rise in the number of cases of skin cancer.

Uncertainty about the amount of damage being inflicted upon the earth's atmosphere does not mean that man should blithely continue with the activities that might prove so calamitous. The risk is not worth taking. It should not be assumed that scientists will be able to devise the means to rectify the situation, once it is clearly seen to be critical.

The situation in West Africa and the future

Air pollution depends upon the quantities and types of pollutants, their rate of dispersion and the diffusion properties of the atmosphere. Wind speed and atmospheric stability are important factors here. The higher the wind speeds the faster the removal of the effluents. Higher mixing heights and turbulence levels enhance the vertical mixing of pollutants and so give lower pollution potentials. Inversions of temperature and anticyclonic conditions will produce high potentials.

Studies of mixing heights and the ventilation factor have permitted atmospheric pollution potentials for Africa to be calculated (Patnaik et al, 1980). The maximum mixing height (MMH) is the maximum height (in km) above the surface through which vigorous vertical mixing occurs. The ventilation factor (VF) (ms^{-1}) is a product of the mixing height (km) and the mean wind speed within the height. The data analyzed came from only 48 radiosonde stations throughout the continent and its offshore islands. For West Africa data was supplied by only eight stations (Nouadhibou, Dakar, Bamako, Niamey, Tamanrasset, Abidjan, Lagos, Douala), so only a very generalized picture was obtained. Temperatures and pressures at the surface, at 850, 700 and 500 mb were studied with wind speeds. The outcome of the research is summarized in Figure 137. Setting aside the perhaps anomalous high values at Abidjan in April and Niamey in October for the ventilation factor, seasonal variation in this factor is more apparent than for the MMH. It is interesting to note that once again it is the Sahel, that transition region between the central Sahara, with its overlying subsiding air and its common inversions of temperature, and the humid Guinea coastlands, that has the highest MMH values. This zone of convection, of squall lines and, of upper air jet streams, has mean MMH values of between 1.5 and 2.5 km. It has the highest VF values too, 3 to 8 ms^{-1}. This is just as well given the prevalence of the dust-laden harmattan in this region for many months of the year. The dispersion of dust across the area has been noted in Figure 59 (p. 80). Perhaps 10 per cent of uplifted dust may remain in the atmosphere for 12 months or more, but because of the VF and MMH, the storms and the jet in particular, the dust does not remain in West Africa.

Air pollution in West Africa is not yet so serious that the levels of mixing and ventilation cannot cope, except locally in some urban areas at times of temperature inversion. There are sound meteorological as well as ecological reasons for ensuring that economic development in the region does not involve the problems of pollution that are the price paid by industrial nations for their 'advancement'. Above all it is now time to cease destruction of the little forest that remains. The change in surface albedos as a result of forest removal has already been noted in Chapter 1 and possible meteorological consequences considered at the end of Chapter 10. Charney et al (1975) suggest that depletion of the vegetation cover increases albedo, decreases net radiation thus leading to a cooling of the atmospheric column from below, discouraging convectional air ascent thus diminishing rainfall. Less rainfall will deplete the natural or semi-natural vegetation thus producing self-perpetuating damage to the environment, to agriculture and to human well-being. It is suggested that the change in albedo in the Sahel in recent years from 14 per cent to 30 per cent or higher, due particularly to overgrazing, has made a significant contribution to the droughts there in the 1970s and 1980s, and to desertification in West Africa.

Grove (1977) firmly considers overgrazing and timber destruction as prime agents of desertification. Devegetated areas can lose their topsoil to winds of only 4 to 5 ms^{-1}. In 1965 Kano in northern Nigeria imported some 75,000 tonnes of wood for fuel, mostly brought in on donkeys from within a radius of 20 km. Since then the population of Kano has grown annually by 5 to 10 per cent and alternatives for wood fuel have not been adopted. The area within 20 km radius has now ceased to be adequate. In the area of Niamey it is thought that the vegetation zones have shifted south by 150 km in the last two centuries, and the pace of movement is quicker each year.

By AD 2000, 70 per cent of the world's population will live in cities. When the major conurbations of West Africa attain the present size of Cairo or Bombay or Mexico City, or come to compare with the great cities of Europe, Asia, North or South America, what will be their impact on the local climate? Little research has been undertaken in the tropics into urban climatology, but if the changes are to compare with those in temperate latitudes then the growth of the cities will mean annual temperatures rising by 1 or 2°C, cloud cover increasing by 5 per cent and rainfall accordingly, relative humidity declining by 4 to 6 per cent, evapotranspiration rising by 30 to 60 per cent, wind speeds diminishing by 40 per cent and the number of thunderstorms climbing by 5 per cent.

The problems of recent years imposed by a decline in precipitation in West Africa are not entirely of man's making, however, as Chapter 10 has pointed out. Years of disastrously low rainfall have been known in the past. Ojo (1982) has noted negative deviations in rainfall at Zinder, Niamey and Maiduguri in the years 1913–17, 1924–7, 1940–5, 1948–53 and 1968–74. Landsberg (1975) has described sudden changes in annual precipitation at Dakar, illustrated in Figure 138 from his data.

But the droughts of the past did not lead to the extent of desertification that is such a feature of the latter part of the twentieth century, nor to such losses of life. (In 1973 alone perhaps 100,000 people died from famine in the West African Sahelian states.) In the years gone by there were not so many people, the pressures on the land were not so intense nor the restrictions imposed by national boundaries so influential. (Although in 1975 an estimated 11 per cent of the population of Bourkina Faso of 6.4 million, and 6 per cent of

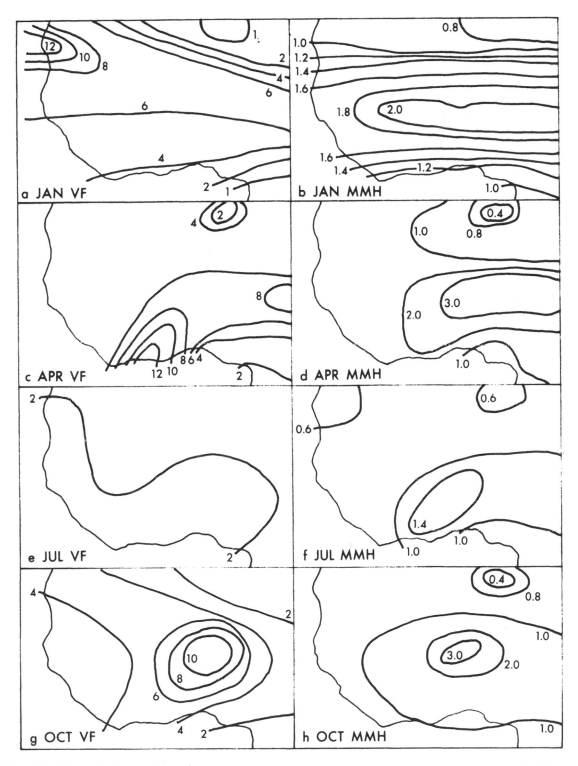

Figure 137 Ventilation factors (km sec⁻¹) and maximum mixing heights (km) for four months (after Patnaik et al. 1980)

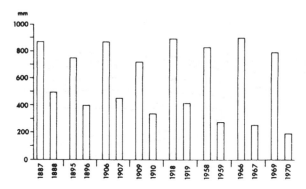

Figure 138 Changes in annual precipitation at Dakar (based on data in Landsberg 1975)

the population of Mali of 6 million, migrated into Ivory Coast, constituting one-fifth of the population of that state in that year (Zachariah and Condé, 1981).)

What is to be done? The grandiose proposals to alter global climates by exploding H-bombs in the Arctic Ocean or on the edge of the Antarctic ice sheet, to cover the polar ice caps with soot to aid absorption of radiation, to dam the Bering Straits to control the flow of cold and warm waters in and out of the Arctic and Pacific Oceans, are so fraught with danger and disadvantage that international agreement for such expensive projects is unlikely ever to be achieved.

The local schemes for weather modification such as cloud seeding with silver iodide or dry ice to produce raindrops in clouds, have had very limited success and then only with certain clouds and conditions. Certainly the method is not promising for the alleviation of drought. In experiments over Senegal (Seck, 1974) it was claimed that 23 per cent of seedings of clouds produced positive results, 33 per cent doubtful results and 44 per cent negative results. In some of the positive cases too the precipitation failed to reach the ground!

Perhaps the most immediate steps that can be taken might involve afforestation, the planting of windbreaks and the cessation of forest destruction. But trees need time to grow and time is short for millions of starving, struggling and destitute people. A major part of the answer to the problem of man's murder of his own environment must lie with education, with an awareness of the basics of ecosystem dynamics, of the importance of ecology and of the role of climatology and meteorology. Then the necessary changes in society that are also required, leading on to the essential lowering of birth rates, may also be achieved.

If this book contributes to that education, then its production will have been worth all the time and effort expended on it.

Appendices

Appendix 1   Conversion Factors

Energy

1 langley (ly)	= 1 calorie (cal)/cm^2 (cal cm^{-2})
	= 4.18 joules/cm^2 (jl cm^{-2})
	= 4.18 × 10^7 ergs/cm^2 (erg cm^{-2})
	= 69.75 milliwatt (mW)/cm^2 (mW cm^{-2})

Temperature

1°C	= 1.8°F
$\dfrac{5 \, (°F - 32)}{9}$	= °C
$\dfrac{9°C}{5} + 32$	= °F

Pressure

1 millibar (mb)	= 1000 dyne/cm^2 (dynes cm^{-2})
	= 100 Newtons/m^2 (N m^{-2})
1 atmosphere (atm)	= approx. 1013 mb
	= approx. 1.1 kg/cm^2 (kg cm^{-2})
	= 29.91 inches of mercury

Velocity

1 metre per second (m/s) (ms^{-1})	= 3.6 km/hr (km/hr^{-1})
	= 2.24 miles per hour (miles/hr) (miles/hr^{-1})
1 km/hr	= 0.28 m/s (ms^{-1})
	= 0.91 ft/s (ft s^{-1})
1 knot (kt)	= 0.51 m/s (ms^{-1})
	= 1.85 km/hr (km hr^{-1})
	= 1.69 ft/s (ft s^{-1})
	= 1.15 miles/hr (miles hr^{-1})
1 mile/hr	= 0.447 m/s (ms^{-1})
	= 1.5 ft/s (ft s^{-1})

Length

1 millimetre (mm)	= 0.1 centimetre (cm)
	= 0.0394 inches (in)
1 metre (m)	= 100 cm
	= 39.37 in
	= 3.28 feet (ft)
	= 1,000,000 microns (μ)
1 kilometre (km)	= 1000 m
	= 0.62 miles
1 mi	= 1609 m
	= 1.609 km
1 micron (μ)	= 0.000001 m
	= 10,000 Angstrom units (Å)

Area

1 cm^2	= 100 mm^2
	= 0.15 in^2
1 m^2	= 10,000 cm^2
	= 10.76 ft^2
1 hectare (ha)	= 10,000 m^2
	= 0.01 km^2
	= 2.47 acres (ac)
	= 0.004 miles^2
1 km^2	= 1,000,000 m^2
	= 247 acres
	= 0.39 miles^2
1 mile^2	= 640 acres
	= 259 ha
	= 2.59 km^2
1 in^2	= 6.45 cm^2

Appendix 2 Instruments for the measurement of radiation and albedo

Radiation

The most commonly used instrument for measuring global or diffuse radiation is the *Solarimeter* (Figure A1). This contains a thermopile under two hemispherical glass domes which protect the thermopile from wind and rain. The surface between the domes is highly polished to reduce radiation absorption and the remainder of the instrument is protected from direct radiation by a guard plate. A drying cartridge of silica gel is incorporated in the housing to keep the air under the glass domes free of moisture.

The thermo-electric electromotive force (EMF) produced by the thermopile may be recorded instantaneously on (1) a direct-reading millivoltmeter, or the instrument can be wired to (2) an electronic integrator which may totalize the solar energy received over a set period of time or, if a printing integrator, can record all measurements continuously on a paper strip, or (3) flatbed recorders which will graph the data instantaneously. The integrators and flatbed recorders may automatically register the radiation in the standard unit of calorie per square centimetre per minute (cal cm^{-2} min^{-1}). Otherwise the readings in millivolts (or milliwatts) cm^{-2} need to be translated into cal cm^{-2} min^{-1}, the solarimeter having been carefully calibrated. (1 watt per square centimetre (1 W cm^{-2}) = 14.33 cal cm^{-2} min^{-1} or 60 joules (J) cm^{-2} min^{-1}.)

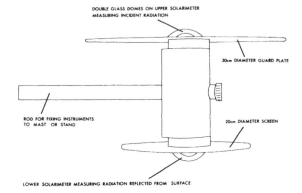

Figure A2 An albedometer

Albedo

An albedometer (Figure A2), for the measurement of albedo, comprises two identical solarimeters, one facing upwards to measure the incident (global) radiation, and a second facing downwards to measure the radiation reflected from the surface, this requiring a smaller screen than the upward facing instrument. Each solarimeter will be wired to a millivoltmeter or separate integrator. (In his studies of albedo in West Africa, Oguntoyinbo set the instrument 1.2 to 1.5 metres above the vegetation. At this level radiation reflected from the canopy with a radius of about 3 metres was recorded. In areas of scanty vegetation the instrument was raised to 2.5 metres above the vegetation to view a wider area.)

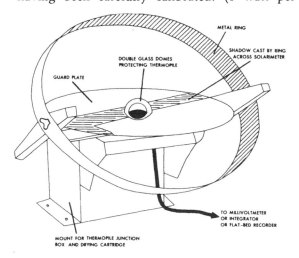

Figure A1 A solarimeter

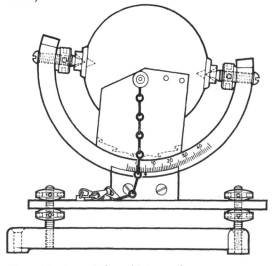

Figure A3 A tropical sunshine recorder

Appendix 3 Tropical sunshine recorder

The recorder illustrated is used to measure the duration of bright sunshine. The glass sphere concentrates the sun's rays on to a specially prepared card graduated in hours and held in a bowl which is half a belt cut from a sphere and concentric to the sphere. The instrument is usually mounted on a concrete pillar about 1.5 metres above the ground surface. In bright sunshine the card will be scorched. The total length of scorch mark permits the duration of sunshine for the day to be calculated. The card is changed daily. The times during the day when the sun shone brightly can also be ascertained. The instrument can be adjusted for latitude. In the diagram it is set for latitude 5°.

Appendix 4 Sunshine data for 70 stations in West Africa

Country, name of station and years of records (where known)		Jan.	Feb.	Mar.	Apr.	May	Jun.	Jul.	Aug.	Sep.	Oct.	Nov.	Dec.	Range
Mauritania														
F'Derik	1	266	243	301	321	338	308	275	296	283	272	241	230	108
(1953–8)	2	8.6	8.8	9.7	10.7	10.9	10.3	8.9	9.5	9.4	8.8	8.0	7.4	
	3	79	77	81	85	83	77	67	74	76	76	72	69	
Nouadhibou	1	253	254	302	305	329	309	287	288	258	250	232	245	97
(1950–8)	2	8.2	9.1	9.7	10.2	10.6	10.3	9.3	9.3	8.6	8.1	7.7	7.9	
	3	74	79	81	81	81	77	70	73	70	69	69	73	
Atar	1	253	244	311	308	319	285	279	275	234	254	234	236	85
(1950–8)	2	8.2	8.7	10.0	10.3	10.3	9.5	9.0	8.9	7.8	8.2	7.8	7.6	
	3	74	76	83	82	79	71	68	69	63	70	70	70	
Nouakchott	1	217	252	310	326	315	303	277	276	254	272	247	221	109
(1953–8)	2	7.0	9.0	10.0	10.9	10.2	10.1	8.9	8.9	8.5	8.5	8.2	7.1	
	3	62	78	83	86	78	76	68	70	69	75	73	64	
Aioun	1	275	265	314	308	300	259	274	256	250	267	261	236	78
(1953–8)	2	8.9	9.5	10.1	10.3	9.7	8.6	8.8	8.3	8.3	8.6	8.7	7.6	
	3	79	82	84	82	75	66	68	65	68	73	77	68	
Néma	1	280	259	303	287	283	267	264	258	263	271	262	219	84
(1953–8)	2	9.0	9.25	9.8	9.6	9.1	8.9	8.5	8.3	8.8	8.7	8.7	7.1	
	3	80	80	82	77	70	68	65	65	72	74	77	64	
Senegal														
St Louis	1	194	199	271	274	260	221	209	203	206	214	205	188	86
(1950–8)	2	6.2	7.1	8.7	9.1	8.4	7.4	6.7	6.5	6.9	6.9	6.8	6.1	
	3	55	61	72	73	65	56	51	52	57	58	60	55	
Thiès	1	266	264	323	328	313	270	234	183	217	258	265	237	145
(1953–8)	2	8.6	9.4	10.4	10.9	10.1	9.0	7.5	5.9	7.2	8.3	8.8	7.6	
	3	76	81	87	87	79	69	58	47	59	70	77	68	
Dakar	1	250	242	292	297	290	257	216	186	216	236	245	232	111
(1950–8)	2	8.1	8.6	9.4	9.9	9.3	8.6	7.0	6.0	7.2	7.6	8.2	7.5	
	3	72	74	78	79	73	66	54	48	59	64	72	67	

Notes: 1: Mean monthly sunshine hours;   2: Mean daily sunshine hours;   3: % of possible maximum.

Country, name of station and years of records (where known)		Jan.	Feb.	Mar.	Apr.	May	Jun.	Jul.	Aug.	Sep.	Oct.	Nov.	Dec.	Range
Tambacounda	1	201	214	273	272	248	179	163	152	178	202	222	160	121
(1950–8)	2	6.5	7.6	8.8	9.1	8.0	6.0	5.3	4.9	5.9	6.5	7.4	5.2	
	3	57	65	73	73	62	46	41	39	48	55	64	46	
Zinguinchor	1	194	218	275	272	263	169	125	104	142	180	214	159	171
(1950–8)	2	6.3	7.8	8.9	9.1	8.5	5.6	4.0	3.3	4.7	5.8	7.1	5.1	
	3	55	67	74	73	67	44	31	26	38	49	61	45	
Mali														
Kayes	1	249	225	298	274	246	201	185	177	204	235	241	195	121
(1950–8)	2	8.0	8.0	9.6	9.1	7.9	6.7	6.0	5.7	6.8	7.6	8.0	6.3	
	3	70	69	80	73	62	52	47	45	56	64	70	56	
Bamako	1	255	240	262	217	207	219	182	153	187	234	234	213	109
(1950–8)	2	8.2	8.6	8.4	7.2	6.7	7.3	5.9	4.9	6.2	7.5	7.8	6.9	
	3	72	73	70	58	53	57	46	39	51	64	68	61	
Ségou	1	299	270	293	261	262	256	233	222	235	281	274	260	77
(1953–8)	2	9.6	9.6	9.4	8.7	8.4	8.5	7.5	7.2	7.8	9.1	9.1	8.4	
	3	84	82	78	70	66	66	59	57	64	77	79	74	
Mopti	1	227	217	250	227	219	230	204	199	220	250	242	184	66
(1950–8)	2	7.3	7.7	8.1	7.6	7.1	7.7	6.6	6.4	7.3	8.1	8.1	5.9	
	3	64	66	67	61	55	60	52	51	60	69	70	52	
Gao	1	271	261	289	291	288	271	268	262	268	288	289	271	30
(1951–8)	2	8.7	9.3	9.3	9.7	9.3	9.0	8.6	8.4	8.9	9.3	9.6	8.7	
	3	78	80	77	78	72	69	66	66	73	79	85	78	
Tessalit	1	301	317	314	300	300	267	294	272	264	293	263	255	62
(June 1952–8)	2	9.7	11.3	10.1	10.0	9.7	8.9	9.5	8.8	8.8	9.4	8.8	8.2	
	3	87	98	84	79	74	67	72	69	71	80	79	75	
Bourkina Faso														
Bobo Dioulasso	1	206	224	246	225	224	201	178	150	188	237	226	208	96
(1950–8)	2	6.6	8.0	7.9	7.5	7.2	6.7	5.7	4.8	6.3	7.6	7.5	6.7	
	3	57	68	65	61	57	52	45	38	52	64	65	59	
Ouagadougou	1	238	212	251	224	234	196	185	158	195	244	252	268	110
(1950–8)	2	7.7	7.6	8.1	7.5	7.5	6.5	6.0	5.1	6.5	7.9	8.4	8.6	
	3	67	65	67	60	59	51	47	41	53	67	73	76	
Algeria														
Tamanrasset	1	284	269	314	298	320	276	319	313	270	243	238	239	82
	2	9.2	9.6	10.1	9.9	10.3	9.2	10.3	10.1	9.0	7.8	7.9	7.7	
	3	84	84	84	78	78	68	77	78	73	67	72	72	
Niger														
Bilma	1	279	275	313	295	307	320	332	309	286	315	300	306	57
(1954–8)	2	9.0	9.8	10.1	9.8	9.9	10.7	10.7	10.0	9.5	10.2	10.0	9.9	
	3	80	85	84	78	76	81	82	79	77	87	88	90	
Agadez	1	291	274	292	297	291	283	275	256	270	311	290	278	55
(1953–8)	2	9.4	9.8	9.4	9.9	9.4	9.4	8.9	8.3	9.0	10.0	9.7	9.0	
	3	84	84	78	79	73	72	68	65	74	85	86	81	

Notes: 1: Mean monthly sunshine hours; 2: Mean daily sunshine hours; 3: % of possible maximum.

Country, name of station and years of records (where known)		Jan.	Feb.	Mar.	Apr.	May	Jun.	Jul.	Aug.	Sep.	Oct.	Nov.	Dec.	Range
Niamey	1	266	239	234	243	235	239	216	179	217	274	272	266	95
(1950–8)	2	8.6	8.5	7.5	8.1	7.6	8.0	7.0	5.8	7.2	8.8	9.1	8.6	
	3	75	73	62	65	59	62	55	46	59	75	79	76	
Birni	1	301	267	274	271	275	278	235	213	250	296	298	296	88
(1952–8)	2	9.7	9.5	8.8	9.0	8.9	9.3	7.6	6.9	8.3	9.5	9.9	9.5	
	3	85	82	73	72	69	72	59	55	68	80	86	84	
Zinder	1	235	218	232	246	245	223	211	194	230	274	249	235	80
(1950–8)	2	7.6	7.8	7.5	8.2	7.9	7.4	6.8	6.3	7.7	8.8	8.3	7.6	
	3	67	67	62	66	62	57	53	50	63	75	72	67	
Guinea-Bissau														
Bafata	1	238	238	297	293	272	208	170	135	158	217	238	240	162
(1959–69 and 1972)	2	7.7	8.5	9.6	9.8	8.8	6.9	5.5	4.3	5.3	7.0	7.9	7.8	
	3	67	73	80	79	69	54	43	34	43	59	68	68	
Bissau	1	241	241	294	292	275	221	166	130	150	215	236	236	164
(1959–69 and 1972)	2	7.8	8.6	9.5	9.7	8.9	7.4	5.3	4.2	5.0	6.9	7.9	7.6	
	3	68	73	79	78	70	58	42	34	41	58	68	67	
Bolama	1	257	253	306	300	274	204	149	122	155	227	248	263	184
(1959–69 and 1972)	2	8.3	9.0	9.9	10.0	8.8	6.8	4.8	3.9	5.2	7.3	8.3	8.5	
	3	72	77	82	81	69	53	38	31	43	61	71	75	
Sierra Leone														
Freetown	1	245	221	245	219	192	153	93	77	120	192	201	223	168
(1949–68)	2	7.9	7.9	7.9	7.3	6.2	5.1	3.0	2.5	4.0	6.2	6.7	7.2	
	3	67	66	65	59	50	40	24	20	33	52	57	62	
Kabala	1	251	238	263	234	214	165	139	93	141	195	207	236	170
(1959–68)	2	8.1	8.5	8.5	7.8	6.9	5.5	4.5	3.0	4.7	6.3	6.9	7.6	
	3	70	72	70	63	55	43	36	24	38	53	59	66	
Njala	1	223	207	211	180	186	150	96	62	96	170	186	195	161
(1949–68	2	7.2	7.4	6.8	6.0	6.0	5.0	3.1	2.0	3.2	5.5	6.2	6.3	
excluding 1967)	3	61	62	56	49	48	40	25	16	26	46	52	54	
Bonthe	1	192	176	192	171	152	114	65	46	84	155	162	179	146
(1949–68)	2	6.2	6.3	6.2	5.7	4.9	3.8	2.1	1.5	2.8	5.0	5.4	5.8	
	3	53	53	51	46	39	30	17	12	23	42	46	50	
Daru	1	189	179	201	174	167	132	93	74	105	164	171	175	127
(1952–68)	2	6.1	6.4	6.5	5.8	5.4	4.4	3.0	2.4	3.5	5.3	5.7	5.6	
	3	52	54	54	47	43	35	24	19	29	44	48	48	
Liberia														
Monrovia	1	170	143	170	162	142	103	87	49	59	132	155	160	121
(1932–9)	2	5.5	5.1	5.5	5.4	4.6	3.4	2.8	1.6	1.8	4.2	5.2	5.2	
	3	47	42	45	44	37	27	23	13	15	35	44	44	
Ivory Coast														
Gagnoa	1	182	178	190	187	182	113	91	82	113	153	178	173	108
(1953–8)	2	5.9	6.3	6.1	6.2	5.9	3.8	2.9	2.6	3.8	4.9	5.9	5.6	
	3	50	52	50	51	48	30	23	21	31	41	50	48	
Bouaké	1	141	164	190	179	166	99	70	68	113	160	151	120	122
(1950–8)	2	4.5	5.8	6.1	6.0	5.3	3.3	2.3	2.2	3.8	5.2	5.0	3.9	
	3	38	49	50	49	43	26	18	18	31	44	42	33	

Notes: 1: Mean monthly sunshine hours;   2: Mean daily sunshine hours;   3: % of possible maximum.

Country, name of station and years of records (where known)		Jan.	Feb.	Mar.	Apr.	May	Jun.	Jul.	Aug.	Sep.	Oct.	Nov.	Dec.	Range
Abidjan	1	207	205	235	206	191	107	148	118	139	192	228	203	128
(1950–70)	2	6.7	7.3	7.6	6.9	6.2	3.6	4.8	3.8	4.6	6.2	7.6	6.6	
	3	56	61	63	57	50	29	39	31	38	52	64	56	
Odiénné	1	264	225	196	214	223	215	176	132	177	226	233	256	132
(1956–8)	2	8.5	8.0	6.3	7.1	7.2	7.2	5.7	4.3	5.9	7.3	7.8	8.3	
	3	73	68	52	58	57	57	45	34	48	61	67	72	
Ferkéssédougou	1	275	254	251	234	236	209	162	148	166	240	265	266	127
(1953–8)	2	8.9	9.1	8.1	7.8	7.6	7.0	5.2	4.8	5.5	7.7	8.8	8.6	
	3	77	77	67	63	60	55	41	38	45	65	75	75	
Man	1	235	197	220	197	176	115	62	88	130	184	175	200	173
(1956–8)	2	7.6	7.0	7.1	6.6	5.7	3.8	2.0	2.8	4.3	5.9	5.8	6.4	
	3	64	59	59	54	46	30	16	22	35	49	49	55	
Tabou	1	203	215	226	202	128	63	77	100	79	144	187	205	163
(1956–8)	2	6.5	7.7	7.3	6.7	4.1	2.1	2.5	3.2	2.6	4.6	6.2	6.6	
	3	55	64	59	55	33	17	20	26	21	38	52	55	
Ghana														
Tamale	1	270	232	242	243	245	213	177	124	165	245	282	276	158
(13 years)	2	8.7	8.3	7.8	8.1	7.9	7.1	5.7	4.0	5.5	7.9	9.4	8.9	
	3	75	70	65	66	63	56	45	32	45	66	80	77	
Kumasi	1	146	143	170	174	149	90	62	40	87	136	162	158	134
(14 years)	2	4.7	5.1	5.5	5.8	4.8	3.0	2.0	1.3	2.9	4.4	5.4	5.1	
	3	40	43	45	47	39	24	16	10	24	37	46	44	
Takoradi	1	208	202	220	210	186	129	139	118	123	189	234	226	116
(16 years)	2	6.7	7.2	7.1	7.0	6.0	4.3	4.5	3.8	4.1	6.1	7.8	7.3	
	3	56	60	59	57	49	35	36	31	34	51	65	62	
Tafo	1	198	179	192	198	180	123	87	71	93	155	198	214	143
(16 years)	2	6.4	6.4	6.2	6.6	5.8	4.1	2.8	2.3	3.1	5.0	6.6	6.9	
	3	54	53	51	54	51	33	23	19	25	42	55	59	
Accra	1	211	199	214	216	211	156	139	139	171	217	240	236	101
(16 years)	2	6.8	7.1	6.9	7.2	6.8	5.2	4.5	4.5	5.7	7.0	8.0	7.6	
	3	57	59	57	59	55	42	36	37	47	59	67	64	
Togo														
Sokodé	1	263	251	238	224	233	178	123	115	130	200	236	273	158
(1955–8)	2	8.5	9.0	7.7	7.5	7.5	5.9	4.0	3.7	4.3	6.4	7.9	8.8	
	3	73	76	64	61	60	47	32	30	35	54	67	76	
Lomé	1	192	200	211	204	194	153	139	148	151	198	234	226	95
(1952–8)	2	6.2	7.1	6.8	6.8	6.2	5.1	4.5	4.8	5.0	6.4	7.8	7.3	
	3	52	60	56	56	50	41	36	39	41	53	66	62	
Benin														
Kandi	1	288	258	268	251	255	233	183	160	199	273	282	282	128
(1952–8)	2	9.3	9.2	8.6	8.4	8.2	7.8	5.9	5.2	6.6	8.8	9.4	9.1	
	3	81	79	71	68	65	61	46	42	54	74	81	80	
Tchaourou	1	235	224	233	218	211	171	100	86	123	183	217	239	153
	2	7.6	8.0	7.5	7.3	6.8	5.7	3.2	2.8	4.1	5.9	7.2	7.7	
	3	65	68	62	59	54	45	25	22	34	50	61	66	

Notes: 1: Mean monthly sunshine hours; 2: Mean daily sunshine hours; 3: % of possible maximum.

Country, name of station and years of records (where known)		Jan.	Feb.	Mar.	Apr.	May	Jun.	Jul.	Aug.	Sep.	Oct.	Nov.	Dec.	Range
Cotonou	1	205	201	212	213	202	152	132	149	153	193	241	232	100
(1953–8)	2	6.6	7.2	6.8	7.1	6.2	5.1	4.3	4.8	5.1	6.2	8.0	7.5	
	3	56	60	56	58	50	41	35	39	42	52	67	64	
Nigeria														
Sokoto	1	279	269	282	255	279	282	229	198	243	307	297	298	109
(1956–60)	2	9.0	9.6	9.1	8.5	9.0	9.4	7.4	6.4	8.1	9.9	9.9	9.6	
	3	79	82	76	69	71	73	58	51	66	83	86	84	
Kano	1	276	255	267	252	273	261	232	186	237	294	294	285	108
(1951–60)	2	8.9	9.1	8.6	8.4	8.8	8.7	7.5	6.0	7.9	9.5	9.8	9.2	
	3	77	78	72	68	70	68	59	48	65	80	84	80	
Maiduguri	1	291	280	285	273	276	264	214	189	222	291	300	298	111
(1956–60)	2	9.4	10.0	9.2	8.8	8.9	8.8	6.9	6.1	7.4	9.4	10.0	9.6	
	3	82	85	77	71	71	69	54	48	60	79	86	84	
Jos	1	307	274	260	213	208	201	152	127	171	242	294	313	186
(1951–60)	2	9.9	9.8	8.4	7.1	6.7	6.7	4.9	4.1	5.7	7.8	9.8	10.1	
	3	85	83	70	58	54	53	39	33	47	66	84	88	
Minna	1	263	244	263	234	242	201	130	121	168	263	279	279	158
(1956–60)	2	8.5	8.7	8.5	7.8	7.8	6.7	4.2	3.9	5.6	8.5	9.3	9.0	
	3	73	74	71	63	62	53	33	31	46	71	79	78	
Ibadan	1	211	188	205	192	205	159	113	87	101	176	217	223	136
(1956–60)	2	6.8	6.7	6.6	6.4	6.6	5.3	3.7	2.8	3.4	5.7	7.2	7.2	
(1968–75)	3	58	56	54	52	53	42	30	23	29	47	61	61	
Makurdi	1	242	235	232	207	220	174	149	121	147	201	216	232	121
	2	7.8	8.4	7.5	6.9	7.1	5.8	4.8	3.9	4.9	6.5	7.2	7.5	
	3	67	71	62	56	57	46	38	31	40	54	61	64	
Yola	1	282	272	267	231	267	237	198	167	189	270	297	301	134
	2	9.1	9.7	8.6	7.7	8.6	7.9	6.4	5.4	6.3	8.7	9.9	9.7	
	3	78	82	72	63	69	62	'51	43	52	73	85	84	
Lagos	1	183	190	198	186	174	120	90	93	93	152	195	205	115
(1959–60)	2	5.9	6.8	6.4	6.2	5.6	4.0	2.9	3.0	3.1	4.9	6.5	6.6	
	3	50	57	53	51	45	32	23	24	25	41	55	56	
Benin	1	180	176	174	177	176	144	99	90	81	149	192	214	133
(1956–60)	2	5.8	6.3	5.6	5.9	5.7	4.8	3.2	2.9	2.7	4.8	6.4	6.9	
	3	49	52	46	48	46	38	26	24	22	40	54	59	
Enugu	1	208	204	192	189	198	159	124	105	117	174	213	229	124
(1951–60)	2	6.7	7.3	6.2	6.3	6.4	5.3	4.0	3.4	3.9	5.6	7.1	7.4	
	3	57	61	51	52	52	42	32	28	32	47	60	63	
Port Harcourt	1	164	171	146	147	152	99	77	77	60	108	141	174	114
(1956–60)	2	5.3	6.1	4.7	4.9	4.9	3.3	2.5	2.5	2.0	3.5	4.7	5.6	
	3	44	51	39	40	40	27	20	20	16	29	39	47	
Cameroon														
Garoua	1	274	253	233	220	239	226	183	163	190	257	265	288	125
(1952–7)	2	8.8	9.0	7.5	7.3	7.7	7.5	5.9	5.2	6.3	8.3	8.8	9.3	
	3	76	76	62	59	61	59	47	42	52	70	75	81	
N'Gaoundéré	1	256	231	190	193	170	133	111	121	126	169	238	280	169
(1952–7)	2	8.3	8.2	6.1	6.4	5.5	4.4	3.6	3.9	4.2	5.4	7.9	9.0	
	3	70	69	50	52	44	35	29	31	34	45	67	77	

Notes: 1: Mean monthly sunshine hours; 2: Mean daily sunshine hours; 3: % of possible maximum.

Country, name of station and years of records (where known)		Jan.	Feb.	Mar.	Apr.	May	Jun.	Jul.	Aug.	Sep.	Oct.	Nov.	Dec.	Range
Yoko	1	233	211	170	203	204	169	107	116	124	159	228	267	160
(1954–7)	2	7.5	7.5	5.5	6.8	6.6	5.6	3.4	3.7	4.1	5.1	7.6	8.6	
	3	63	62	45	56	54	45	27	30	34	42	64	73	
Batouri	1	160	163	150	174	184	140	108	94	112	138	179	189	95
(1944–57)	2	5.2	5.8	4.8	5.8	5.9	4.7	3.5	3.0	3.7	4.4	6.0	6.1	
	3	43	48	40	47	48	38	28	24	30	37	50	51	
Yaoundé	1	175	168	151	152	151	120	87	85	103	124	150	179	94
(1941–6)	2	5.6	6.0	4.9	5.1	4.9	4.0	2.8	2.7	3.4	4.0	5.0	5.8	
(1948–57)	3	47	50	40	42	40	32	23	22	28	33	42	49	
Douala	1	92	118	112	134	123	80	37	34	66	110	109	111	100
(1941–57)	2	3.0	4.2	3.6	4.5	4.0	2.7	1.2	1.1	2.2	3.5	3.6	3.6	
	3	25	35	30	37	32	22	10	9	18	29	30	30	
Equatorial Guinea														
Malabo	1	102	100	112	93	81	56	36	50	40	72	89	91	76
(1935–49)	2	3.3	3.5	3.6	3.1	2.6	1.9	1.2	1.6	1.3	2.3	3.0	3.0	
	3	28	29	30	25	21	15	10	13	11	19	25	25	

Notes: 1: Mean monthly sunshine hours;   2: Mean daily sunshine hours;   3: % of possible maximum.

Appendix 5   Cup–counter anemometer

In the example illustrated (British Meteorological Office anemometer Mark II) the instrument has a cup wheel with three conical cups. The spindle is connected by worm gearing to a revolution counter, with a gear ratio that can indicate directly the run of the wind in kilometres or miles, with tenths and hundreds. If the reading of the counter is taken at two different times the mean wind speed of the period can be determined. Cup-counter anemometers of this type normally begin to rotate at wind speeds of the order of 2.7 to 3.7 km hr^{-1} (1.8 to 2.3 miles hr^{-1}) have errors of less than 1.8 km hr^{-1} at normal speeds above 9 km hr^{-1}, but are unreliable at speeds below 9 km hr^{-1} (5 knots).

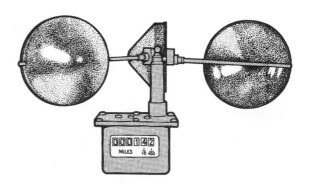

Figure A4   A British Meteorological Office anemometer Mark II

Appendix 6 Instruments for measuring evaporation and evapotranspiration

The Piché evaporimeter

This consists of a narrow glass tube, closed at one end, open at the other and graduated in cubic centimetres. The open flat end is covered with a circular filter paper held in position with a disc and clip. The tube is filled with distilled water, closed with the filter paper and clip and then inverted. The water soaks the filter paper and then evaporates from it. The amount of water that evaporates in a given time can be measured by making two consecutive readings of the level of the water in the tube. The difference is the volume of water evaporated. If it is desired to express the evaporation in centimetres the following equation can be used:

$$E = \frac{V}{2\,\pi(r_1^{\,2} - r_2^{\,2})}$$

where E is the evaporation in centimetres (expressed as the height of water column evaporated from an area of 1 cm^2), and V is the recorded evaporation in cc. $2\pi(r_1^{\,2} - r_2^{\,2})$ is the area of the evaporating surface; r_1 is the radius of the filter paper and r_2 is the radius of the base of the clip in centimetres.

Tank evaporimeters

These may be raised or sunken, square, rectangular or round. In common use is the Class 'A' pan, circular with a diameter of 121 cm and 25.5 cm deep. Made of galvanized iron it is normally set on a platform 15 cm above the ground. With reference to a fixed point in the tank or by means of a hook gauge, the tank is filled to a depth of 19 cm. Evaporation can be measured by noting the amount of water required to refill the tank to the standard depth, using a graduated measuring jar.

Lysimeters

Lysimeters are essentially tanks buried in the ground through which water percolation is measured. Sophisticated models can be expensive to install but a simple evapotranspirometer designed by Garnier (1952a) is illustrated here. It comprises an oil-drum cut to a height of about 55 cm, containing a layer of gravel, a filling of light soil which will permit free drainage, and topped with short-cut grass. A pipe of about 2 cm diameter leads from the base of the tank (the outlet protected by mesh) to a measuring cylinder in a covered chamber.

A known amount of water is applied to the tank and the drainage is collected through the pipe. Potential evapotranspiration equals the amount of water applied, plus rainfall, minus the amount collected, converted into millimetres of equivalent precipitation.

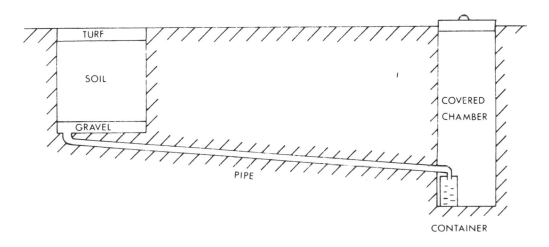

Figure A5 A simple lysimeter

Appendix 7 An instrument for measuring humidity – the hair hygrograph

Elevation and plan of a typical instrument are in Figure A6. The hygrograph uses a type of human hair which is very sensitive to changes in atmospheric moisture, the length of the hair increasing by 2 to 2½ per cent when relative humidity changes from 0 to 100 per cent. In the instrument a bundle of hairs (H) is held between two points and caught up near the centre by a hook (K) joined to a lever (L). Quadrants A and B are attached to the lever, a spring (S) keeping them in contact. To quadrant B a pen arm is linked, which rests against a rotating drum that may be fitted with a clock turning the drum once in 24 hours or once in a week. On the drum is placed a chart with a scale graduated from zero to 100 (per cent relative humidity). Totally dry hair will contract, pulling on the lever (L), so pivoting the quadrants and bringing the pen to the zero position on the chart. With increasing humidity the hair lengthens and the pen rises up the chart to the highest position of 100 per cent when relative humidity attains this maximum level.

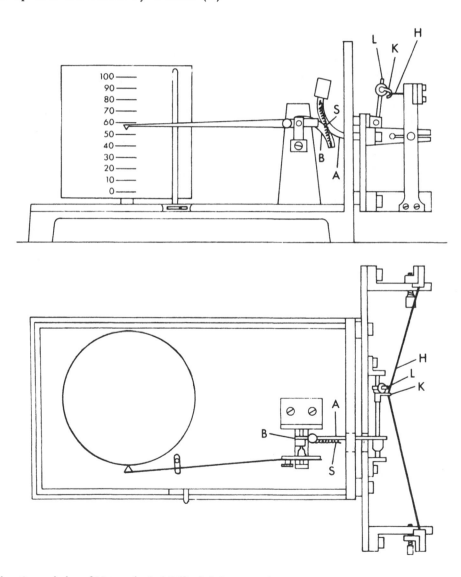

Figure A6 Elevation and plan of Meteorological Office hair hygrograph

Appendix 8   A psychrometric chart

Figure A7

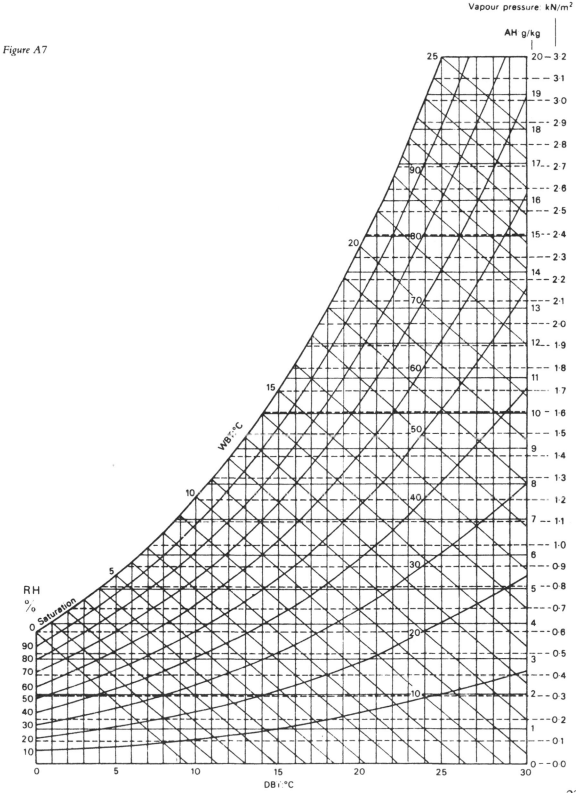

Appendix 9   Climatic data for 38 stations

		J	F	M	A	M	J	J	A	S	O	N	D	Year mean or total
Mauritania														
F'Derik	1	25	27	30	33	36	39	42	42	39	35	30	25	
	2	11.5	13	15	17	19	22	24	25	24	21	17	13	
	3	55	54	52	54	55	50	47	49	50	50	51	58	52
	4	42	42	36	32	32	27	26	29	31	34	37	44	34
	5	1	3.5	2.5	1	Tr	1	3	10	15	15	9	2	62
Nouadhibou	1	25	26	26	26	26	28	27	28.5	31	30	27	25	
	2	13	14	14	15	16	17	18.5	20	20	19	17	15	
	3	69	71	80	87	83	80	85	84	80	79	71	69	78
	4	38	40	43	50	51	51	61	64	53	50	42	48	49
	5	2	1	1.5	1	Tr	1	Tr	2.5	6.5	6	5	1.5	28
Nouakchott	1	29	30	32.5	33	34.5	34	32	32	34	35	33	29	
	2	13	14	16	18	20	22	24	24	24	22	18	14	
	3	51	54	65	68	73	79	85	88	85	73	64	58	70
	4	31	30	30	32	35	48	63	69	59	41	35	34	42
	5	1	2	1	0.5	1	1	12	62	40	10.5	2	6	137
Tidjikja	1	28	30	34	37	40	41	39	37	38	37	32	27	
	2	12	13.5	16	19	23	26	25	24	24	21	17	12	
	3	55	54	44	41	34	54	65	74	70	58	57	62	55
	4	34	33	27	24	26	34	46	52	45	32	35	40	36
	5	1	3	1.5	Tr	4	10.5	24	59	33	8	5	1	151
Néma	1	30	33.5	37	40	42	41.5	38	35	37	39	36	31	
	2	17	19	23	26	29	28	26	24	25	26	23	18	
	3	33	32	26	24	48	50	72	82	79	47	35	39	47
	4	24	24	19	18	19	31	49	57	49	29	25	28	31
	5	1	Tr	Tr	2	10	34	63	114	90	14	1	2	331
Algeria														
Tamanrasset	1	19	22	26	30	33	35	35	34	33	29	26	21	
	2	4	6	9	13	17	21	22	21	19	15	11	6	
	3	36	30	24	22	25	22	23	26	25	28	32	35	27
	4	24	20	20	17	19	17	19	16	18	22	22	24	20
	5	3	1	1	4	7	4	3	10	12	2	1	3	51
Senegal														
Thiès	1	31	32	33	33	32.5	33	32	30	31	32	33	31	
	2	15	16	18	18	19.5	22	22	23	23	22	19	16	
	3	72	81	80	83	89	92	92	93	97	97	92	72	86
	4	28	39	31	36	44	54	66	74	74	62	44	31	48
	5	Tr	1.5	Tr	Tr	1	27	113	277	202	52	3	5	683

1: Mean monthly maximum temperatures (°C);
2: Mean monthly minimum temperatures (°C);
3: Mean monthly maximum relative humidity (%);
4: Mean monthly minimum relative humidity (%);
5: Mean monthly rainfall (mm), (Tr = Trace, <0.5 mm).

		J	F	M	A	M	J	J	A	S	O	N	D	Year mean or total
Gambia														
Banjul	1	31	32	34	33	32	32	30	29	31	32	32	31	
	2	15	16	17	18	19	23	23	23	23	22	18	16	
	3	67	66	76	82	88	91	94	95	95	95	90	77	85
	4	27	26	29	41	49	61	72	78	73	65	47	36	50
	5	Tr	1	Tr	Tr	4	73	270	502	275	90	7	2	1224
Mali														
Kéniéba	1	35	37	40	41	39.5	35	31	30	31.5	33	34.5	34	
	2	17	20	22	24	26	24	23	22	22	22	18	16	
	3	52	47	40	48	68	86	94	95	96	96	92	71	74
	4	24	24	23	28	46	65	76	79	76	69	53	33	50
	5	Tr	Tr	Tr	7.5	54	150	259	443	305.5	120	14	2	1356
Bamako	1	33	36	38.5	39.5	38.5	35	31	30	32	34	35	33	
	2	17	19.5	23	26	26	24	23	23	21	21.5	19	17	
	3	38	33	41	63	70	74	91	94	93	73	70	69	67
	4	19	18	23	36	40	49	70	73	68	41	34	40	43
	5	1	Tr	4	16.5	67.5	141	234.5	338.5	207	61	10	Tr	1081.5
Tessalit	1	27	30	33.5	37	40.5	43	42	40	40	37	33	27	
	2	13	14	18	22	26	28	27	26	26	24	19	14	
	3	31	27	27	22	22	30	43	58	46	31	29	30	33
	4	18	16	17	14	13	16	22	32	24	18	17	17	18
	5	Tr	Tr	1	Tr	1.5	8	21	51	23	1	Tr	Tr	107
Ménaka	1	32	33	39	42	43	42	38	35	38	40	37	33	
	2	15	16	21	24	28	27	25	24	24	22	20	16	
	3	33	28	24	24	40	58	72	85	87	57	38	33	48
	4	15	12	11	12	20	32	43	56	42	23	13	15	25
	5	Tr	Tr	1	Tr	7	21	67	112	44	4	Tr	Tr	256
Niger														
Agadez	1	30	33	38	41	43	42	41	38	40	39	35	31	
	2	10	13	17	21	25	24	23.5	23	23	20	15.5	12	
	3	38	33	29	23	36	42	67	80	68	41	40	41	45
	4	17	14	14	10	18	20	33	45	30	17	16	17	21
	5	Tr	Tr	Tr	1.5	5	9	45	91	16	Tr	Tr	Tr	168
Birni	1	33	36	39	41	40	38	34	32	34	37.5	37	34	
	2	16	18	22	24	26	25	23	22	22.5	21	18	16	
	3	35	31	29	45	61	74	85	92	92	80	50	38	59
	4	16	14	14	24	33	45	58	69	62	39	20	18	34
	5	Tr	Tr	Tr	2	30.5	71	137	231	109	15	Tr	Tr	599
N'guigmi	1	29	31	35	37	39	38	36	34	35	36	33	30	
	2	13.5	15	20	22	24.5	25	25	23	23	21	18	14	
	3	36	32	30	32	54	64	74	83	76	53	39	43	54
	4	21	21	21	24	30	34	43	53	42	27	24	23	30
	5	Tr	Tr	Tr	Tr	5	6	55.5	131	19	1	Tr	Tr	217

1: Mean monthly maximum temperatures (°C);
2: Mean monthly minimum temperatures (°C);
3: Mean monthly maximum relative humidity (%);
4: Mean monthly minimum relative humidity (%);
5: Mean monthly rainfall (mm), (Tr = Trace, <0.5 mm).

		J	F	M	A	M	J	J	A	S	O	N	D	Year mean or total
Guinea-Bissau														
Bolama	1	31	32	33	33	32	31	29	28	29	31	31	31	
	2	17	20	22	23	24	24	23	23	23	23	23	21	
	3	51	57	63	64	67	79	84	86	84	81	78	58	71
	4	38	39	46	50	56	69	78	82	77	71	60	45	59
	5	1	0.5	Tr	0.5	2	229	582	632	444	223	43	Tr	2184
Bourkina-Faso														
Ouagadougou	1	34	37	39.5	40	38	35	32	31	32	36	37	35	
	2	16	18	22	25.5	27	24	23	22	22	23	20	16.5	
	3	42	38	39	51	65	73	78	84	79	72	58	46	60
	4	19	19	20	28	40	49	62	67	60	44	30	23	38
	5	Tr	2	6	21	87	118	187	269	150	38	1	Tr	879
Sierra Leone														
Freetown	1	29	30	30	31	30	29	28	28	28	29	29	29	
	2	24	24	25	25	25	24	23	23	23	23	24	24	
	3	82	80	81	81	83	86	89	91	90	87	85	82	85
	4	67	67	69	71	74	76	81	82	81	77	75	71	74
	5	10	6	27	81	229	433	869	872	652	288	138	34	3639
Liberia														
Monrovia	1	30	29	31	31	30	27	27	27	27	28	29	30	
	2	23	23	23	23	22	23	22	23	22	22	23	23	
	3	95	94	92	91	89	89	88	87	92	92	91	93	91
	4	78	76	77	80	79	82	83	84	86	84	80	79	81
(Robertsfield)	5	45	82	114	157	356	630	698	508	715	472	198	73	4048
Ivory Coast														
Odienné	1	34	36	36	35	34	32	30	29	31	32	33	33	
	2	17	19	22	23	23	21	21	21	21	20.5	20	17	
	3	67	70	79	87	90	93	94	95	95	94	92	84	86
	4	33	39	48	58	63	69	74	77	73	68	59	44	59
	5	4	16	37	74	120	176	296	395	294	162	54	9	1637
Bouaké	1	33	35	35	34	32	31	29	29	30	31	32	32.5	
	2	20	21.5	22	22	22	21	21	20	21	21	21	20	
	3	89	87	92	93	96	96	97	97	98	97	96	92	94
	4	55	57	62	64	69	74	78	79	77	72	67	56	67
	5	13	43	93	143	149	134	95.5	108	220	142	37	22	1200
Tabou	1	30	30	30	31	29	28	27	26	27	28	29	29.5	
	2	23	23	23	24	24	24	23	22	23	23	24	23	
	3	94	95	95	94	94	89	84	88	94	94	94	95	93
	4	79	78	78	77	81	83	76	80	87	84	80	79	80
	5	55	51	92	143.5	444	513	130	82	221	221	200	151	2304.5

1 Mean monthly maximum temperatures (°C);
2 Mean monthly minimum temperatures (°C);
3 Mean monthly maximum relative humidity (%);
4 Mean monthly minimum relative humidity (%);
5 Mean monthly rainfall (mm), (Tr = Trace, <0.5 mm).

		J	F	M	A	M	J	J	A	S	O	N	D	Year mean or total
Ivory Coast—*continued*														
Abidjan	1	31	31	32	32	31	29	28	27	28	29	31	31	
	2	23	24	25	25	24	23	23	22	22.5	23	24	24	
	3	96	96	97	94	95	96	94	94	95	94	95	96	95
	4	72	68	69	69	72	81	78	76	79	76	72	73	74
	5	32	49	115	140	358	554	217	39.5	62	206	182	91	2045
Ghana														
Tamale	1	36	37	37	36	33	31	29	29	30	32	34	35	
	2	21	23	24	24	24	22	22	22	22	22	22	20	
	3	86	56	62	80	88	92	94	95	95	94	78	54	77
	4	20	33	37	52	62	69	72	74	74	66	42	27	52
	5	2	7	51.5	82.5	118.5	144	144.5	196	223	95.5	14	4	1082
Accra	1	31	31	31	31	31	29	27	27	27	29	31	31	
	2	23	24	24	24	24	23	23	22	23	23	24	24	
	3	95	96	95	96	96	97	97	97	96	97	97	97	96
	4	61	61	63	65	68	74	76	77	72	71	66	64	68
	5	16	32	59	86	136	182	47	15	38	63	36	22	732
Togo														
Atakpame	1	34	35	35	34	33	31	29	28.5	30	31	33	33.5	
	2	20.5	21	22	22	21	21	21	20	21	21	21	21	
	3	77	82	86	83	84	89	92	93	93	92	86	76	86
	4	42	48	59	62	64	73	75	76	73	69	58	49	63
	5	19	46	105	132	155	182	194	164	189	144	43	22	1395
Benin														
Kandi	1	34	37	39	38	36	33	31	29.5	31	34	36	34.5	
	2	16	19	23	25	24	23	22	22	21	22	18.5	16	
	3	48	50	66	70	83	92	95	97	97	93	80	58	76
	4	23	24	35	39	54	67	74	76	72	58	32	24	48
	5	1	1	8	31	98	144	194	299	222	54	3	Tr	1055
Nigeria														
Kano	1	30	33	37	38	37	34	31	29	31	34	33	31	
	2	13	15	19	24	24	23	22	21	21	19	16	13	
	3	40	36	33	47	72	81	90	94	93	82	52	45	64
	4	13	13	11	14	33	43	59	68	57	32	16	14	31
	5	Tr	Tr	2	8	69	114	208	312	140	13	Tr	Tr	864
Maiduguri	1	32	34	38	40	38	36	32	30	33	36	35	32	
	2	12	14	18	22	25	24	23	22	22	20	15	12	
	3	49	43	35	36	61	74	87	94	93	84	61	58	65
	4	18	15	12	14	30	40	59	68	59	37	20	21	33
	5	Tr	Tr	Tr	7	38	66	186	221	102	19	Tr	Tr	640
Jos	1	28	30	32	31	29	27	25	24	27	29	29	28	
	2	14	16	18	19	18	18	17	17	17	17	16	14	
	3	31	35	38	74	90	94	97	98	96	83	49	37	69
	4	13	14	17	34	60	64	78	81	68	47	22	17	43
	5	2	8	28	84	203	226	325	279	216	43	2	2	1420

1   Mean monthly maximum temperatures (°C);
2   Mean monthly minimum temperatures (°C);
3   Mean monthly maximum relative humidity (%);
4   Mean monthly minimum relative humidity (%);
5   Mean monthly rainfall (mm), (Tr = Trace, <0.5 mm).

		J	F	M	A	M	J	J	A	S	O	N	D	Year mean or total
Lokoja	1	33	35	36	34	33	32	31	31	31	32	33	33	
	2	19	22	24	24	23	22	22	22	22	22	22	19	
	3	87	81	83	82	88	91	92	92	94	94	93	90	89
	4	54	50	53	55	64	70	72	72	75	73	62	58	63
	5	5	5	48	94	155	160	173	165	211	137	13	2	1168
Lagos	1	31	32	32	32	31	29	28	28	28	29	31	31	
	2	23	25	26	25	24	23	23	23	23	23	24	24	
	3	84	83	82	81	83	87	87	85	86	86	85	86	85
	4	65	69	72	72	76	80	80	76	77	76	72	68	74
	5	28	47	102	144	254	437	248	62	144	200	68	27	1761
Calabar	1	30	32	32	31	31	30	29	28	29	29	31	30	
	2	22	22	23	22	22	22	22	22	22	22	22	22	
	3	88	87	88	88	88	91	92	91	91	89	89	88	89
	4	67	67	70	73	76	80	84	85	83	78	75	68	75
	5	38	76	157	218	312	411	455	419	422	328	190	48	3073

Cameroon

		J	F	M	A	M	J	J	A	S	O	N	D	Year mean or total
Maroua	1	34	36	39	41	38	35	32	30	32	35	36	33.5	
	2	19	21	24	26	24	23	22	21	21	21	21	19	
	3	35	30	30	40	71	81	89	91	91	78	40	37	59
	4	22	20	19	22	40	51	63	68	63	41	23	26	38
	5	Tr	Tr	2	12	69	108	181	265	158	28	Tr	Tr	823
N'Gaoundéré	1	30	31	31	30	28	27	26	26	26	28	29	30	
	2	13	14	17	18	17	17	17	17	16	16	14	13	
	3	53	60	80	87	95	95	98	98	98	97	85	61	84
	4	16	19	36	46	61	63	68	69	67	59	32	19	46
	5	4	2	44	145	205	241	271	275	241	153	12	3	1596
Yoko	1	30	30	29.5	28	27	26	25	25	26	27	28	29	
	2	18	18	19	19	18	17	17	17	17	17	18	18	
	3	75	86	91	93	94	96	99	99	98	95	79	62	89
	4	35	46	62	67	72	76	80	81	77	72	50	32	63
	5	16	25	85	134	195	170	148	178	311	302	72	12	1651
Kríbi	1	29.5	30	30	30	29	28	27	27	27	28	29	29	
	2	24	23.5	23	23	23	23	22	22	22	22.5	23	23	
	3	95	95	95	95	96	95	98	95	96	97	96	95	95
	4	78	77	77	77	80	81	80	82	86	86	82	77	80
	5	104	141	213	261	376	285	112	244	517	534	199	38	3024
Batouri	1	30	31	31	31	30	29	27	27	27	29	30	30	
	2	17	18	19	19	19	19	18	18.5	19	18.5	18	17	
	3	97	97	97	97	98	99	99	99	99	99	98	97	98
	4	58	63	65	66	69	74	76	76	72	72	63	54	67
	5	34	55	114	150	202	170	102	157	221	281	127	35	1648.5

1: Mean monthly maximum temperatures (°C);
2: Mean monthly minimum temperatures (°C);
3: Mean monthly maximum relative humidity (%);
4: Mean monthly minimum relative humidity (%);
5: Mean monthly rainfall (mm), (Tr = Trace, <0.5 mm).

Glossary

Adiabatic   Changes in the temperature, pressure and volume of air without transfer of heat. In thermodynamics a process in which heat does not leave or enter the system. An ascending body of air will expand and so cool, a descending body will be compressed and so become warmer, but there will be no net loss or gain of heat. The dry adiabatic lapse rate is 3°C per 305 m of change. The saturated adiabatic lapse rate is not constant and, near the earth's surface, is less at about 0.5°C per 100 m, due to the latent heat of condensation.

Adobe   Unburnt brick, dried in the sun.

Advection   The transfer of the properties of air masses essentially in a horizontal direction.

Aerosol   Minute liquid or solid particles in the atmosphere.

Airmass   A body of air throughout which there is little horizontal variation in humidity or temperature.

Albedo   The proportion of the incident radiation that is reflected from a surface.

Albedometer   An instrument for measuring albedo (see Appendix 2).

Ambient   Relating to the immediate surroundings.

Anemograph   A self-recording instrument for measuring the direction and strength of the wind, usually by means of a pen on a rotating chart.

Anemometer   An instrument for measuring wind strength and direction, but lacking the means for permanently recording these on time-charts (see Appendix 5).

Angstrom Unit   A unit of length, especially used for the measurement of wave-lengths of light. It is 1/10,000th of a micron.

Anticyclone   An atmospheric high-pressure system in which barometric pressure is centrally high, diminishing outwards in all directions. Thus the related isobaric pattern is commonly near circular around the central high pressure. Winds flow from such a system in a clockwise direction in the northern hemisphere, anticlockwise in the southern.

Astronomical twilight   *See* twilight

Baroclinic   Applied to the atmosphere in which there is a marked horizontal temperature gradient.

Barograph   A self-recording barometer.

Barometer   An instrument for measuring atmospheric pressure.

Barotropic   A term describing the atmosphere in which there is no marked horizontal temperature gradient, so that the vertical gradient is significant.

Bowen Ratio   The ratio of the sensible heat to latent heat lost by a surface due to conduction.

Calorie   A unit of heat. Gram–calorie (gcal or cal), the heat required to raise the temperature of 1 gram of water by 1°C. Kilogram–calorie (kcal) is 1000 cal.

Civil twilight   *See* twilight.

Climate   A synthesis of the day-to-day values of the meteorological elements of a region or locality.

Climatological station   One from which climatological data are received concerning wind, cloud, temperature, humidity, pressure, precipitation and sunshine. *See also* principal climatological station.

Climatology   The science which considers the earth's climates and their influence on natural global processes and life.

Clo-unit   A measurement of the thermal resistance from the skin to the outer surface of a clothed body.

Conduction   (thermal) The transmission of heat from points of higher to points of lower temperature by the imparting, on collision, of energy from molecules or electrons with greater levels of kinetic energy to those with less.

Convection   In the atmosphere relates to the localized vertical movement of heat and other airmass properties.

Coriolis effect   The impression, gained by earth-bound observers, that airflow in the northern hemisphere is deflected to the right, in the southern to the left; the result of the earth's rotation.

Coriolis Force   The apparent force resulting from the earth's rotation which brings about the Coriolis effect.

Daylight   The interval between sunrise and sunset, these taken to occur when the upper edge of the disk of the sun appears to be exactly on the horizon.

Dew-point   The air temperature at which water vapour saturation is achieved and dew deposited.

Dicotyledon   Having two cotyledons or leaves that form from the embryo of plants.

Diffuse radiation   *See* radiation.

Dropwindsondes   Devices released by high-flying aircraft to descend on parachutes at a constant rate (e.g. 25 mb per min) to measure temperature, humidity and pressure. Wind direction and speed can also be measured using methods to track the sondes.

Dyne   A unit of force. The force which, acting upon a mass of 1 gram, will impart to it an acceleration of 1 cm per second per second.

Electromagnetic waves   Wave motions not requiring any material medium for their propagation, ranging from gamma rays to radio waves (see Figure 3) and travelling at 299,300 km per second.

Emissive power   The total energy emitted from a

unit area of a surface of a body per second, depending upon the temperature of the body and the nature of its surface.

Energy   The capacity for doing work.

Kinetic energy   Possessed by a body by virtue of its motion.

Potential energy   Possessed by a body by virtue of its position. Measured by the amount of work performed in passing from that position to a standard position in which the potential energy is deemed to be zero.

Static energy   The energy content of the atmosphere, i.e. latent heat plus sensible heat plus potential energy, but excluding kinetic energy.

Enthalpy   Heat content per unit mass.

Equivalent temperature   *See* temperature.

Evaporation   Conversion of liquid into vapour.

Potential evaporation   That evaporation which could occur from a surface given the climatic conditions, were the water to be available for evaporation.

Evaporimeter   (or atmometer) An instrument for measuring evaporation (see Appendix 6).

Evapotranspiration   Evaporation from the earth's surface combined with transpiration (water loss) from plants.

Potential evapotranspiration   The addition of water vapour to the atmosphere from the earth's surface and from plants to a given climatic environment if there were no lack of available water.

Fadama   A Hausa word for valley-land that may be cultivated in the dry season.

Geostrophic wind   A wind which blows parallel to the isobars due to the balance between the horizontal pressure force (at right angles to the isobars) and the Coriolis force.

Global radiation   *See* radiation.

Greenhouse effect   The re-radiation from the atmosphere back to the earth of infra-red radiation emitted from the earth and absorbed by the water vapour and carbon dioxide of the atmosphere, the effect being to maintain the temperature of the earth's surface and lower atmosphere at a higher level than would otherwise exist.

Hadley Cell   A simple thermal circulation, if the earth's rotation is ignored, most commonly applied to low latitudes where it involves air ascent at the equator, thence high-altitude poleward airflow, descent in the subtropics and a returning equatorward flow at low altitude.

Heat   Energy possessed by a substance in the form of kinetic energy of its molecules. Usually measured in joules or calories and transmitted by conduction, convection and radiation.

Latent heat   The quantity of heat required to bring about a change of state of 1 gram of a substance from solid to liquid, or from liquid to vapour without change of temperature.

Sensible heat   That which is felt or experienced in contradistinction to latent heat. The transfer of sensible heat causes an increase in the temperature of the receiving body.

Specific heat   The quantity of heat required to raise the temperature of 1 gram of a substance through 1°C.

Humidity   Measure of water vapour content of the air.

Absolute humidity   The mass of water vapour present in a cubic metre of air.

Relative humidity   Either the ratio of the actual water vapour pressure to the vapour pressure that would be present if the vapour were saturated at the same temperature, or the ratio of the mass of water vapour per unit volume of air to the mass per unit volume of saturated air at the same temperature.

Hydrological cycle   The cycling of water from earth sources to atmosphere and back again.

Hygrometer   An instrument for measuring the relative humidity of the atmosphere.

Hygrometric tables or **charts**   (also psychrometric). Tables or charts to permit the computation of relative humidity (see Appendix 8).

Isobars   Lines on a chart joining points of equal barometric pressure.

Isohels   Lines on a chart joining points of equal amounts of sunshine.

Isohyets   Lines on a chart joining points of equal amounts of rainfall.

Isotachs   Lines on a chart joining points of equal wind speed.

Isotherms   Lines on a chart joining points of equal temperature.

Joule   A unit of work; that done in 1 second by a current of 1 ampere flowing through a resistance of 1 ohm.

Kelvin wave   A tidal flow in which, in the northern hemisphere, the tidal range is increased on the right hand side, decreased on the left of the direction of flow, due to the earth's rotation.

Langley (ly)   A measurement of energy per unit area. 1 ly = 1 cal/cm^2.

Latent heat   *See* heat.

Lysimeter   An instrument for measuring evapotranspiration (see Appendix 6).

Metabolism   The chemical processes occurring within an organism.

Meteorology   The science concerned with the physics and phenomena of the atmosphere.

Micron (μ)   One millionth of a metre. 10,000 Angstrom units.

Millibar   A unit of atmospheric pressure, used in meteorology. 1000 dynes per cm^2.

Monocotyledon   Having only one leaf forming from the embryo of a plant.

Normals   Period averages computed for a uniform period comprising at least three consecutive 10-year time-spans.

Oktas   Eighths. Used to measure the proportion of

the sky covered by cloud, e.g. 4 oktas means half the sky is covered, 8 oktas means total cover.

Pentad   A period of five consecutive days.

Photoperiodism   The effect of light and dark periods on the growth and development of flowers and fruits.

Photosensitive   Sensitive to light.

Photosynthesis   The development in the green cells of plants of carbohydrates and thence proteins from carbon dioxide and water in conditions of sufficient light.

Potential energy   *See* energy.

Potential evaporation   *See* evaporation.

Potential evapotranspiration   *See* evapotranspiration.

Potential temperature   *See* temperature.

Precipitable water   The resulting depth of water that could be condensed from the water vapour of a vertical column of air of unit area. It is a direct indicator of the amount of moisture available for precipitation and an indirect indicator of the net inflow of moisture over an area.

Principal climatological station   One in which readings are taken hourly or at least three times daily, plus autographic recordings.

Principal land station   A surface synoptic station observing specific WMO Technical Regulations and reporting observations for international exchange.

Pyranometer   (pyrheliograph, pyrheliometer, solarimeter, actinometer) Instruments for measuring solar radiation (see Appendix 2).

Radiation   The emission of rays or waves, usually electromagnetic waves.

   Diffuse radiation   Radiation received simultaneously from many different directions.

   Global radiation   The sum of direct and diffuse radiation received by a unit surface.

   Net radiation   The sum of the negative upward and positive downward solar radiation on a horizontal surface.

Radiosonde   Balloon-carried equipment normally measuring air pressure, temperature and humidity, released at ground level or aloft by rockets or aircraft.

Radiowind   The electronic tracking of a free balloon in order to determine upper wind patterns.

Rain day   A day having at least 2.5 mm of rain.

Rawinsonde   A combined radiosonde and radiowind.

Solarimeter   An instrument for measuring solar radiation.

Specific heat   *See* heat.

Standard times   (for meteorological observations). For the upper air: 00.00, 06.00, 12.00 and 18.00 hours Greenwich Mean Time. For the surface: additionally, 03.00, 09.00, 15.00 and 21.00 hours GMT.

Streamlines   Lines drawn on weather charts to show the essential direction of air flow.

Sublimation   Without melting, the conversion of a solid direct into vapour and subsequent condensation.

Synoptic   Pertaining to local or regional climates or climatology in terms of their relationship with the atmospheric circulation.

Synoptic observations   Those made at the same time all over the world to obtain the state of the atmosphere at that time.

Synoptic stations   Those recording at the specified observing times, which may be hourly, a comprehensive list of climatic parameters which, in a full station, will include present and past weather, through wind, cloud, visibility, temperature, air pressure, rainfall and sunshine.

Temperature   A measure of the 'hotness' or 'coldness' of a body; measured commonly, as in this book, in degrees Celsius or Fahrenheit.

   Equivalent temperature   The sum of the actual air temperature plus that due to the latent heat of the water vapour it contains.

   Potential temperature   The temperature that the air could attain if brought adiabatically to a standard pressure, usually 1000 mb.

Temperature inversion   The condition in which air temperature increases, rather than decreases, with altitude.

Thermal capacity   The quantity of heat required to raise the temperature of a body 1°C.

Tropopause   The upper limit of the troposphere, where temperature ceases to fall with increasing height.

Troposphere   The lower part of the atmosphere characterized by temperature decreases with altitude.

Turbidity   The property of the cloudless atmosphere which causes a diminution of incoming solar radiation. Dust and smoke are the main agents increasing turbidity.

Twilight

   Astronomical twilight   The interval between sunrise or sunset and the time when the true position of the centre of the sun is 18° below the horizon, at which time stars of the sixth magnitude are visible and twilight glow has disappeared.

   Civil twilight   The interval between sunrise or sunset and the time when the true position of the centre of the sun is 6° below the horizon, at which time stars and planets of the first magnitude become visible and normal outdoor activities will be suspended because of the darkness.

Volt   A unit of electromotive force and potential difference.

Vorticity   Rotation imparted to 'parcels' of air by the rotating earth.

Watt   Unit of power. The rate of work done in joules per second. 1000 watts = 1 kilowatt; 746 watts = 1 horsepower.

Weather   The state of the atmosphere with respect to meteorological phenomena at a given place and time or over a short period of time.

Zenith   Directly overhead.

Bibliography

Abbreviations

The following abbreviations have been used in the bibliography.

Acad.	Académie
Afr.	Africa, African, Afrique, africain(e)(s)
Agric.	Agricultural, Agriculture
Am.	America, American
Ann.	Annual
Annls.	Annals, Annales
Appl.	Applied
ASECNA	Agence pour la Securité de la Navigation Aérienne en Afrique et à Madagascar
Ass.	Association
Atmos.	Atmosphere, Atmospheric, Atmosphérique(s)
Bioklim.	Bioklimatologie
Biomet.	Biometeorology
Bldg.	Building
BRAB	Building Research Advisory Board
BRE	Building Research Establishment
Brit.	British
BRS	Building Research Station
Bull.	Bulletin
Clim.	Climate, Climatology
Conf.	Conference
Contrib.	Contributions
C.R.	Compte rendu
CSIRO	Commonwealth Scientific and Industrial Research Organization
DEM	Direction d'Exploitation de la Météorologie
Dep.	Department
E.	East
EDSTM	Environmental Data Service Technical Memorandum
Envir.	Environment, Environmental
ERL	Environmental Research Laboratories
ERS	Economic Research Service
ESA	European Space Agency
ESSA	Environmental Science Services Administration
FAO	Food and Agriculture Organization
Fed.	Federation
FGGE	First GARP Global Experiment
Fr., fr.	France, français(e)
GARP	Global Atmospheric Research Programme
GATE	GARP Atmospheric Tropical Experiment
Geog., Géog.	Geography, Géographie
Geogr.	Geographical
Geophys.	Geophysics, Geophysik, Geophysical
HMSO	Her (His) Majesty's Stationery Office
ICSU	International Council for Scientific Unions
IEEE	Institute of Electrical and Electronics Engineers
IITA	International Institute of Tropical Agriculture
Inst.	Institute, Institut
Int.	International
J.	Journal
Lab.	Laboratory
Mag.	Magazine
Med.	Medical, Medicine
Mem.	Memoirs
Met.	Meteorology, Meteorologie, Météorologie, Meteorological
Mon.	Monthly
NHRL	National Hurricane Research Laboratory
Nig.	Nigerian
NOAA	National Oceanic and Atmospheric Administration
ORSTOM	Office de la Recherche Scientifique et Technique Outre–Mer
Phys.	Physics, Physical, Physique
Proc.	Proceedings
Publ.	Publication(s)
Q.	Quarterly
Rech.	Recherches
Rep.	Report
Res.	Research
Rev.	Review
Roy.	Royal
Sci.	Science, Scientific

SCOPE	Scientific Committee on Problems of the Environment
Serv.	Services
Soc.	Society
Symp.	Symposium
Tech.	Technical
Trans.	Transactions
Trop.	Tropical
UGGI	Union de Géodésique et Géophysique Internationale (Union for Geodesy and Geophysics International Association)

UN	United Nations
UNESCO	United Nations Educational Scientific and Cultural Organization
Univ.	University, Université
US	United States
W.	West, Western
Weath.	Weather
WMO	World Meteorological Organization
WMPO	Weather Modification Program Office

Acheampong, P. K. (1982). 'Rainfall anomaly along the coast of Ghana – its nature and causes'. *Geografiska Annaler*, **64A** (3–4), 199–211

Adams, W. M. (1986). 'Traditional agriculture and water use in the Sokoto Valley, Nigeria'. *Geogr.J.* **152**(1), 30–43.

Adedokun, J. A. (1978), 'West African precipitation and dominant atmospheric mechanisms'. *Archiv für Met. Geophys. Bioklim.*, **27A**, 289–310.

Adedokun, J. A. (1983), 'Intralayer (low level/mid tropospheric) precipitable water vapour stations and precipitation in West Africa'. *Archiv für Met. Geophys. Bioklim.*, **33B**, 117–130.

Adedokun, J. A. (1986). 'On a relationship for estimating precipitable water vapour aloft from surface humidity over West Africa'. *J. of Climatology*, **6**(2), 161–72.

Adefolalu, D. O. (1972). 'On the mean equivalent potential temperature of the tropical atmosphere and the "little dry season" of West Africa'. *Nig. Met. Serv. Q. Met. Mag.* **2**(1), 15–40.

Adefolalu, D. O. (1973a). 'The mean troposphere over West Africa'. *Nig. Met. Serv. Q. Met. Mag.* **3**(2), 31–72.

Adefolalu, D. O. (1973b). 'On the mean low-level synoptic features over West Africa during northern summer'. *Nig. Met. Serv. Q. Met. Mag.* **3**(3), 83–94.

Adefolalu, D. O. (1973c). 'Composite models of lower tropospheric synoptic-scale wave disturbances at 700mb over West Africa'. *Nig. Met. Serv. Q. Met. Mag.* **3**(3), 95–106.

Adefolalu, D. O. (1973d). 'Further studies on the mid-summer drought in West Africa'. *Nig. Met. Serv. Q. Met. Mag.* **3**(4), 110–20.

Adefolalu, D. O. (1977). 'A study of synoptic types of the African easterly perturbation'. In *Lectures on Forecasting of Tropical Weather*. WMO 492, 255–71.

Adefolalu, D. O. (1981), *The Weather and Climate of Calabar*, Senate Res. Project, Rep. on the Weather Patterns of Calabar.

Adefolalu, D.O. (1984a). 'Weather hazards in Calabar – Nigeria' *Geogr.J., Dordrecht*, **9**(4), 359–68.

Adefolalu, D. O. (1984b). 'On bioclimatological aspects of harmattan dust haze in Nigeria'. *Archiv. für Met. Geophys. Bioklim.*, **33(B)**, 387–404.

Adefolalu, D. O. (1985). 'On transport of zonal momentum and energy exchange processes in West Africa'. *Archiv. für Met. Geophys. Bioklim.* **33**, 4A, 277–87.

Adefolalu, D. O. (1986), 'Wamex 79: its significant contributions to the understanding of monsoonal circulations in West Africa during the northern summer', ICCU and WMO, *GARP Publ. Ser.* No. 26, 429–60.

Adefolalu, D. O., Senouci, M., Bounoua, A. and Boukri, A. (1985). 'Mean state during the onset phase of the West African monsoon in 1979'. *Archiv. für Met. Geophys. Bioklim.* **33**, 4A, 327–43.

Adejokun, J. A. (1964). 'The three-dimensional structure of the inter-tropical discontinuity over Nigeria'. *Nig. Met. Serv. Tech. Note* 17.

Adejokun, J. A. (1966). 'A three-dimensional study of the ITD'. *Tech. Note* 30, Met. Dept., Lagos.

Adejokun, J. A. (1979). 'Towards achieving an in-season forecast of the West African precipitation'. *Archiv. für Met. Geophys. Bioklim.* **28A**, 19–38.

Adejuwon, J. O. (1983) 'Pests and diseases'. Chapter 7 in Oguntoyinbo, J. S. et al. (eds.). *A Geography of Nigerian Development*, 117–30.

Adeoye, K. B. (1973). 'An assessment of aridity and

the severity of the 1972 drought in northern Nigeria and neighbouring countries'. *Savanna*, **2**, 145–58.

Adesina, H. O. (1981). 'A statistical analysis of distribution characteristics of cholera within Ibadan City, Nigeria (1971)'. *Soc. Sci. Med.*, **15D**, 121–32.

Adetunji, J., McGregor, J. and Ong, C. K. (1979). 'Harmattan haze'. *Weather*, **34**(11), 430–6.

Adetunji, J. and Ong, C. K. (1980). 'Qualitative analysis of the harmattan haze by x-ray diffraction'. *Atmos. Environ.* **14**, 857–8.

Adolph, E. F. et al. (1947). *Physiology of Man in the Desert*. Inter Science Publishers, Wiley.

Adu, V. S. (1972). 'Eroded savanna soils of the Navrongo–Bawku area, Northern Ghana'. *Ghana J. Agric. Sci.* **5**, 3–12

Agboola, J. A. (1979). *An Agroclimatological Atlas of Nigeria*. Oxford Univ. Press.

Aina, J. O. (1969). 'Wind flow in the lower troposphere over West Africa and associated weather'. *Nig. Met. Serv. Tech. Note* 30.

Aina, J. O. (1974). 'A synoptic study of the harmattan dust haze'. Am. Met. Soc., E. Afr. Met. Dep. and WMO. *Int. Trop. Met. Meeting* 31 Jan.– 7 Feb. 1974, Nairobi.

Air Ministry, Met. Office, (1958). *Tables of Temperature, Relative Humidity and Precipitation for the World*, 6 parts. Brit. Information Serv., New York.

Air Ministry, Met. Office, (1962). *Weather in the Mediterranean. Vol. 1. General Meteorology. M.O. 391.* HMSO.

Ajayi, G. O. and Olsen, R. L. (1983). 'Measurements and analysis of raindrop size distribution in south-western Nigeria'. *ESA SP – 194*, 173–84. European Space Agency.

Akinola, R. A. (1966). 'Floods in Ibadan, western Nigeria'. *Nig. Geogr. J.* **9**(2), 101–13.

Ali, F. M. (1969). 'Effects of rainfall on yield of cocoa in Ghana'. *Experimental Agric.* **5**, 209–13.

Alvim, P. de T. and Kozlowski, T. T. (eds.) (1977). *Ecophysiology of Tropical Crops*. Academic Press.

Ambler, H. R. (1955). 'Notes on the climate of Nigeria with reference to personnel'. *J. Trop. Med. Hygiene* **58**, 99–112.

American Society of Heating and Ventilating Engineers, (1962). *Heating, Ventilating and Air-Conditioning Guide*, Waverly Press, Baltimore.

Ananaba, S. E. (1983). 'A comparison of ground based and sunshine derived estimates of cloud cover in Nigeria'. *Nig. Met. J.* **1**(1), 33–42.

Angstrom, A. (1925). 'The albedo of various surfaces'. *Geografiska Annaler* **7**, 323–42.

Anon, (1973). 'Wind generators use sun's energy more effectively than solar cells'. *Electrical Rev.* 782–3.

Anyadike, R. N. C. (1977). *West African Rainfall: Its Synoptic and Dynamic Climatology*. Unpublished Ph.D thesis, Univ. Sheffield.

Anyadike, R. N. C. (1979a). 'The content of water vapour in the atmosphere over West Africa'. *Archiv.*

für Met. Geophys. Bioklim. **28A**, 245–54. Anyadike, R. N. C. (1979b). 'The field of mean low-level divergence and rainfall over the West African region'. *Nig. Geogr. J.* **22**(2), 121–33.

Anyadike, R. N. C. (1981). 'On the relative contributions of atmospheric parameters to the rainfall of West Africa'. *Archiv. für Met. Geophys. Bioklim.* **30A**, 87–98.

Anyamba, E. K., Kiangi, P. M. R. and Patnaik, J. K. (1982). 'The mean horizontal motions over the African atmosphere during the southern summer'. *WMO* 596, 158–70.

ASECNA, (1973). 'La structure continue de l'équateur météorologique sur l'Afrique intertropicale'. *ASECNA Publ. Séries I* (29).

ASECNA, Bureau d'Etudes, (1976). 'Analyse de lignes de grains dans la boucle du Niger'. *La Météorologie VI*(6), 183–89.

ASECNA, Bureau d'Etude de la Direction de L'Exploitation Météorologique, (1980). 'Fréquences horaires et journalières des valeurs de la visibilité en Afrique centrale et occidentale'. *D.E.M. Publ.* 43, Dakar.

Aspliden, C. I. H. (1974). 'The low-level wind field and associated perturbations over tropical Africa during northern summer'. *Preprints, Int. Trop. Met. Meeting, Am. Met. Soc., Boston*, 218–23.

Aspliden, C. I. and Adefolalu, D. (1976). 'The mean troposphere of West Africa'. *J. Appl. Met.* **15**, 705–16.

Aspliden, C. I., Tourre, Y., and Sabine, J. B. (1976). 'Some climatological aspects of West African disturbance lines during GATE'. *Mon. Weath. Rev.* **104**, 1029–35.

Assamoi, P. and Planton, S. (1984), 'Etude numérique des perturbations sur la région Ouest Africaine en période de mousson'. *J. Rech Atmos.*, **18** (2), 81–94.

Atkinson, G. A. (1950a). 'African housing'. *African Affairs* **49**(196), 228–37.

Atkinson, G. A. (1950b). 'Building in the tropics'.*Roy. Inst. Brit. Architects J.* **57**, 313–20.

Atkinson, G. A. (1952). 'Building in warm climates'.*BRAB Res. Conf. Rep.* **5**, 66–74.

Aubreville, A. (1949). *Climates, Forêts et Désertification de l'Afrique Tropicale*. Soc. d'Editions Géographiques, Maritimes et Coloniales, Paris.

Ausubel, J. and Biswas, A. K. (eds.), (1980). *Climate Constraints and Human Activities*. Pergamon.

Ayoade, J. O. (1970). 'The seasonal incidence of rainfall'. *Weather*, **25**, 414–18.

Ayoade, J. O. (1971). *Rainfall, Evapotranspiration and the Water Balance in Nigeria*. Unpublished Ph.D thesis, Univ. London.

Ayoade, J. O. (1973). 'Annual rainfall trends and periodicities in Nigeria'. *Nig. Geog. J.* **16**, 167–76.

Ayoade, J. O. (1974). 'A statistical analysis of rainfall over Nigeria'. *J. Trop. Geog.* **39**, 11–23.

Ayoade, J. O. (1976a). 'Evaporation and evapo-transpiration in Nigeria'. *J. Trop. Geog.***43**, 9–19.

Ayoade, J. O. (1976b). 'A preliminary study of the magnitude, frequency and distribution of intense rainfall in Nigeria'. *Bull. Int. Ass. Sci. Hydrology*, **21**(3), 419–29.

Ayoade, J. O. (1977a). 'On the use of multivariate techniques in climatic classification and regionalisation'. *Archiv. für Met. Geophys. Bioklim.* **24B**, 257–67.

Ayoade, J. O. (1977b). 'Perspectives on the recent drought in the Sudano–Sahelian region of West Africa with particular reference to Nigeria'. *Archiv. für Met. Geophys. Bioklim.* **25B**, 67–77.

Ayoade, J. O. (1978). 'Spatial and seasonal patterns of physiologic comfort in Nigeria'. *Archiv. für Met. Geophys. Bioklim.* **26B**, 319–37.

Ayoade, J. O. (1980). 'A note on spatial and seasonal variations in wind speed over Nigeria'. *J. Trop. Geog.* **1**(2), 11–15.

Ayoade, J. O. (1983). *Introduction to Climatology for the Tropics*. Wiley.

Ayoade, J. O. and Akintola, F. O. (1982).'A note on some characteristics of rainstorms in Ibadan, Nigeria'. *Weather*, **37**(2), 56–8.

Ayoade, J. O. and Oyebande, B. C. (1983). 'Water Resources'. Chapter 4 in Oguntoyinbo, J. S. et al. (eds.). *A Geography of Nigerian Development*.

Bach, W., Pankrath, J. and Kellogg, W. W. (eds.), (1979). *Man's Impact on Climate*. Elsevier.

Bakun, A. (1978). 'Guinea current upwelling'. *Nature*, (London) **271**, 147–50.

Baldy, Ch., Delecolle, R. and Kontongomde, H. (1982). 'Study of surface winds in Upper Volta', in WMO *Proc. of Tech. Conf. on Climate – Africa.* WMO 596, 325–32.

Balek, J. (1977). 'Hydrology and water resources in tropical Africa'. Elsevier. *Developments in Water Science*, 8.

Balek, J. (1983). *Hydrology and Water Resources in Tropical Regions*. Elsevier.

Balogun, A. A. (1971). 'Seasonal variations in the conception rate of beef cattle in the seasonal equatorial climate of southern Nigeria'. *Int. J. Biomet.* **15**, 71–9.

Balogun, C. (1972). 'The variability of rainfall in Nigeria'. *Nig. J. Sci.* **6**(1), 87–92.

Balogun, E. E. (1981a) 'Convective activity over Nigeria during the monsoon season'. I.C.S.U., WMO, GARP *Int. Conf. on Early Results of FGGE and Large-Scale Aspects of its Monsoon Experiments.* Tallahassee, Florida, 12–17 Jan. 1981, 8/22–8/27.

Balogun, E. E. (1981b). 'Seasonal and spatial variations in thunderstorm activity over Nigeria'. *Weather*, **36**(7), 192–97.

Balogun, E. E. (1982), *West African Monsoon Windsets from Geostationary Satellite*. Space and Science Engineering Center, Univ. of Wisconsin, Madison. App. II.

Barberon, J. (1935). 'La visibilité en Afrique occidentale française. *Annls. Phys. Globe Fr. Outre-Mer.* **2**, 136.

Bargman, D. J. (1960). *Tropical Meteorology in Africa*. Munitalp Foundation, Nairobi.

Barker, D., Oguntoyinbo, J. S. and Richards, P. (1977). 'The utility of Nigerian peasant farmers' knowledge in the monitoring of agricultural resources'. *Monitoring and Assessment Res. Centre, Chelsea College. General Report Series*, 4.

Barrefors, B. B. (1966). 'Disturbances in West Africa as gravity waves in the inter-tropical discontinuity surface'. *Nig. Met. Serv. Tech. Note* 29.

Barry, R. C. (1969). *Introduction to Physical Hydrology*. Methuen.

Barry, R. G. and Chorley, R. J. (1982). *Atmosphere, Weather and Climate*. Methuen.

Bartrum, P. C. (1952). *Statistics Concerning the Climate of Sierra Leone, 1951*. Met. Office, Freetown.

Batten, A. (1967). 'Seasonal movements of swarms of *Locusta migratoria* 1928 to 1931'.*Bull. Entomological Res.* **57**, 357–80.

Bauchardeau, M. A. (1958). 'Etudes d'évaporation dans les régions Sahelo–Soudaniennes'. U.G.G.I. *Ass. Int. Hydrologie Sci. Ass. Gen.,* Toronto, 1957. Publ. 45, 407–20.

Baudin, de Thé, B. M. S. (1960). *Essais de Bibliographie du Sahara Français et des Régions Avoisinantes*. Arts et Métiers graphiques, Paris.

Baumgartner, A. and Reichel, E. (1975). *The World Water Balance*. Elsevier.

Beckinsale, R. P. (1957). 'The nature of tropical rainfall'. *Trop. Agric., Trinidad*, **34**, 77.

Bedford, T. (1936). 'Warmth factor in comfort at work' *Med. Res. Council, Ind. Health Res. Board. Rep. No. 76*. HMSO.

Beer, T., Greenhut, G. K. and Tandoh, S. E. (1977). 'Relations between the Z criterion for the sub-tropical high, Hadley cell parameters, and the rainfall in northern Ghana'. *Mon. Weath. Rev.* **105**, 849–55.

Belding, H. S. and Hatch, T. F. (1955). 'Index for evaluating heat stress in terms of resulting physiological strains'. *Heating, Piping and Air Conditioning*. August, 129–36.

Benoit, P. (1977). 'The start of the growing season in northern Nigeria'. *Agric. Met., Int. J.* **18**, 91–9.

Berger, J. M. (1964). 'Profits culturaux dans le centre de la Côte d'Ivoire'. *Cahiers d'ORSTOM, Sér Pédologie*, **2**(1), 41–69.

Bernet, G. (1968). 'Note sur la structure du front intertropical'. *ASECNA, Sér. 11*, No. 16.

Bernet, G. (1969). 'Recherche d'un mode de formation des lignes de grains en Afrique centrale'. *ASECNA Publ.* No. 5.

Berrit, G. R. (1965). 'Les conditions de saison chaude dans la région du Golfe de Guinée'. *Progress in Oceanography*, **3**, 31–47.

Berry, F. A., Bollay, E. and Beers, N. R. (eds.) (1945). *Handbook of Meteorology*. McGraw-Hill.

Berry, L., Campbell, D. J. and Emker, I. (1977). 'Trends in mainland interaction in the West African

Sahel'. Chapter 9, in Dalby, D. et al. *Drought in Africa.*

Bertrand, J. (1976). 'Visibilité et brume sèche en Afrique'. *La Météorologie.* **VI**(6), 201-11.

Bertrand, J. and Baudet, J. (1976). 'Concentration et propriétés des noyaux glacogènes naturels en Afrique de l'ouest'. *J. Rech. Atmos.* **10**, 25-43.

Bertrand, J., Baudet, J. and Drochon, A. (1974). 'Importance des aérosels naturels en Afrique de l'ouest'. *J. Rech. Atmos.* **8**, 845-60.

Bertrand, J., Cerf, A. and Domergue, I. L. (1979). 'Repartition in space and time of dust haze south of the Sahara'. *WMO* **538**, 409-15.

Bigelstone, H. J. (1958a). 'Diurnal variations in surface wind speed at Kano'. *Nig. Met. Serv. Tech. Note* 8.

Bigelstone, H. J. (1958b). 'Harmattan haze at Kano'. *Nig. Met. Serv. Tech. Note* 10.

Bisseck, H. (1968). 'Etude des lignes de grains au Cameroun'. *C. R. Acad. Sci., Paris* (B), **266**, 1295-6.

Bisseck, H. (1977). 'Phenomène des perturbations d'est dans l'aérojet intertropical Africain'. *J. Rech. Atmos.* **11**, 141-4.

Biswas, A. K. (1984). *Climate and Development.* Tycooly Int. Pub. Ltd, Dublin.

Biswas, M. R. and A. K. (1979). *Food, Climate and Man.* Wiley.

Blaxter, K. and Fowden, L. (1982). *Food, Nutrition and Climate.* Applied Science Publishers.

Bolin, B. (1980). *Man's Impact on Climate.* Int. Conf. Clim. and Offshore Energy Res., Roy. Soc., London, 21-23 Oct.

Bonfils, J. (1962). 'Etudes lysimétriques au Sénégal'. *L'Agronomie Tropicale*, Paris.

Booker, R. H. (1963). 'The effect of sowing date and spacing on the rosette disease of groundnut in Nigeria with observation on the vector *Aphis craccivora*'. *Annls. Appl. Biology*, **52**, 125-31.

Booker, R. H. (1965). 'Pests of cowpeas and their control in northern Nigeria'. *Bull. Entomological Res.* **55**, 663-72.

Bougnol, M. (1936). 'La grêle en Afrique occidentale française'. *Annls. Phys. Globe Fr. Outre-Mer.* **3**, 70-1.

Bougnol, M. (1937a). 'Note sur le régime du vent à Port Etienne (Dakar)'. *Publ. du Comité d'Etudes Historiques et Sci. de l'Afr. Occidentale Fr.* (B) **3**, 33-4.

Bougnol, M. (1937b). 'Note sur la grêle en Afrique occidentale française'. *Publ. du Comité d'Etudes Historiques et Sci. de l'Afr. Occidentale Fr.* (B) **3**, 63-8.

Bourke, P. A. M. (1963). 'Agricultural biometeorology'. *Int. J. Biomet.* **7**, 121-5.

Bourn, D. (1978). 'Cattle, rainfall and tsetse in Africa'. *J. of Arid Envir.* **1**(1), 49-61.

Bowden, D. J. (1980). 'Rainfall in Sierra Leone'. *J. Trop. Geogr.* **1**(2), 31-9.

Bradley, P. N. (1977). 'Vegetation and environmental change in the West African Sahel'. pp. 35-54 in O'Keefe, P. and Wisner, B. (eds.). *Land Use and Development.* Afr. Inst., London.

Brazol, D. (1954). 'Bosquejo bioclimático de la República Argentina'. *Meteoros,* **4**(4), 381-94.

Briggs, J. (1972). 'Probabilities of aircraft encounters with heavy rain'. *Met. Mag.* **101**, 8-13.

Brinkman, A. W. and McGregor, J. (1983). 'Solar radiation in dense Saharan aerosol in northern Nigeria'. *Q. J. Roy. Met. Soc.* **109**(462), 831-47.

British West African Meteorological Services, Lagos, (1954). *Bibliography of Publications Relating to the Meteorology of British West Africa.*

British West African Meteorological Services, Lagos, (1959 and 1960). *Tech. Notes* 16 and 17.

Brooks, C. E. P. (1925). 'The distribution of thunderstorms over the earth'. *Geophys. Mem.* **24**.

Brooks, C. E. P. (1946). 'Climate and the deterioration of materials'. *Q. J. Roy. Met Soc.* **72**(311), 87-91.

Brooks, C. E. P. (1950). *Climate in Everyday Life.* E. Benn., London.

Bryson, R. A. (1973a). 'Climatic modification by air pollution II: The Sahelian effect'. *Wisconsin Univ. Inst. for Environmental Studies. Rep* 9.

Bryson, R. A. (1973b). 'Drought in Sahelia, who or what is to blame?' *Ecologist,* **3**, 366-71.

Buchanan, R. O. and Pugh, J. C. (1955). *Land and People in Nigeria.* London Univ. Press.

Budyko, M. I. (1971). 'The influence of man's activity on climate'. pp. 17 and 18 in *WMO Special Environmental Rep. No. 2., Met. as Related to the Human Environment.* WMO 312, Genève.

Buettner, K. J. K. (1951). 'Physical aspects of human bioclimatology', pp. 1112-25 in Malone, T. F. (ed.). *Compendium of Meteorology.*

Building Research Advisory Board (1952). 'Housing and building in hot-humid and hot-dry climates'. *BRAB Res. Conf. Rep.*5. National Academy of Sciences, Washington.

Building Research Advisory Board (1953). *Preliminary BRAB Bibliography of Housing and Building in Hot-Humid and Hot-Dry Climates.* National Academy of Sciences, Washington.

Building Research Establishment (1982). 'Thermal comfort in hot climates'. *BRE News No.* 57, 13.

Building Research Station (1949). 'The deterioration and preservation of book material stored in libraries'. *Note No.* D62. Garston, Watford, UK.

Building Research Station (1960-1963). *Tropical Building Studies* 1-5. HMSO.

Building Research Station (1967). 'Building for warm climates - a bibliography'. *Overseas Building Notes,* **118**, 2-12.

Building Research Station (1972). A select bibliography of publications of interest to physical planners in developing countries'. *Overseas Building Notes,* **147**.

Bultot, F. (1964). 'Sur la délimitation de la zone tropicale humide'. *Bull. Séan. Acad. Roy. Sci. Outre-Mer. (Bruxelles).* **2**, 406-12.

Bunting, A. H. (1968). 'Agroclimatology and

agriculture'. *Unesco, Natural Resources Res.* 7. In *Agrometeorological Methods Proceedings of Reading Symposium* 1968, 361–6.

Bunting, A. H. and Curtis, D. C. (1968). 'Local adaptation of sorghum varieties in northern Nigeria'. pp. 101–5 in *Agrometeorological Methods Proceedings of Reading Symposium*. UNESCO, Paris.

Bunting, A. H., Dennett, M. D., Elston, J. and Milford, J. R. (1975). 'Seasonal rainfall forecasting in West Africa'. *Nature (London)*, **253**, 622–3.

Bunting, A. H., Dennett, M. D., Elston, J. and Milford, J. R. (1976). 'Rainfall trends in the West African Sahel'. *Q. J. Roy. Met. Soc.* **102**, 59–64.

Burdon, D. J. et al. (1973). *Water and Sahelian Zone Drought Problems in the Republic of Senegal.* FAO, WS/D9635.

Burnham, J. (1969). 'Atmospheric turbulence at supersonic transport cruise altitudes'. *Q. J. Roy. Met. Soc.* **95**, 782–3.

Burns, F. (1961). 'Dust haze in relation to pressure gradients'. *Met. Mag.* **90**, 223–6.

Buroleau, M. (1937). 'Note sur la visibilité à Dakar'. *Publ. du Comité d'Etudes Historiques et Scientifiques de l'Afr. Occidentale Fr.* Sér. B. No. 3, 69–70.

Burpee, R. W. (1972). 'The origin and structure of easterly waves in the lower troposphere of North Africa'. *J. Atmos. Sci.* **29**, 77–90.

Burpee, R. W. (1974). 'Characteristics of North African easterly waves during the summers of 1968 and 1969'. *J. Atmos. Sci.* **31**(6), 1556–70.

Burpee, R. W. (1976a). 'The influence of easterly waves on the patterns of precipitation in tropical North Africa'. *NOAA, Rep. National Hurricane and Experimental Laboratory, Miami* also in *WMO 492*, 41–71.

Burpee, R. W. (1976b). 'Structure énergétique des ondes d'est.' *La Météorologie* VI Sér. No. 6, 137–48.

Burpee, R. W. and Dugdale, G. (1975) 'A summary of weather systems affecting western Africa and the eastern Atlantic during GATE'. *GATE Rep.* 16. WMO.

Burroughs, W. (1981). 'Climate and the earth's albedo'. *New Scientist,* **91**(1262), 144–6.

Burton, A. C. and Edholm, O. G. (1955). *Man in a Cold Environment.* Arnold.

Cardinall, A. W. (1932). *A Bibliography of the Gold Coast.* Government Printer, Accra.

Carlson, T. N. (1969a). 'Synoptic histories of three African disturbances that developed into Atlantic hurricanes'. *Mon. Weath. Rev.* **97**, 256–76.

Carlson, T. N. (1969b). 'Some remarks on African disturbances and their progress over the tropical Atlantic'. *Mon. Weath. Rev.* **97**, 716–26.

Carlson, T. N. (1971). 'A detailed analysis of some African disturbances'. *N.O.A.A. Tech. Memo. ERL NHRL* 90.

Carlson, T. N. and Prospero, J. M. (1972). 'The large-scale movement of Saharan air outbreaks over the northern equatorial Atlantic'. *J. Appl. Met.* **11**(2), 283–97.

Carlson, T. N., Prospero, J. M. and Hanson, K. J. (1973). 'Attenuation of solar radiation by windborne Saharan dust off the west coast of Africa'. *N.O.A.A. Tech. Mem. ERL WMPO–7, NHRL–106.* Boulder, Col.

Carvalho Guerra, A. de. (1947). 'Subsidios para o estudo do clima da Guiné Portuguesa'. *Boletin Cultural da Guiné Portuguesa, Bissau,* **2**(5), 3–75.

Casanova, H. (1967). 'Principaux types de temps en Afrique occidentale illustrés par des situations météorologiques réelles'. *Bull. de l'Inst. Fr. d'Afr. Noire,* **29**(A) 1, 383–408.

Casanova, H. (1968). 'Principaux types de temps en Afrique occidentale'. *ASECNA Met. Publs.* 6, Dakar.

Castet, J. (1942), (1948). 'Sur l'intensité de la radiation solaire directe et ses causes d'affaiblissement à Tamanrasset'. *Annls. de l'Inst. de la Mét. et Phys. du Globe du Sahara,* 37–9, Alger.

Cauvet-Duhamel, P. (1936). 'La température de la mer à Lomé (Togo)'. *Annls. de Phys. du Globe de la Fr. d'Outre-Mer, Paris,* **3**, 54.

Cena, K. and Clark, J. A. (1981). *Bioengineering, Thermal Physiology and Comfort.* Studies in Environmental Science. Elsevier.

Cerf, A. (1980). 'Atmospheric turbidity over West Africa'. *Contributions to Atmos. Physics,* **53**, 414–29.

Cerf, A. (1981). 'Caractéristiques du trouble atmosphérique en Afrique de l'ouest'. *La Météorologie* (VI Sér.) **23**, 5–24.

Chaggar, T. S. (1985). 'Debundscha, the wettest place in Africa'. *Weather,* **40**(6), 192–3.

Chalmers, J. (1941). 'Cloud and earth lightning flashes'. *Philosophical Mag.* **32**, 77.

Chamney, N. P. (1930). 'The hourly distribution of rainfall in the Gold Coast'. *Bull. Dep. Agric. Gold Coast, Accra,* **22**, 360–3.

Chang, C-B. (1979). 'On the influences of solar radiation and diurnal variation of surface temperature on African disturbances'. *Rep. 79–3, Dept. Met. Florida State Univ.,* Tallahassee.

Chang, Jen-Hu (1968). *Climate and Agriculture, an Ecological Survey.* Aldine, Chicago.

Chapas, L. C. and Rees, H. R. (1964). 'Evaporation and evapotranspiration in southern Nigeria'. *Q. J. Roy. Met. Soc.* **90**, 313–19.

Charney, J. G. (1975). 'Dynamics of deserts and droughts in the Sahel'. *Q. J. Roy. Met. Soc.* **101**, 193–202.

Charney, J. G., Quirk, W. J., Chow, S. and Kornfield, J. (1977). 'A comparative study of the effects of albedo change on drought in semi-arid regions'. *J. Atmos. Sci.* **34**, 1366–85.

Charney, J. G., Stone, P. H. and Quirk, W. J. (1975). 'Drought in the Sahara; a biogeophysical feedback mechanism'. *Science,* **187**, 434–5.

Charney, J. G., Stone, P. H. and Quirk, W. J. (1976).

'Reply to "Drought in the Sahara: insufficient biogeophysical feedback?" by Ripley, E. A.'. *Science*, **191**, 100–2.

Charre, J. (1973). 'La variabilité du rythme annuel des pluies au Niger'. *Rév. Géog. Alpine* **61**, 411–26.

Charreau, C. (1969). 'Influence des techniques culturales sur le dévelopement du ruissellement et de l'érosion en Casamance'. *Agron. Trop.* **24**, 836–42.

Charreau, C. (1974). *Soils of Tropical Dry and Dry-Wet Climate Areas and their Use and Management.* Cornell Univ., New York.

Charreau, C. and Nicou, R. (1971). 'L'amélioration du profit cultural dans les sols sableux et sablo-argileux de la zone tropicale sèche de l'ouest Africain'. *Bull. Agron.* **23**.

Chelam, E. V. (1977). 'Forecasting techniques for West Africa' in *Lectures on Forecasting of Tropical Weather*. WMO 492.

Chester, R. and Johnson, L. R. (1971). 'Atmospheric dusts collected off the West African coast'. *Nature* (London), **229**(5280), 105–7.

Clackson, J. R. (1952). 'Gusts in Nigeria'. *Brit. W. Afr. Met. Serv. Met. Note* 1. Lagos.

Clackson, J. R. (1957). 'The seasonal movement of the boundary of northern air'. *Nig. Met. Serv. Tech. Note* 5. Lagos. Addenda, 1958 and 1959.

Cloudsley-Thompson, J. L. (1966). 'Biometeorological problems in the ecology of animals in the tropics'. *Int. J. Biomet.* **10**, 253–71.

Cloudsley-Thompson, J. L. (ed.), (1984). *Key Environments: Sahara Desert*. Pergamon Press.

Cochemé, J. (1968). 'FAO/UNESCO/WMO agroclimatological survey of a semi-arid area in West Africa, south of the Sahara'. In *Agroclimatological Methods*. Natural Resources Res. Series VII, UNESCO, Paris.

Cochemé, J. and Franquin, P. (1967). *An Agroclimatology Survey of a Semi-Arid Area in Africa, South of the Sahara*. WMO 86(210), TP 110, 136.

Comité Interafricain d'Etudes Hydrauliques. *Precipitations Journalières de l'Origine des Stations à 1965* (for Senegal, Ivory Coast, Togo, Cameroon, Niger, Mauritania, Dahomey, Mali, Upper Volta). Office de la Recherche Scientifique et Technique Outre-Mer. Ministère de la Coopération. République Française.

Constantin, Le Frère (1901). 'Observatoire de Saint-Louis du Sénégal (Ecole Secondaire). Observations Météorologiques. Moyennes conclus de 23 anneés d'observations'. *Bull. du Comité d'Etudes Historiques et Scientifiques*, **13**, 437–73.

Courel, M. F., Kandel, R. S. and Rasool, S. I. (1984). 'Surface albedo and the Sahel drought'. *Nature*, **307**(5951), 528–31.

Coursey, P. R. (1944). 'Extreme climatic conditions. Resumé of climatic factors, their influence on radio equipment and components, and of methods of testing'. *Wireless Engineer*, **21**, 412–20.

Crane, A. J. (1982). 'Man's impact on climate'.

pp. 313–54 in Blaxter, K., and Fowden, L. *Food, Nutrition and Climate*.

Cross, M. (1985). 'Africa's drought may last many years'. *New Scientist*, **105**(1439), 9.

Crozat, G., Domergue, J. L., Baudet, J. and Bogui, V. (1978). 'Influence des feux de brousse sur la composition chimique des aérosols atmosphériques en Afrique de l'ouest'. *Atmos. Envir.* **12**, 1917–20.

Cunnington, W. M. and Rowntree, P. R. (1985). 'The sensitivity of the Saharan region in a general circulation model'. *Dynamical Clim. Tech. Note* 19, Met. Office, London.

Curry, L. (1952). 'Climate and economic life: a new approach'. *Geogr. Rev.* **42**, 367–83.

Curtis, D. C. (1968). 'The relationship between the date of heading of Nigerian sorghums and the duration of the growing season'. *J. Appl. Ecol.* **4**, 215–6.

Daily Times, Nigeria Ltd. (1985). *Ravages of Erosion in Imo State* I and II, Aug. 23 and 24.

Dalby, D. and Harrison Church, R. J. (1973). *Report of the 1973 Symposium Drought in Africa*. Centre for African Studies, School of Oriental and African Studies, Univ. London.

Dalby, D., Harrison Church, R. J. and Bezzaz, F. (eds.), (1977). *Drought in Africa. African Environment Special Rep.* 6. Int. Afr. Inst.

Dankwa, J. B. (1974). 'Maximum rainfall intensity-duration frequencies in Ghana'. *Ghana Met. Serv. Dep. Note* 23.

Davey, J. T. (1957). 'The African migratory locust in the central Niger Delta. I. Climate and vegetation'. *Locusta*, **5**.

Davies, J. A. (1965a). 'Evaporation and potential evapotranspiration at Ibadan'. *Nig. Geogr. J.* **8**, 17–31.

Davies, J. A. (1965b). 'Estimation of insolation for West Africa'. *Q. J. Roy. Met. Soc.* **91**, 359–63.

Davies, J. A. (1966a). 'The assessment of evapotranspiration for Nigeria'. *Geografiska Annaler*, **48**(3), 139–56.

Davies, J. A. (1966b). 'Solar radiation estimates for Nigeria'. *Nig. Geogr. J.* **9**, 2.

Davies, J. A. (1967). 'A note of the relationship between net radiation and solar radiation'. *Q. J. Roy. Met. Soc.* **93**, 109–10.

Davies, J. A. (1973). 'Albedo investigation in Labrador-Ungava' *Archiv. für Met. Geophys. Bioklim.* (B), **13**, 137–51.

Davy, E. G. (1974a). 'Drought in West Africa'. *WMO Bull.* **23**(1), 18–23.

Davy, E. G. (1974b). *A Survey of Meteorological and Hydrological Data Available in Six Sahelian Countries of West Africa*. WMO No. 379.

Davy, E. G., Mattei, F. and Solomon, S. I. (1976). 'An evaluation of climate and water resources for development of agriculture in the Sudano–Sahelian zone of West Africa'. *WMO Special Envir. Rep.* **9**, 1–286. WMO 459.

Defant, F. and Taba, H. (1957). 'The threefold structure of the atmosphere and the characteristics of the tropopause'. *Tellus*, **9**, 259–74.

Delacroix, G. (1911). *Maladies des Plantes Cultivées dans les Pays Chauds*. Paris.

Delormé, G. A. (1963). 'Répartition et durée des précipitations en Afrique occidentale'. *Met. Nationale Monographies* 28. Paris.

Demmler, K. (1969). 'Characteristic of the North African jet stream' *Meteorologische Abhandlungen*, **101**(H.2), Beil 37. Berlin.

Deneau, V. (1956). 'Singularités climatiques du bas Togo'. *Mét. Nationale. Mémorial* No. 42. Paris.

Dennett, M. D., Elston, J. and Rodgers, J. A. (1985). 'A reappraisal of rainfall trends in the Sahel'. *J. Clim.* **5**(4), 353–61.

Dennett, M. D., Elston, J. and Speed, C. B. (1981). *Climate and Cropping Systems in West Africa.* Geoforum.

Dennett, M. D., Rodgers, J. A. and Stern, R. D. (1983). 'Independence of rainfalls through the rainy season and the implications for the estimation of rainfall probabilities'. *J. Clim.* **3**(4), 375–84.

Department of Demography, Australia National University, (1974). *African Drought Bibliography*, 1973, 1974. 2 vols.

Department of Hydrometeorological Services, Gambia, (1977). 'Analyses of intense rainfall'. *Tech. Rep.* 2.

Dettwiller, J. (1962). 'Sur les vents forts dans la haute troposphère au dessus du Sahara' Mét. Nationale. *Notice d'Informations Techniques*, Section X. Pièce 17. Paris.

Dettwiller, J. (1967). 'Caractéristiques du jet subtropical sur le nord–ouest de l'Afrique' Mét. Nationale. *Monographies* 58.

Devynck, J. L. (1981). 'Study of some Sahelian disturbances during WAMEX' *GARP Special Report* No. 37. Appendix 15, 16–26.

Devynck, J. L. and Ago. J. (1985). 'Cloud clusters over West Africa during the Wamex experiment'. *WMO Train. Semin. Use Met. Data Implications Forecast. Res. Trop. Countries, 1981*, 271–303.

De Wit, C. T. (1965). 'Photosynthesis of leaf canopies'. *Agric. Res. Rep.* 663, Inst. for Biological and Chemical Res. on Field Crops and Herbage, Wageningen, Netherlands.

Dhonneur, G. (1970). 'Essai de synthèse sur les théories des lignes de grains en Afrique occidentale et centrale'. *ASECNA Pub.* 20. Dakar.

Dhonneur, G. (1981a). 'Sécheresse et/ou désertification au Sahel.' *La Météorologie.* VIᵉ Sér. 24, 119–23.

Dhonneur, G. (1981b). 'Moving cloud clusters – main component of Sahel meteorology'. *Progress in Res. in Trop. Met. App.* D., 17–25. WMO.

Dhonneur, G. (1985). 'Intertropical convergence and intertropical discontinuity'. '*WMO Train. Semin. Use Met. Data Implications Forecast. Res. Trop. Countries 1981*, 1–26.

Dhonneur, G., Finaud, L., Garnier, R., Gaucher, L. and Rossignol, D. (1974a). 'Analyse de deux perturbations ayant évolué en dépressions tropicales au large de Dakar'. *Int. Trop. Met. Meeting*, 1974, Nairobi. Pt II, 10–15.

Dhonneur, G., Finaud, L., Garnier, R., Gaucher, L. and Rossignol, D. (1974b). 'Etude d'une ligne de grains sur la boucle du Niger'. In *Int. Trop. Met. Meeting* 1974, Nairobi, Pt II, 16–23.

Dhonneur, G., Finaud, L., Garnier, R., Gaucher, L. and Rossignol, D. (1974c). 'Structure continue de l'équateur météorologique sur l'Afrique intertropicale'. In *Int. Trop. Met. Meeting* 1974, Nairobi, Pt II, 29–33.

Dhonneur, G. and Tourre, Y. (1981). 'Easterly waves and squall lines – two different phenomena'. *GARP Special Report No. 37.* Appendix 15, 9–15.

Dines, J. S. (1922). 'Note on the effect of a coastline on precipitation'. *Q. J. Roy. Met. Soc.* **48**, 357–62.

Dittberner, G. J. (1978). 'Climatic change: volcanoes, man–made pollution and carbon dioxide'. *IEEE Trans. on GeoScience Electronics.* GE 16, 50–61. New York.

Division de la Météorologie du Sénégal, (1972). *Approche d'une Analyse de l'Evolution de la Pluviométrie au Sénégal depuis plus d'un Siècle.* Dakar.

Doggett, H. (1970). *Sorghum.* Longmans.

Dorot, G. (1973). 'Contribution a l'étude des interactions océan–atmosphère sur les côtes de l'Afrique occidentale'. *ASECNA,* Sér. 1 (24). Dakar.

Dorrell, A. T. (1947). 'Severe squall at Kano, northern Nigeria'. *Met. Mag.* **76**, 206–7.

Drew, J. B. and Fry, E. M. (1947). *Village Housing in the Tropics with Special Reference to West Africa.* Lund Humphries.

Drochon, A. (1976). 'Donneés climatologiques au sol et en altitude pour la station d'Abidjan'. *Notes, Traductives et Informatives Sélectionnés de la D.E.M.* 55. ASECNA, Dakar.

Dubief, J. (1947). 'Les pluies au Sahara central'. *Travaux de l'Inst. de la Rech. Sahariennes,* **4**, 7–24.

Dubief, J. (1979). 'Review of the North African climate with particular emphasis on the production of eolian dust in the Sahel Zone and in the Sahara'. In Morales, C. (ed.). *Saharan Dust.*

Dugdale, G. and Pearce, R. P. (1977). 'Practical examples, including interpretation of satellite data'. Chapter 3 in Part I in *WMO Lectures on Forecasting of Tropical Weather*. WMO 492.

Durand, J. H. (1977). 'A propos de la sécheresse et de ses conséquences au Sahel'. *L'Université de Bordeaux. Inst. Géog. Cahiers d'Outre-Mer*, **30**, 383–403.

Ehrenkrantz, E. D. (1962). *Traditional Building Forms for Design of Buildings for Hot Climates*. New Building Res. Conf, Building Research Institute, Washington.

Eldridge, R. H. (1957). 'A synoptic study of West African disturbance lines'. *Q. J. Roy. Met. Soc.* **83**, 303–14 and discussion in 1958, **84**, 468–9.

Electricity Council, Intelligence Section, (1973). *Select List of References on Wind Power, 1949–Nov. 1973*. Bibliographies B. 70. London.

Erickson, C. O. (1963). 'An incipient hurricane near the West African coast'. *Mon. Weath. Rev.* **91**, 61–8.

Ette, A. I. I. (1971). 'The effect of harmattan dust on atmospheric electrical parameters'. *J. Atmos. Terrestrial Phys.* **33**(2), 295–300.

Ette, A. I. I. and Adewolu, D. O. (1980). '*Atmospheric Dust Content and Associated Electric and Meteorological Effects during the Harmattan*'. Manchester Univ. Inst. Sci. and Technology. Abstract VI, Int. Conf. Atmospheric Electricity, 28 July–1 Aug. 1980. Sess. IV, A–9, 4.

Ezekwe, C. I. and Ezeilo, C. C. O. (1981). 'Measured solar radiation in a Nigerian environment compared with predicted data'. *Solar Energy*, **26**, 181–6.

Fajemisin, J. M. and Olaniyan, G. O. (1977). 'Exploiting weather variability for the improvement in the level and efficiency of maize production in Nigeria'. *WMO* 481, 350–74.

Fanger, P. O. (1970). *Thermal Comfort*. Danish Technical Press, Copenhagen. Reprinted by McGraw-Hill Book Co. New York, 1973. Reprinted by Robert E. Krieger Publishing Company, Malabar, Florida, 1982.

Faniran, J. (1977). 'The use of drainage basins in development planning in West Africa'. *Nig. Geogr. J.* **20**, 189–97.

FAO (1966). *Agricultural Development in Nigeria 1965–1980*. FAO, Rome.

FAO/UNESCO/WMO, (1967). *A study of the Agroclimatology of the Semi-arid Area South of the Sahara in West Africa*. FAO/UNESCO/WMO Interagency Project. Tech. Rep. by FAO.

Faria, J. M. R. (1971). 'Frequency analysis of the annual highest values of daily precipitation in some Portuguese overseas sites'. *Fomento, (Lisbon)*, **9**, 237–70.

Farrow, R. A. (1975). 'The African migratory locust in its main outbreak area of the middle Niger: quantitative studies of solitary populations in relation to environmental factors'. Int. Af. Migratory Locust Org. *Locusta* **11**, 1–198. Bamako.

Faure, H. and Gac, J–Y. (1981). 'Will the Sahelian drought end in 1985?' *Nature*, **291**, 475–8.

Fitch, J. M. and Branch, D. P. (1960). 'Primitive architecture and climate'. *Sci. Am.* **207**, 134–44.

Flohn, H. (1966). 'Warum ist die Sahara trocken?' *Zeitschrift für Met.* **17**, 316–20.

Font, T. I. (1951). *El Clima de las Posesiones Españolas del Golfa de Guinea*. Consejo Superior de Investigaciones Científicas, Instituto Estudios Africanos, Madrid.

Fortune, M. (1980). 'Properties of African squall lines inferred from time-lapse satellite imagery'. *Mon. Weath. Rev.* **108**, 153–68.

Fosberg, F. R., Garnier, B. J. and Kuchler, A. W. (1961). 'Delimitation of the humid tropics'. *Geogr. Rev.* **51**, 333–47.

Foyle, A. M. (1950). 'Nigerian architecture'. *Geogr. Mag.* **23**(5), 173.

Franquin, P. (1973). 'Agroclimatic analysis in tropical areas. (Model for defining the growing season)'. *Agronomie Tropicale*, **28**(6/7), 665–82.

Franquin, P. (1985). 'Météorologie et agrostratégies pour le Scahel'. *La Météorologie* (VII sér), **8**, 1–7.

Fry, M. and Drew, J. (1956). *Tropical Architecture in the Humid Zone*. Batsford.

Gallardo, Y. (1978). 'Hydrometeorological relationships on the coast of the Gulf of Guinea – influence of the orientation of the coast'. Annex (Section 2) in *GARP Publications, Series No. 21. The West African Monsoon Experiment (WAMEX)*, WMO.

Gamble, D. P. (1952). 'Infant mortality rates in rural areas in the Gambia Protectorate'. *J. Trop. Med.* 145–9.

Garbutt, D. J., Stern, R. D. , Dennett, M. D. and Elston, J. (1981). 'A comparison of the rainfall climate of eleven places in West Africa using a 2-part model for daily rainfall'. *Archiv. für Met. Geophys. Bioklim.* **29B**, 137–55.

Garnier, B. J. (1952a). 'A simple apparatus for measuring potential evapotranspiration'. *Nature* (London), **170**, 4320.

Garnier, B. J. (1952b). 'A preliminary experiment to measure potential evapotranspiration at University College, Ibadan'. *Res. Notes Dep. Geog. Univ. College, Ibadan*, **1**, 4–20.

Garnier, B. J. (1953a). 'Some comments on measurements of potential evapotranspiration in Nigeria'. *Res. Notes Dep. Geog. Univ. College, Ibadan*, **2**, 11–20.

Garnier, B. J. (1953b). 'Some comments on rainfall at University College, Ibadan, during 1952'. *Res. Notes Dep. Geog. Univ. College, Ibadan*, **3**.

Garnier, B. J. (1953c). 'The incidence and intensity of rainfall at Ibadan, Nigeria'. *C. R. IVᵉ Réunion Conf. Int. des Africanistes de l'Ouest, Abidjan*, 87.

Garnier, B. J. (1954). 'Measuring potential evapotranspiration in Nigeria'. pp. 140–8 in Mather, J. R. (ed.). *The Measurement of Potential Evapotranspiration*.

Garnier, B. J. (1956a). 'Report on experiments to measure potential evapotranspiration in Nigeria'. *Res. Notes Dep. Geog. Univ. College, Ibadan,* **8**, 2–10.

Garnier, B. J. (1956b). 'A method of computing potential evapotranspiration in West Africa'. *Bull. Inst. Fr. Afr. Noire 18,* Sér. A, **3**, Dakar, 665–76.

Garnier, B. J. (1960a). 'Maps of the water balance in West Africa'. *Bull. Inst. Fr. Afr. Noire,* Sér. A, **XXII**, 709–22.

Garnier, B. J. (1960b). 'Delimiting the humid tropics'. *I.C.S.U. Rev.* 2, 210–8.

Garnier, B. J. (1961). 'Maps of the water balance of West Africa'. *Bull. Inst. Fr. Afr. Noire,* Sér. A, **22**, 709–22.

Garnier, B. J. (1967). 'Weather conditions in Nigeria'. *Climatological Res. Series* No. 2, McGill Univ., Montreal.

Garnier, P. (1951). 'Le déficit de saturation absolue en Afrique occidentale française' *Bull. Inst. Fr. Afr. Noire* **XIII**, 3, Dakar, 734–48.

Garnier, P. (1954). 'Quelques mesures de luminosité à Bamako'. *Notes Afr.* **61**, 2–4.

Garnier, R. (1976). 'Situations à ondes d'est sur l'Afrique occidentale'. *ASECNA Publ.* 38.

Gates, D. M. (1972). *Man and his Environment: Climate.* Harper and Row.

Gaucher, L. R. (1976). 'Origine Africaine du cyclone Fifi' *La Météorologie* VI Sér., 6, 159–66.

Geckor-Kove, N. A. and Dankwa, J. B. (1966). 'Diurnal variation of rainfall in Ghana'. *Ghana Met. Dep.* Dep. Note 15.

Genève, R. (1957). *Météorologie Tropicale.* Météorologie Nationale, Paris.

Gentilli, J. (1958). *A Geography of Climate.* Univ. W. Australia Press.

Germain, H. (1968). 'Situation typique de petit hivernage'. *ASECNA Publ.* 9.

Ghana Meteorological Department (1959). 'Number of days of lightning in Ghana'. *Climatological Note* 2.

Ghana, Ministry of Construction and Commerce (1962). *Handbook on Rural Housing, Village and Peri-urban Development in Ghana.* Accra.

Giessner, K. (1985). ('Hydrological aspects of the Sahel problem'.) *Die Erde,* **116** (H. 2–3), 137–57 (in German).

Gifford, P. L. (1973). 'L'importance du climat dans l'écologie des arbres forestiers au Sénégal'. *WMO* No. 340, 158–69.

Gillett, J. D. (1974). 'The mosquito: still man's worst enemy'. *Am. Scientist,* **61**, 430–6.

Giraud, J. M. and Grégoire, P. (1976). 'Le climat Soudano–Sahélien. Anneé sèche – anneé pluvieuse'. *La Météorologie* VI Sér., 6, 69–81.

Givoni, B. (1969). *Man, Climate and Architecture.* Applied Sci. Publishers.

Giwa, F. B. A. (1972). 'A note on the diurnal variation of surface temperature and pressure at some Nigerian meteorological stations'. *Planetary and Space Science* 20, 1941–8.

Glantz, M. H. (1977). 'The value of a long-range weather forecast for the West African Sahel'. *Bull. Am. Met. Soc.* **58**(2), 150–8.

Glantz, M. H. and Parton, W. (1976). 'Weather and climate modification and the future of the Sahara'. pp. 303–24 in Glantz, M. H. (ed.). *Politics and Natural Disaster. Case of Sahelian Drought,* Praeger, New York.

Glavnaya Geofizicheskaya Observatoriya (1978). *Climatic Atlas of Africa.* Pts 1 and 2. Leningrad.

Glover, J. and McCullough, J. S. G. (1958). 'The empirical relationship between solar radiation and hours of sunshine'. *Q. J. Roy. Met. Soc.* **84**, 172–5.

Glover, J. and Robinson, P. (1953). 'A simple method of calculating the reliability of rainfall'. *East Af. Agric. J.* **19**, 11–13.

Gold Coast, Agricultural Dept., (1930). 'Tables of mean annual rainfall and rainfall chart of West Africa'. *Agric. Dep. Bull.* 23, 324–9.

Golding, E. W. (1955). *The Generation of Electricity by Wind Power.* E. and F. N. Spon Ltd.

Golding, E. W. (1962). 'Energy from wind and local fuels'. *Arid Zone Res.* 18, UNESCO.

Goldman, R. F., Green, E. and Iampietro, P. F. (1965). 'Tolerance of hot-wet environments by resting men'. *J. Appl. Physiology,* **20**, 271–7.

Gornitz, V. (1985). 'A survey of anthropogenic vegetation changes in West Africa during the last century – climatic implications'. *Climatic Change,* 7(3), 285–325.

Gourou, P. (1966). *The Tropical World.* Longman.

Grandidier, G. (1934). *Atlas des Colonies Françaises, Protectorats et Territoires sous Mandat de la France* (Text on climate and climatic zones of West Africa by H. Hubert). Académie des Sciences, Paris.

Greathouse, G. A. and Wessel, C. J. (1954). *Deterioration of Materials.* Reinhold, New York.

Greenhut, G. K. (1978). 'Correlations between rainfall and sea surface temperature during GATE'. *J. Phys. Oceanography,* **8**, 1135–8.

Greenhut, G. K. (1981). 'Comparison of temperature gradient model predictions with recent rainfall trends in the Sahel'. *Mon. Weath. Rev.* **109**, 137–47.

Gregorczuk, M. (1967). *Bioclimates of the World Under Air Enthalpy Point of View.* Typescript 4 pp. (Quoted in Landsberg, 1972).

Gregory, S. (1964). 'A multiple regression analysis of average rainfall patterns over Sierra Leone' (Abstract). *20th Int. Geog. Congress Abstracts of Papers,* 63–4. Nelson.

Gregory, S. (1965). 'Rainfall over Sierra Leone'. Department of Geography, *University of Liverpool, Research Paper* 2.

Gregory, S. (1967). 'The orographic component in rainfall distribution patterns'. In *Mélanges de Géographie, Physique, Humaine, Economique,*

Appliquée, offerts M. Omer Tulippe. Gembloux, Editions Duculot, S.A.

Gregory, S. (1982). 'Spatial patterns of Sahelian annual rainfall, 1961–80' *Archiv. für Met. Geophys. Bioklim.* **31B**, No. 3, 273–86.

Gregory, S. (1983). 'A note on mean seasonal rainfall in the Sahel, 1931–60 and 1961–80'. *Geography*, **68**, 31–6.

Gribbin, J. (1985). 'The drying of the Sahel'. *New Scientist*, **105**, (1447), 8–9.

Griffiths, J. F. (1960). 'Bioclimatology and the meteorological services'. pp. 283–300 in Bargman, D. J. (ed.). *Proceedings of the Symposium on Tropical Meteorology in Africa*. Munitalp Foundation and WMO, Nairobi.

Griffiths, J. F. (1968). *Applied Climatology*. Oxford Univ. Press.

Griffiths, J. F. (ed.) (1972). *Climate of Africa. World Survey of Climatology*. Vol. 10. Elsevier.

Griffiths, J. F. (1976). *Climate and the Environment*. Elek Books.

Griffiths, J. F. and M. J. (1969). 'A bibliography of weather and architecture'. US Dep. Commerce., *ESSA Tech. Memorandum EDSTM 9*, Silver Spring, Md.

Grosrey, A. (1974). 'Rélation entre la hauteur moyenne des précipitations et la distance à la mer. Etude mathématique de la répartition des pluies en Afrique occidentale'. *La Météorologie*, **30**, 31–52.

Grove, A. T. (1973). 'Desertification in the African environment'. pp. 33–45 in Dalby, D. and Harrison Church, R. J. *Drought in Africa*.

Grove, A. T. (1977). 'Climate for deserts' *Geog. Mag.* **49**, 634–8.

Grove, A. T. (ed.) (1985). *The Niger and its Neighbours: Environmental History and Hydrobiology, Human Use and Health Hazards of the Major West African Rivers*. Balkema.

Grundke, G. (1955/56). 'Uber die Bedeutung Technoklimatischer Forschungen' *Wissenschaftliche Zeitschrift der Hochschule für Binnenhandel*, **1**(3), 19–38.

Guimaraes, C. (1957). ('The rains of Portuguese Guinea'). *Boletin Cultural da Guiné Portuguesa*, **12**, 315–32. Lisbon.

Guimaraes, C. (1959). 'O clima da Guiné Portuguesa'. *Boletin Cultural da Guiné Portuguesa*, **14**, 295–357.

Hales, W. B. (1949). 'Micrometeorology in the tropics'. *Bull. Am. Met. Soc.* **30**, 124–37.

Hall, N. et al., (1958). 'Trees and microclimate'. *Climatology and Microclimatology*. Proc. of Canberra Symp., UNESCO, 259–63.

Hamilton, R. A. (1943a). 'Line squalls in Nigeria'. *Met. Mag. Lagos*, May, 1–2.

Hamilton, R. A. (1943b). 'An interesting squall'. *Met. Mag. Lagos*, October, 3–4.

Hamilton, R. A. (1943c). 'Morning coastal fog'. *Met. Mag. Lagos*, December, 4.

Hamilton, R. A. and Archbold, J. W. (1945).

'Meteorology of Nigeria and adjacent territory'. *Q. J. Roy. Met. Soc.* **71**, 231–64.

Hamilton, R. A. and Fisher, J. F. (1943). 'Squall statistics, Lagos'. *Met. Mag.* Lagos, July, 4–5.

Hanson, K. J. (1976). 'A new estimate of solar irradiance at the earth's surface on zonal scales'. *J. Geophys. Res.* **81**, 4435–43.

Hardy, F. (1958). 'The effects of air temperature on growth and production in cacao'. *Cacao*, **3**(17), 1–14.

Hare, F. K. (1977). 'Connections between climate and desertification'. *Envir. Conservation*, **4**, 81–90.

Hare, F. K., Kates, R. W. and Warren, A. (1977). 'The making of deserts: climate, ecology and society'. *Economic Geography*, **53**, 332–46.

Harris, O. (1957). *Notes on Book Preservation in West Africa*. Ibadan Univ.

Harrison Church, R. J. (1961). *West Africa*. Longmans.

Harrower, T. N. S. and Evans, D. C. (1956). 'Damage to aircraft by heavy hail at high altitude'. *Met. Mag.* **85**, 330–8.

Haufe, W. O. (1963). 'Entomological biometeorology' *Int. J. Biomet.* **7**, 129–36.

Hay, W. M. (1957). 'Effects of weather on railroad operation, maintenance and construction in industrial operation under extremes of weather'. *Met. Monographs*, **2**(9), 10–36.

Hayward, D. F. (1968). 'Dynamic climatology and the climate of West Africa'. *Sierra Leone Geogr. J.* **12**, 3–16.

Hayward, D. F. and Addae, E. (1978). *Meteorological Station, Summary of Records, 1965–1976*. Dept. Geog., Univ. Cape Coast, Ghana.

Heinricy, D. J. and Young, J. A. (1974). 'Long term variations in cloud activity over the east Atlantic–West Africa region'. *Univ. Wisconsin., Space Sci. and Engineering Center. Annual Rep. 1973*, 28–82.

Henderson-Sellers, A. and Gornitz, V. (1984). 'Possible climatic aspects of land cover transformations, with particular emphasis on tropical deforestation'. *Clim. Change*, **6**(3), 231–57.

Her Majesty's Stationery Office (1958). *Tables of Temperature, Relative Humidity and Precipitation for the World. Vol. IV. Africa*. HMSO.

Hidore, J. J. (1978). 'Acceleration of desiccation and population trauma in sub-Saharan Africa'. *UN Economic and Social Commission for Asia and the Pacific. Water Resources*, **119**, 8–16.

Hiernaux, J., Rudan, P. and Brambati, A. (1975). 'Climate and the weight/height relationship in sub-Saharan Africa'. *Annls. Human Biology*, **2**, 3–12.

High, C. and Oguntoyinbo, J. S. (1974). 'Ground level raingauge catch at Ibadan'. *Weather*, **29**, 384–7.

Hillier, J. (1707). 'Part of two letters dated Cape Corse, Jan. 3, 1687 and April 25, 1688 . . . giving . . . an account of the weather there from Nov. 24, 1686 to the same day 1687'. *Miscellanea Curiosa*. Vol. III. London.

Hills, E. S. (1966). *Arid Lands*. UNESCO.

Hisdal, V. (1953). 'The influence of weather on mortality'. *Meteorologiske Annaler* (Oslo), **3**(10), 257–83.

Hobbs, J. E. (1981). *Applied Climatology*. Butterworths.

Hobeniche, P. (1936). 'La grêle en Guinée française en 1932–1933–1934'. *Annls. de Phys. du Globe de la Fr. d'Outre-Mer., Paris*, **3**, 72.

Hodder, B. W. (1957). 'A note on delimiting the humid tropics: the case of Nigeria in West Africa.' *Res. Notes Dep. Geog. Univ. College, Ibadan*, **10**, 1–7.

Hogg, W. H. (1950). 'Diurnal variation of upper wind at Freetown'. *Met. Mag.* **79**, 133–7.

Höller, E. (1959). 'Cases of damage to cargoes and their meteorological significance'. *Marine Observations*, **29**, 19–25.

Höller, E. and Kerner, G. (undated). *Africa. Climate Almanac for Business and Travel*. Afrika–Verein Hamburg–Bremen E. V. Uebersee–Verlag, Hamburg.

Holmes, B. M. (1951). *Weathering in the Tropics*. Division of Building Research, CSIRO. Melbourne, Australia.

Hookey, P. (1970). 'Revenge of the Gods'. *Weather*, **25**(9), 425–8.

Hoppen, C. E. (1958). *The Pastoral Fulbe Family in Gwandie*. Unpublished MS.

Houghton, F. C. and Yaglou, C. R. (1923). 'Determining lines of equal comfort'. *Trans. Am. Soc. Heating and Ventilating Engineers*, **28**, 163–76, and 361–84.

Houghton, R. W. (1973). 'Evaporation during upwelling in Ghanaian coastal waters'. *J. Phys. Oceanography*, **3**(4), 487–9.

Houghton, R. W. and Beer, T. (1976). 'Wave propagation during the Ghana upwelling'. *J. Geophys. Res.* **81**(24), 4423–9.

Housing and Home Finance Agency, (1951). *Climate and Architecture*. Bibliography. 175 references.

Houze, R. A. (1977). 'Structure and dynamics of a tropical squall-line system'. *Mon. Weath. Rev.* **105**, 1540–67.

Howe, G. M. (ed.), (1977). *A World Geography of Human Diseases*. Academic Press.

Hubert, H. (1922). *Nouvelles Recherches sur les Grains Orageux et les Pluies en Afrique Occidentale*. Publ. du Gouvernement Générale de l'Afr. Occidentale Fr.

Hubert, H. (1934). 'L'organisation du service météorologique colonial'. *Annls. du Phys. du Globe de la Fr. d'Outre-Mer*. **1**, 2–9.

Hubert, H. (1938). 'Les masses d'air de l'ouest Africain'. *Annls. du Phys. du Globe de la Fr. d'Outre-Mer*. **5**, 33–64.

Hubert, H. (1939a). 'Origine Africaine d'un cyclone tropical Atlantique'. *Annls. du Phys. du Globe de la Fr. d'Outre-Mer*. **6**, 97–115.

Hubert, H. (1939b). 'Les pluies torrentielles en Afrique occidentale française'. *Annls. du Phys. du Globe de la Fr. d'Outre-Mer*. **6**, 161–7.

Hubert, H. (1943). 'A summary of knowledge of the sand and dust storms of French West Africa'. *Bull. Am. Met. Soc.* **24**, 243–6.

Hudson, J. C. and Stanners, J. F. (1953). 'The effect of climate and atmospheric pollution on corrosion'. *J. Appl. Chemistry*, **3**, 86–96.

Humphreys, M. A. (1971). 'A simple theoretical derivation of thermal comfort conditions', in Humphreys, M. A. and Nicol, J. F. *Theoretical and Practical Aspects of Thermal Comfort*. BRS, Watford, UK.

Humphreys, M. A. (1981). 'The dependence of comfortable temperatures upon indoor and outdoor climates'. Chapter 15 in Cena, K. and Clark, J. A. (eds.). *Bioengineering, Thermal Physiology and Comfort*.

Hunter, J. M. and Thomas, M. O. (1984). 'Hypothesis of leprosy, tuberculosis and urbanization in Africa'. *Soc. Sci. Med.*, **19**(1), 27–57.

Hutchings, J. W. (ed.), (1964). *Proceedings of the Symposium on Tropical Meteorology*. New Zealand Met. Serv., Wellington.

Hutchinson, P. (1985). 'Rainfall analysis of the Sahelian drought in the Gambia', *J. Clim.*, **5**(6), 665–72.

Huxley, P. A. and Summerfield, R. J. (1974). 'Effects of night temperature and photoperiod on reproductive ontogeny of cultivars of cowpea and soyabean selected for the wet tropics'. *Plant Science Letters*, **3**, 11–17.

Ibukun, O. (1964). 'Occurrences of thunderstorm and lightning in the tropics'. *Nig. Engineer*, **2**.

Ilesanmi, O. O. (1970). *A Study of Nigerian Rainfall Patterns from the Viewpoint of Precipitation Dynamics*. Ph.D. thesis, Univ. Wisconsin.

Ilesanmi, O. O. (1971). 'An empirical formulation of an ITD rainfall model for the tropics: a case study of Nigeria'. *J. Appl. Met.* **10**, 882–91.

Ilesanmi, O. O. (1972a). 'Rainfall amounts and duration characteristics in Nigeria'. *Nig. Agric. J.* **9**(1), 25–37.

Ilesanmi, O. O. (1972b). 'An empirical formulation of the onset, advance and retreat of rainfall in Nigeria'. *J. Trop. Geog.* **34**, 17–24.

Ilesanmi, O. O. (1972c). 'Aspects of the precipitation climatology of the July–August rainfall minimum of southern Nigeria'. *J. Trop. Geog.* **35**, 51–9.

Ilesanmi, O. O. (1972d). 'The diurnal variation of rainfall in Nigeria'. *Nig. Geogr. J.* **15**(1), 25–34.

Ingram, D. L. and Mount, L. E. (1975). *Man and Animals in Hot Environments*. Springer-Verlag.

International Institute of Tropical Agriculture. *Annual Reports*. IITA, Ibadan, Nigeria.

Ireland, D. H. (1962). 'The little dry season of southern Nigeria'. *Nig. Met. Serv. Tech. Note* 24.

Jackson, I. J. (1977). *Climate, Water and Agriculture in the Tropics*. Longman.

Jackson, I. J. (1986). 'Relationships between raindays, mean daily intensity and monthly rainfall in the tropics'. *J. Clim.* **6**(2), 117–34.

Jackson, S. P. (Director), (1961). *Climatological Atlas of Africa*. Commission for Tech. Cooperation in Africa, South of the Sahara and Sci. Council for Afr., South of the Sahara. Joint Project No. 1. Lagos/Nairobi.

Jeffreys, M. D. W. (1952). 'Lightning'. *Nigeria* **39**, 228–36.

Johnson, A. F. (1964). *A Bibliography of Ghana, 1930–61*. Longmans.

Johnson, D. H. (1965). 'African synoptic meteorology'. WMO 171. Tech. Note 69, 48–90.

Johnstone, D. R. and Huntington, K. A. (1977). 'Weather and crop spraying in northern Nigeria'. *Weather*, **32**(7), 248–57.

Joucla, E. A. (1937). *Bibliographie de l'Afrique Occidentale Française*. Paris.

Junge, C. (1979). 'The importance of mineral dust as an atmospheric constituent'. In Morales, C. (ed.) *Saharan Dust*.

Kalu, A. E. (1979). 'The African dust plume: its characteristics and propagation across West Africa in winter'. pp. 95–118 in Morales, C. (ed.) *Saharan Dust*.

Kamara, S. I. (1986). 'The origins and types of rainfall in West Africa'. *Weather* **41**(2), 48–56.

Kamark, A. (1976). *The Tropics and Economic Development*. Johns Hopkins Univ. Press, Baltimore.

Kassam, A. H. and Kowal, J. M. (1973). 'Productivity of crops in the savanna and rain forest zones in Nigeria'. *Savanna*, **2**(1), 39–49.

Kassam, A. H. and Kowal, J. M. (1975). 'Water use, energy balance and growth of Gero millet at Samaru, northern Nigeria'. *Agric. Met., Inst. J.* **15**, 333–42.

Kates, R. W. (1981). *Drought Impact in the Sahelian–Sudanic Zone of West Africa: A Comparative Analysis of 1910–15 and 1968–74*. Centre for Technology, Environment and Development, Clark Univ., Worcester, Mass.

Keeling, C. D., Bacastow, R. D. and Whors, T. P. (1976). 'Measurements of the concentration of carbon dioxide at Mauna Loa Observatory, Hawaii'. pp. 377–85 in Clark, W. C. (ed.) *Carbon Dioxide Review*. Oxford Univ. Press.

Kellogg, W. W. (1977). 'Effects of human activities on global climate'. *WMO Tech. Note* 156. WMO 486.

Kellogg, W. W. (1979). 'Influences of mankind on climate' *Annual Rev. of Earth and Planetary Sci.* **7**, 63–92.

Kelly, T. J. (1975). 'Climate and the West African drought' pp. 14–31 in Newman, J. L. (ed.). *Drought, Famine and Population Movements in Africa*. Maxwell School of Citizenship and Public Affairs, Syracuse Univ. New York.

Kidson, J. W. (1977). 'African rainfall and its relation to the upper air circulation'. *Q. J. Roy. Met. Soc.* **103**, 441–56.

Kimble, G. H. T. (1946). 'Tropical land and sea breezes, with special reference to the East Indies'. *Bull. Am. Met. Soc.* **27**, 99–113.

King, H. E. (1957). 'Cotton yields and weather in northern Nigeria'. *Empire Cotton Growers Rev.* **34**, 153–4.

Klaus, D. (1978). 'Spatial distribution and periodicity of mean annual precipitation south of the Sahara'. *Archiv. für Met. Geophys. Bioklim.* **26B**, 17–27.

Knoch, K. and Schulze, A. (1956). *Precipitation, Temperature and Sultriness in Africa*. II/151–II/156. Falk-Verlag, Hamburg.

Koenigsberger, O. H., Ingersoll, T. G., Mayhew, A. and Szokolay, S. V. (1974). *Manual of Tropical Housing and Building*. Longman.

Koteswaram, P. (1958). 'The easterly jet stream in the tropics'. *Tellus*, **10**, 43–57.

Kowal, J. M and Kassam, A. H. (1973). 'Water use, energy balance and growth of maize at Samaru, northern Nigeria'. *Agric. Met.* **12**, 391–406.

Kowal, J. M. and Kassam, A. H. (1975). 'Rainfall in the Sudan savanna region of Nigeria'. *Weather*, **30**, 24–8.

Kowal, J. M. and Kassam, A. H. (1976). 'Energy load and instantaneous intensity of rainstorms at Samaru, northern Nigeria'. *Tropic. Agric. Trinidad*, **53**, 185–97.

Kowal, J. M. and Kassam, A. H. (1978). *Agricultural Ecology of Savanna: a Study of West Africa*. Clarendon Press.

Kowal, J. M. and Knabe, D. T. (1972). *An Agroclimatological Atlas of the Northern States of Nigeria*. Ahmadu Bello Univ. Press, Samaru, Nigeria.

Kramer, M. P. (1952). 'Selective annotated bibliography in the climate of Central and West Africa including the Belgian Congo'. *Met. Abstracts*. **3**, 837–73.

Kravcova, L. M. and Makarova, N. M. (eds.), (1965). *Climatic Characteristics of Cloudiness for the IGY and its Extension. Northern Hemisphere. Vyp. 4, Africa*. Naucno–Issledovatel 'Skij Institut. Aeroklimatologii, Moscow.

Krishnamurti, T. N. (1977). 'Primitive equation models and their application to the tropics', in WMO 492.

Krishnamurti, T. N., Pasch, R. J. and Ardanuy, P. (1980). 'Prediction of African waves and specification of squall lines'. *Florida State Univ., Dept. Met. Rep.* 79–82 and *Tellus*, **32**(3), 215–31.

Kullenberg, B. (1955). 'Quelques observations microclimatologiques en Côt d'Ivoire et Guinée française'. *Bull. de l'Inst. Fr. d'Afr. Noire*, (A)17, 755–68.

Kung, E. L., Bryson, R. A. and Lenschow, D. H. (1964). 'Study of a continental surface albedo on the basis of flight measurements and structure of the earth's surface cover over North America'. *Mon. Weath. Rev.* **92**, 543–64.

Ladell, W. S. S. (1949). 'Physiological classification of climates: illustrated by reference to Nigeria'. *Proc. Inst. W. Af. Conf., Ibadan*, 4–41.

Ladell, W. S. S. (1953). 'The physiology of life and work in high ambient temperatures'. *Res. Council Israel Special Publ. 2. Desert Research*, 187–204. Jerusalem.

Lal, R. (1973). 'Soil erosion and shifting agriculture'. Paper presented at the *FAO. Regional Seminar on Shifting Cultivation and Soil Conservation in Africa*, 2–25 July, 1973, Univ. Ibadan.

Lal, R. (1974). 'Soil temperature relations in tropical Africa and their effects on crop yield'. Paper presented at the *International Expert Consultation on the Use of Improved Technology for Food Production in Rainfed Areas in Tropical Asia*. Kuala Lumpur, 1974.

Lamb, H. H. (1977). 'Some comments on the drought in recent years in the Sahel-Ethiopian zone of North Africa'. Chapter 3 in Dalby, D. et al. (eds.) *Drought in Africa*.

Lamb, P. J. (1983a). 'Sub-Saharan rainfall update for 1982: continued drought'. *J. Clim.* **3**, 419.

Lamb, P. J. (1983b). 'West African water vapor variations between recent contrasting sub-Saharan rainy seasons'. *Tellus*, **35A**, (3), 198–212.

Lamb, P. J. (1985). 'Rainfall in sub-Saharan West Africa during 1941–83'. pp. 64–7 in *Am. Met. Soc. Third Conf. on Climatic Variations and Symp. on Contemporary Climate 1850–2100*. Los Angeles, Cal.

Lamb, P. J. (1986), 'Rainfall to subsaharan West Africa during 1982–84'. *WMO Prog. Long-range Forecast Res. Rep. Ser.* no. 1986, no. 6, vol. 1, 86.

Lambergeon, D. (1977). 'Relation entre les pluies et les pressions en Afrique occidentale'. *ASECNA.* No. 57. Dakar.

Lambergeon, D., Dzietara, S. and Janicot, S. (1981). 'Comportement du champ de vent sur L'Afrique occidentale'. *La Météorologie*, **25**, 69–82.

Landsberg, H. E. (1964). *Physical Climatology*. Gray Printing C. Inc., Du Bois, Pennsylvania.

Landsberg, H. E. (1969). *Weather and Health, an Introduction to Biometeorology*. Doubleday.

Landsberg, H. E. (1972). *The Assessment of Human Bioclimate, a Limited Review of Physical Parameters.* WMO 331, *Tech. Note* 123.

Landsberg, H. E. (1975). 'Sahel drought: change of climate or part of climate?' *Archiv. für Met. Geophys. Bioklim.* **23B**, 193–200.

Landsberg, H. E. (1979). 'The effect of man's activities on climate'. pp. 187–236 in Biswas, M. R. and A. K. (eds.). *Food, Climate, Man.*

Landsberg, H. E. (1984). 'Climate and Health'. pp. 26–64 in Biswas, A. K. *Climate and Development.*

Landsberg, H. E. and Jacobs, W. C. (1951). 'Applied climatology'. pp. 976–92 in Malone, T. F. (ed.). *Compendium of Meteorology.*

Lapeyssonnie, L. (1963). *La Méningite Cérébro-spinale en Afrique.* WMO 28.

Larmie, G. A. K. (1965). 'Multiple tropopauses over Ikeja'. *Nig. Met. Serv. Proc. Symp. Trop. Met., Oshodi-Lagos, 1964*, 54–61.

La Seur, N. E. and Adefolalu, D. (1975). 'A multiple scale research program in tropical meteorology'. *Res. Tech. Rep. No. ECOM – 69 – 0062F.* Dept. Met., Florida State Univ.

Lawson, T. L. (1977). 'Possible impact of agrometorological studies on food production in the humid tropics, with particular reference to West Africa'. Paper presented at the *WMO/FAO Tech. Conf. on Application of Meteorology to Agriculture.* IITA, Ibadan.

Lawson, T. L. (1978). 'Effects of agroclimatological changes in food production in Africa'. Paper presented at the *AASA 3rd General Conf. and 10th Anniversary*. Univ. Ibadan, Nigeria.

Lawson, T. L., Oguntoyinbo, J. S. and Ojo, S. O. (1979). 'Agroclimatic conditions in West Africa'. Paper presented at the *IITA Annual Res. Conf. on Soil and Climatic Resources and Constraints in Relation to Food Crop Production in West Africa.* IITA. Ibadan.

Lean, O. B. (1931). 'On the recent swarming of *Locusta migratoria*'. *Bull. Entomological Res.* **22**, 365–78.

Lebedev, A. N. (ed.), (1967). *The Climate of Africa. Vol. II. Wind, Relative Humidity, Cloud Amount, Fog, Thunder and Aeroclimatic Conditions* (in Russian). Hydrometeorological Publishing House, Leningrad.

Lebedev, A. N. (ed.), (1970). *The Climate of Africa. Vol. I. Air Temperatures, Precipitation.* Israel Program for Scientific Translations, Jerusalem.

Lebedev, A. N. (ed.), (1978). *Climatic Atlas of Africa.* Pts 1 and 2. Glavnaya Geofizicheskaya Observatoriya, Leningrad.

Ledger, D. C. (1964). 'Some hydrological characteristics of West African rivers'. *Trans. Inst. Brit. Geographers*, **35**, 73–90.

Ledger, D. C. (1975). 'The water balance of an exceptionally wet catchment in West Africa'. *J. of Hydrology*, **XXIV**, 207–14.

Lee, D. H. K. (1952). 'Significance of hot environments for man'. *BRAB, Res. Conf. Rep.* **5**, 6–12. Washington.

Lee, D. H. K. (1953). 'Tropical housing'. *Geogr. Rev.* **43**, 571–3.

Lee, D. H. K. (1957). *Climate and Economic Development in the Tropics.* Harper.

Lee, D. H. K. (1958). 'Proprioclimates of man and domestic animals'. *Climatology. Arid Zone Research* **X**, 102–25. UNESCO.

Lee, D. H. K. (1965). 'Climatic stress indices for domestic animals'. *J. Biomet.* **9**(1), 23–35.

Lefèvre, R. (1972). 'Aspect de la pluviométrie dans la région du Mont Caméroun'. *WMO* **326**, (2), 373–82.

Legrand, M., Bertrand, J. J. and Desbois, M. (1982). 'Etude des brumes sèches sur l'Afrique de l'Ouest à l'aide de Météosat I'. *La Météorologie* Nos. 29–30, 153–9.

Leluin, P. (1949). 'Le temps à Port-Etienne'. *La Météorologie*, **IV**(15), 147–56.

Lemon, E. R. et al., (1957). 'Some aspects of the relationship of soil, plant and meteorological factors to evapotranspiration'. *Proc. Soil Sci. Soc.* **21**, 464–8.

Leroux, M. (1972). 'La dynamique des précipitations en Afrique occidentale'. *ASECNA Publ.* 23.

Leroux, M. (1973). 'Les principales discontinuités africaines F.I.T.–C.I.O.' pp. 21–36 in *La Structure Continue de l'Equateur Météorologique sur l'Afrique Intertropicale*, Dakar. Direction de l'Exploitation Météorologique.

Leroux, M. (1974). 'Champ de vent en altitude en Afrique occidentale et centrale'. *ASECNA Publ.* 34.

Leroux, M. (1976). *Processus de Formation et d'Evolution des Lignes de Grains de l'Afrique Tropicale Septentrionale*, Univ. de Dakar, Dép. de Géog. Rech. de Clim. Trop. 1.

Leroux, M. (1983). *Le Climat de l'Afrique Tropicale*. Champion–Slatkine, Paris. 2 Vols. (No. 2 is an atlas).

L'Homme, J. P. (1981). 'L'évolution de la pluviosité annuelle en Côte d'Ivoire au cours des soixantes dernières années'. *La Météorologie* 25, 135–40.

Licht, S. (ed.), (1964). *Medical Climatology*. Waverly Press, Baltimore.

Liss, P. S. and Crane, A. J. (1983). *Man-made Carbon Dioxide and Climatic Change: a Review of Scientific Problems*. Geo Books.

List, R. J. (ed.), (1984). *Smithsonian Meteorological Tables*. Smithsonian Institute, Washington D.C.

Lough, J. M. (1980). 'West African rainfall variations and tropical Atlantic sea surface temperatures', Univ. East Anglia, Climatic Res. Unit. *Clim. Monitor* 9, 150–7.

Lough, J. M. (1986). 'Tropical Atlantic sea-surface temperatures and rainfall variations in subsaharan Africa'. *Mon. Weath. Rev*, **114** (3), 561–70.

Lowry, W. P. (1969). *Weather and Life*. Academic Press.

McArdle, B., Dunham, W., Holling, H. E., Ladell, W. S. S., Scot, J. W., Thomson, M. L. and Winer, J. S. (1947). 'The prediction of the physiological effects of warm and hot environments: the P₄SR index'. *Roy. Naval Pers. Res. Co. Med. Res. Council.* Report 47/391. London.

Mackenzie, A. F. (1947). 'Farmer's calendar'. *Sierra Leone Agric. Note* 17.

McKeown, H. D. J. (1958). 'Dust concentrations in the harmattan'. *Q. J. Roy. Met. Soc.* **84**(361), 280–2.

Macleod, N. H. (1976). 'Dust in the Sahel: cause of drought?' pp. 214–31 in Glantz, M. H. (ed.). *The Politics of Natural Disaster. The Case of the Sahel Drought.*

McQuigg, J. D. (ed.) (1975) *Economic Impacts of Weather Variability*. Univ. Missouri.

McTainsh, G. (1980). 'Harmattan dust deposition in northern Nigeria'. *Nature, (London)*. **286**, 587–8.

Magor, J. I. (1962). *Rainfall as a Factor in the Geographical Distribution of Desert Locust Breeding Areas, with Particular Reference to the Swamp Breeding Area of India and Pakistan*. Unpublished Ph.D. thesis, Edinburgh Univ.

Malone, T. F. (ed.), (1951). *Compendium of Meteorology*. Am. Met. Soc., Boston, Mass.

Mandengue, D. (1965). 'Les perturbations atmosphériques et les précipitations dans la région de Douala'. *Notes de l'Establissement d'Etudes et de Rech. Mét.* 209. Direction de la Mét. Nationale, Paris.

Mangelsdorf, P. C. (1965). 'The evolution of maize'. In Hutchinson, J. (ed.). *Essays on Crop Plant Evolution*. Cambridge Univ. Press.

Manshard, W. (1974). *Tropical Agriculture*. Longman.

Marie-Sainte, Y. (1963). *Influence des pluies sur la date des semis de l'arachide dans diverses régions du Sénégal*. Sécrétariat d'Etat au Plan et au Développement. Aménagement du Territoire, Dakar.

Marinho, T. (1946). 'Esbôco do clima da Guine Portuguesa' *Junta das Missões Geograficas e de Investigacões Coloniais, Anais*. 1, 153–90.

Martin, D. W. (1975). 'Characteristics of West African and Atlantic cloud clusters based on satellite data'. *GATE Rep.* **14**, 182–90. ICSU/WMO.

Mascarenhas, A. (1982). 'Climatological aspects of human settlements in Africa'. pp. 460–77 in WMO *Proceedings of the Technical Conference on Climate in Africa*. WMO 596.

Mason, B. J. (1976). 'Towards the understanding and prediction of climatic variations'. *Q. J. Roy. Met. Soc.* **102**, 473–98.

Mather, J. R. (ed.) (1954). 'The measurement of potential evapotranspiration'. *Publ. in Climatology* 7(1).

Mather, J. R. (1958). 'Potential evapotranspiration and the water balance'. pp. 20–46 in *Inter-Regional Seminar on Hydrologic Forecasting and the Water Balance*. WMO, Belgrade, 1957.

Mather, J. R. (1964). 'Average climatic water balance data of the continents. Africa'. *Publ. in Climatology*, 17(1), Drexel Inst. of Technology, Centerton.

Mather, J. R. (1974). *Climatology: Fundamentals and Applications*. McGraw-Hill.

Mattei, F. (1979). 'Climatic variability and agriculture in the semi-arid tropics'. *WMO* **537**, 475–509.

Matthews, L. S. (1961). 'The unusual storms of Christmas week 1960 in Nigeria'. *Nig. Met. Serv. Tech. Note* 15.

Maunder, W. J. (1970). *The Value of the Weather*. Methuen.

Mbele-Mbong, S. (1972). 'Some characteristics of the rainfall regime at Douala, Cameroun'. *Colorado State Univ., Dep. Atmos. Sci., Paper* 194.

Mbele-Mbong, S. (1974). 'Rainfall in West Central Africa'. *Colorado State Univ., Dep. Atmos. Sci., Paper* 222, 6–20.

Melville, E. H. (1849). *A Residence at Sierra Leone.* John Murray. Republished 1968 by Frank Cass.

Mensching, H. G. (1980). 'The Sahelian zone and the problems of desertification'. *Palaeoecology of Africa* 12, 257–66.

Meteorological Office, (1958). *Tables of Temperature, Relative Humidity, and Precipitation for the World.* 6 Vols. HMSO, London.

Météorologie Nationale, Service Météorologique de la France d'Outre-Mer, (1960). *Annales. Vol. 1. Territoires Français de l'Afrique Noire.* Paris.

Middleton, N. J. (1985). 'Effect of drought on dust production in the Sahel'. *Nature* (London), 316(6027), 431–4.

Migahid, A. M. (1964). 'Vegetation types of Africa as related to climatic conditions'. *WMO Seminar Agrometeorological Problems in Africa. Cairo, 1964.* Cas/Tec 16.

Ministry of Supply (1959). *Comparative Temperatures and Humidities in West Africa.* London; Typescr. in Met. Office, Bracknell.

Ministry of Supply and Department of Scientific and Industrial Research (1949). *Tropic Proofing. Protection Against Deterioration due to Tropical Climates.* HMSO.

Ministry of Supply Tropical Testing Establishment (1954). *Ultra-violet Radiation in Nigeria 1951–1954.* London. Typescr. in Met. Office, Bracknell.

Mkpanmam, E. O. (1965). 'Dust haze in Nigeria'. *Nig. Met. Serv. Proc. Symp. Trop. Met., Oshodi-Lagos, August 1964*, 49–53.

Molga, M. (1962). *Agricultural Meteorology.* Office of Tech. Serv., Washington.

Molski, B. A. (1966). 'Tree belts as a protection against the harmattan in Nigeria'. *Proc. (Pt. 1) Int. Clean Air Congress, London. Paper* 111/6, 54–7.

Monteith, T. L. (1977). 'Soil temperature and crop growth in the tropics'. Paper presented at the *Int. Conf. on the Role of Soil Physical Properties in Maintaining the Productivity of Tropical Soil*, IITA, Ibadan.

Monteny, B. A. and Gosse, G. (1978). 'Trouble (turbidity) atmosphérique et rayonnement solaire en basse Côte d'Ivoire'. *Agric. Met., Int. J.* 19, 121–36.

Monteny, B. A., Humbert, J., L'Homme, J. P. and Kalms, J. M. (1981). 'Le rayonnement net et l'estimation de l'évapotranspiration en Côte d'Ivoire'. *Agric. Met., Int. J.* 23, 45–59.

Moor, H. W. (1939). 'The influence of vegetation on climate in West Africa, with particular reference to the protective aspects of forestry in the Gold Coast'. *Imperial Forestry Institute, Inst. Paper* 17.

Morales, C. (ed.) (1979). *Saharan Dust: Mobilization, Transport, Deposition.* SCOPE 14, ICSU.

More, R. J. (1967). 'Hydrological models and geography'. Ch. 5 in Chorley, R. J. and Haggett, P. (eds.) *Models in Geography.* Methuen.

Morth, H. T. and Jones, P. D. (1977). 'The 1977 summer rains in Sahelian countries'. Univ. East Anglia, Climatic Res. Unit. *Climatic Monitor* 6(3), 98–100.

Mortimore, M. (1983). 'Livestock production'. Chapter 16 in Oguntoyinbo, J. S. et al. (eds.). *A Geography of Nigerian Development.*

Motha, R. P., Leduc, S. K., Steyaert, L. T. and Sakamoto, C. M. (1980). 'Precipitation patterns in West Africa'. *Mon. Weath. Rev.* 108, 1567–78.

Moughtin, J. C. (1985). *Hausa Architecture.* Ethnographica and the Institute of Planning Studies, Univ. Nottingham.

Mukherjee, A. K. (1976). 'Monsoon in Sierra Leone'. *Indian J. Met. Hydrology and Geophys.* 27, 385–90. Met. Dept, Poona, India.

Mukherjee, A. K. and Massaquoi, A. E. (1973). 'Rainfall in Sierra Leone'. *Sierra Leone Met. Dept. Sci. Note* 3.

Mukherjee, A. K. and Moore, H. G. (1972). 'Fog at Freetown airport, Lungi'. *Sierra Leone Met. Dept. Sci. Note* 1.

Mukherjee, A. K. and Moore, H. G. (1973). 'Surface wind at Freetown airport'. *Sierra Leone Met. Dept. Sci. Note* 2.

Mulero, M. A. A. (1973). 'On seasonal distribution of thunderstorm days in Nigeria'. *Nig. Met. Serv. Q. Met. Mag.* 3, 73–8.

Munn, R. E. and Machta, L. (1979), 'Human activities that affect climate'. *WMO* 537, 170–209.

Musk, L. F. (1983). 'Outlook – changeable'. *Geogr. Mag.* LV(10), 532–3.

Namias, J. (1963). 'Surface–atmosphere interactions as fundamental causes of drought and other climatic fluctuations'. pp. 345–59 in *Changes of Climate. Arid Zone Res.* XX. UNESCO.

Navarro, J. (1950). *Les Grains du Nord Est et la Régime des Pluies sur la Côte du Golfe de Guinée.* Duplicated manuscript. Cited in Harrison Church, R. J. *West Africa.*

Newell, R. E. and Kidson, J. W. (1984). 'African mean wind changes between Sahelian wet and dry periods'. *J. Clim.*, 4(1), 27–33.

Nicholl, B. (1965). *A Survey of the Levels of Lightning and Thunderstorm Activity in Sierra Leone, 1964/65.* Unpublished M.Sc. thesis., Univ. Durham.

Nicholson, S. E. (1975). 'Sea-surface temperature variation off the west coast of Africa during Phase 1 of GATE'. Preliminary Scientific Results of the GARP Atlantic Tropical Experiment. Vol. I. *GATE Report* 14, 129–36. WMO.

Nicholson, S. E. (1978). 'Climatic variations in the Sahel and other African regions during the past five centuries'. *J. Arid Envir.* 1, 3–24.

Nicholson, S. E. (1979). 'Saharan climates in historic times' in Williams, M. E. J. and Faure, H. (eds.). *The Sahara and the Nile.*

Nicholson, S. E. (1980). 'The nature of rainfall

fluctuations in subtropical West Africa'. *Mon. Weath. Rev.* **108**, 473–87.

Nicholson, S. E. (1983). 'The climatology of sub-Saharan Africa'. National Res. Council Board on Sci. and Technology for Int. Development, *BOSTID, Rep.* 50, 77–92. National Academy of Sciences, National Academy Press. Washington.

Nicholson, S. E. (1985). 'Sub-saharan rainfall 1981–84'. *J. Clim. Appl. Met.*, **24** (12), 1388–91.

Nicolas, J. P. (1962). 'Etude bioclimatologique du Sénégal et de la Mauritanie'. *Int. Soc. Biomet. Proc., 2nd Int. Bioclim. Congress, London 1960*, 437–60.

Nieuwolt, S. (1977). *Tropical Climatology.* Wiley.

Nigeria Meteorological Services. *Agrometeorological Bulletins*, August 1962 onwards. Lagos.

Nishiyama, I. (1976). 'Effects of temperature on the vegetative growth of rice plants'. pp. 158–83 in *Climate and Rice.* Int. Rice Res. Inst., Los Baños, Philippines.

Norman, R. C. (1959). 'Lightning research in Ghana'. *J. Kumasi College of Technology.* **1**, 31–6.

Norton, C. C., Mosher, F. R. and Hinton, B. (1979). 'An investigation of surface albedo variations during the recent Sahel drought'. *J. Appl. Met.* **18**, 1252–62.

Noyalet, A. (1978). 'Utilisation des images Météosat: genèse et évolution d'une tempête de sable sur l'ouest Africain'. *La Météorologie*, VIᵉ Sér. 14, 113–15.

Nweke, F. I. (1981). 'Weather constraints on the smallholder cropping system of south eastern Nigeria: a case study of two villages in Anambra State'. *Agric. Met., Int. J.* **23**, 309–15.

Oakley, D. J. (1961). *Tropical Houses.* Batsford.

Obasi, G. O. P. (1965a). 'Atmospheric, synoptic and climatological features of the West African region'. *Nig. Met. Serv. Tech. Note* 28.

Obasi, G. O. P. (1965b). 'Thermodynamic and dynamic transformation over Ikeja'. *Nig. Met. Serv. Proc. Symp. Trop. Met., Oshodi-Lagos, 1964*, 1–18.

Obasi, G. O. P. (1972). 'Water balance in Nigeria'. *Nig. Met. Serv. Q. Met. Mag.* **2**(2), 91–127.

Obasi, G. O. P. (1974a). 'The environmental structure of the atmosphere near West African disturbance lines'. *Proc. Inter. Trop. Met. Meeting*, Nairobi, Kenya. Am. Met. Soc.

Obasi, G. O. P. (1974b). 'Some statistics concerning the disturbance lines of West Africa'. *Preprints Symp. Trop. Met. Pt. II.*, Nairobi. Am. Met. Soc. 62–6.

O'Brien, J. J. and Adamec, D. (1977). 'Upwelling in the Gulf of Guinea created by remote winds off Brazil'. Summary in *Bull. Am. Met. Soc.* **58**(10), 1129.

Ochse, J. J. et al. (1961). *Tropical and Sub-tropical Agriculture.* Vols. I and II. Macmillan.

Odumodu, L. O. (1983). 'Rainfall distibution, variability and probability in Plateau State, Nigeria'. *J. Clim.* **3**(4), 385–93.

Office de la Recherche Scientifique et Technique Outre-Mer. Service Hydrologique, Paris.

Précipitations Journalières de l'Origine des Stations à 1965. République du Dahomey, 1973. Côte d'Ivoire, 1973. Mali, 1974. Niger, 1976. Sénégal, 1976. Haute Volta, 1977.

Ofari-Sarpong, E. (1980). 'Impact of drought in Ghana and Upper Volta 1970–1977'. *Clim, Res. Paper* 1, Dep. Geog., Univ, Ghana.

Ofori-Sarpong, E. (1985). 'The nature of rainfall and soil moisture in the northeastern part of Ghana during the 1975–1977 drought'. *Geografiska Annaler*, **67A** (3–4), 177–86.

Ogallo, L. (1978). 'Rainfall in Africa'. *East Af. Inst. Met. Training and Res. Res. Rep. No.* 5/78. Nairobi.

Ogallo, L. (1979). 'Rainfall variability in Africa'. *Mon. Weath. Rev.* **107**, 1133–39.

Ogundipe O. O. (1980). *Conservation of Library Material in Tropical Conditions: the Example of Nigeria.* Int. Fed. Libr. Ass., Ann. Conf., Manila, Philippines. The Hague.

Oguntala, A. B. and Oguntoyinbo, J. S. (1982). 'Urban flooding in Ibadan: a diagnosis of the problem'. *Urban Ecology* 7, 39–46.

Oguntoyinbo, J. S. (1967). 'Rainfall, evaporation and cotton production in Nigeria'. *Nig. Geogr. J.* **14**(2), 185–98.

Oguntoyinbo, J. S. (1970a). 'Reflection coefficient of natural vegetation, crops and urban surfaces in Nigeria'. *Q. J. Roy. Met. Soc.* **96**, 430–41.

Oguntoyinbo, J. S. (1970b). 'Surface measurements of albedo over different agricultural crop surfaces in Nigeria: their spatial and seasonal variability'. *Nig. Geogr. J.* **13**(1), 39–55.

Oguntoyinbo, J. S. (1971a). 'Heat and water balance studies in Nigeria'. *WMO* **301**(2), 1–9.

Oguntoyinbo, J. S. (1971b). 'Seasonal variations of the radiative fluxes over cocoa in Nigeria'. *Nig. Geogr. J.* **14**(2), 185–98.

Oguntoyinbo, J. S. (1973). 'Rainfall, drought and food supply in south-western Nigeria'. *Savanna* **2**, 115–20.

Oguntoyinbo, J. S. (1974). 'Land usc and reflection coefficient (albedo) map for southern parts of Nigeria'. *Agric. Met., Int. J.* **13**, 227–37.

Oguntoyinbo, J. S. (1976). 'Solar energy as a potential environmental resource in sub-Saharan Africa'. *Nig. Geogr. J.* **19**, 165–94.

Oguntoyinbo, J. S. (1979). 'An albedo map of Nigeria'. *Nig. Geogr. J.* **22**, 185–91.

Oguntoyinbo, J. S. (1981). 'Climatic variability and food crop production in West Africa'. *Geojournal* **5**(2), 139–49.

Oguntoyinbo, J. S. (1983). *Climate and Mankind.* Inaugural lecture, Univ. Ibadan, Nigeria, 24 March, 1982.

Oguntoyinbo, J. S., Areola, O. O. and Filani, M. O. (eds.), (1983). *A Geography of Nigerian Development.* Heinemann.

Oguntoyinbo, J. S. and Odingo, R. S. (1979). 'Climatic variability and land use. An African perspective'. *WMO* **537**, 552–80.

Oguntoyinbo, J. S. and Ojo, O. (1972). 'Estimating net radiation in West Africa'. Unpublished monograph.

Oguntoyinbo, J. S. and Richards, P. (1977). 'The extent and intensity of the 1969–73 drought in Nigeria: a provisional analysis' in Dalby, D. et al. (eds.). *Drought in Africa*.

Oguntoyinbo, J. S. and Richards, P. (1978). 'Drought and the Nigerian farmer'. *J. Arid. Envir.* **1**, 165–94.

Ojo, O. (1969). 'Potential evapotranspiration and the water balance in West Africa: an alternative method of Penman'. *Archiv. für Met. Geophys. Bioklim.* **B17**, 239–60.

Ojo, O. (1970a). 'The distribution of mean monthly precipitable water vapour and annual precipitation efficiency in Nigeria'. *Archiv. für Met. Geophys. Bioklim.* **B18**, 221–38.

Ojo, O. (1970b). 'The seasonal march of the spatial patterns of global and net radiation in West Africa'. *J. Trop. Geog.* **30**, 48–62.

Ojo, O. (1971). 'Bovine energy balance climatology and livestock potential in Nigeria'. *Agric. Met.* **8**, 353–69.

Ojo, O. (1972a) 'The distribution of net radiation in Nigeria'. *Nig. Geogr. J.* **15**(2), 115–25.

Ojo, O. (1972b). 'Energy balance climatology of man in Ibadan, Nigeria'. *Int. Geogr. Congress Proc.*, Montreal. 172–4.

Ojo, O. (1973). 'The distribution of solar and net heat load on cattle in Nigeria'. *J. Trop. Geog.* **36**, 50–9.

Ojo, O. (1977). *The Climates of West Africa*. Heinemann.

Ojo, O. and Agun, A. (1972). 'The spatial distribution of rainfall and rainfall variabilities in metropolitan Lagos'. *Nig. Met. Serv. Q. Met. Mag.* 2, 158–77.

Ojo, S. O. (1982). 'Rainfall, drought and the dynamics of population in West Africa'. *Univ. Tsukuba, Inst. Geoscience. Clim. Notes* 29, 105–8.

Okulaja, F. O. (1968). 'The frequency distribution of Lagos/Ikeja wind gusts'. *J. Appl. Met.* **7**, 379–83.

Okulaja, F. O. (1970). 'Synoptic flow perturbations over West Africa'. *Tellus*, **22**(6), 663–79.

Olaniran, O. J. (1981). 'Empirical methods of computing potential maximum evapotranspiration'. *Archiv. für Met. Geophys. Bioklim.* **30A**, 369–81.

Olaniran, O. J. (1982a). 'The physiological climate of Ilorin, Nigeria'. *Archiv. für Met. Geophys. Bioklim.* **31B**(3), 287–99.

Olaniran, O. J. (1982b). 'Problems in the measurement of rainfall – an experiment at Ilorin, Nigeria'. *Weather*, **37**(7), 201–4.

Olaniran, O. J. (1983a). 'Problems in the measurement of pan evaporation in Nigeria'. *Nig. Met. J.* 1 (1), 7–12.

Olaniran, O. J. (1983b). 'Flood generating mechanisms at Ilorin, Nigeria. *Geojournal* (Wiesbaden) 7(3), 271–7.

Olgyay, V. (1952). 'Bioclimatic approach to architecture'. *BRAB Res. Conf. Rep.* 5, 13–23.

Olgyay, V. (1963). *Design with Climate: Bioclimatic Approach to Architectural Regionalism*. Princeton.

Oliver, J. E. (1973). *Climate and Man's Environment: An Introduction to Applied Climatology*. Wiley.

Oliver, J. E. (1981). *Climatology: Selected Applications*. Arnold.

Oliver, P. (ed.) (1971). *Shelter in Africa*. Barrie and Jenkins, London.

Omar, M. H. (1980). *The Economic Value of Agrometeorological Information and Advice*. WMO 526. Tech. Note 164.

Omolayo, S. A. (1983). 'A statistical study of the distribution of the largest daily rainfall in Nigeria'. *Nig. Met. J.* **1**(1), 44–57.

Omorinbola, E. O. (1964). 'Ground water resources in tropical African regoliths'. In Walling, D.E., Forster, S. and Warzel, A. *Challenges in African Hydrology and Water Resources*.

Omotosho, J. B. (1984). 'Spatial and seasonal variation of line squalls over West Africa'. *Archiv. für Met. Geophys. Bioklim.* **33A** (2–3), 143–50.

Omotosho, J. B. (1985). 'The separate contributions of line squalls, thunderstorms and the monsoon to the total rainfall in Nigeria'. *J. Clim.* **5**(5), 543–52.

Oshodi, F. R. (1966). *Biometeorological Studies of Nigerian Crops*. Nig. Met. Serv., Lagos.

Oshodi, F. R. (1971). 'A study of pentade normals of rainfall in Nigeria'. *Nig. Met. Serv. Q. Met. Mag.* 1(4), 44–67.

Ouedraogo, J. P., Ouedraogo, J. B. G. and Baldy, Ch. M. (1981). 'Premières données sur le rayonnement global et la durée d'insolation en Haute-Volta'. *La Météorologie*, **25**, 123–34.

Oyebande, S. L. and Oguntoyinbo, J. S. (1970). 'An analysis of rainfall patterns in the south-western states of Nigeria'. *Nig. Geogr. J.* **13**(2), 141–62.

Oyegun, R. O. (1982a). 'Discomfort in the use of domestic tap water in a developing tropical city'. *Weather*, **37**(2), 41–3.

Oyegun, R. O. (1982b) 'Insolation hazard in unshaded car parks in the tropics: a case study at Ilorin, Nigeria'. *Weather*, **37**(9), 260–2.

Page, J. K. et al. (1964). 'Indoor climate in arid and humid zones'. Symposium during the 3rd Int. Biomet. Congress, Pau (France) 1–7 Sept. 1963. *Int. J. Biomet.* **8**(2), 93–163.

Pagney, P. and Besancenot, J-P. (1982). 'Human bioclimatology in tropical zones. Discussion and practical implications (1)'. pp. 56–63 in Yoshino, M. M. (ed.). Proc. of the Tokyo meetings on tropical climatology and human settlements. *Tsukuba Univ., Inst. Geoscience, Clim. Notes* 29.

Palmen, E. (1951). 'The role of atmospheric disturbances in the general circulation'. *Q. J. Roy. Met. Soc.* **77**, 337–54.

Palmer, W. C. and Denny, L. M. (1971).

Drought Bibliography. *N.O.A.A. Tech. Mem. EDS.* 20.

Palmieri, S. and Stefanelli, R. (1979). ('Extreme precipitations in a region of Nigeria: a comparison between results by Gumbel Method and Meteorological Maximization Method'.) *Rivista di Meteorologia Aeronautica* **39**, 237–45 (in Italian).

Palutikof, J. R., Lough, J. M. and Farmer, G. (1981). 'Senegal river runoff'. *Nature* **293**, 414.

Papadakis, J. (1961). 'Crop ecology in West Africa'. *FAO: UN Publ. MR/16439/1.* Vol. 2.

Papadakis, J. (1964). 'Climatic belts and agriculture in West Africa'. *W.M.O. Seminar. Agrometeorological Problems in Africa.* Cairo 1964. Cas/Tec. 22.

Parker, D. E., Folland, C. K. and Palmer, T. N. (1986). 'African drought and anomalous sea-surface temperatures'. *WMO Prog. Long-range Forecast Res. Rep. Ser.* no. 6, vol. 1, 103–12.

Pastuh, V. P. and Vinnikova, E. B. (1965). ('Some features of the distribution of thunder over the African continent'). *Glavnaja Geofiziceskaja Observatorija, Trudy Vypusk* **182**, 88–93. Leningrad.

Patalagoity, J. (undated). 'Climatologie de la Côte d'Ivoire. Application à la protection aéronautique'. *ASECNA Publ.* 48.

Patnaik, J. K., Nganga, J. K. and Kiangi, P. M. R. (1980). 'Atmospheric pollution potentials over Africa'. *East Af. Inst. for Met. Training and Res. Res. Rep. No.* 1/80, Nairobi.

Payne, S. W. and McGarry, M. M. (1977). 'The relationship of satellite infrared convective activity to easterly waves over West Africa and the adjacent ocean during Phase III of GATE'. *Mon. Weath. Rev.* **105**, 413–20.

Peczely, G. (1984). 'Relationship between the oscillations of precipitation in the Sahel zone in Africa and weather anomalies of the earth'. *Időjárás* (az Országos Meteorológiai Szolgálat Folyóirata, Budapest), **88**(4), 185–92.

Pedgley, D. E. and Krishnamurti, T. N. (1976). 'Structure and behaviour of a monsoon cyclone over West Africa'. *Mon. Weath. Rev.* **104**, 149–67.

Peel, C. (1954). 'Thermal conditions in African dwellings in Sierra Leone'. *J. Inst. Heating and Ventilation Engineers* **22**, 125–43.

Peel, C. (1961). 'Thermal comfort zones in northern Nigeria: an investigation into the physiological reactions of nursing students to their thermal environment'. *J. Trop. Medicine and Hygiene*, May, 113–21.

Penman, H. L. (1948). 'Natural evaporation from open water, bare soil and grass'. *Proc. Roy. Soc.* **A. 193**, 120–45.

Penman, H. L. (1955). *Humidity*. Inst. of Physics, London.

Penman, H. L. (1956). 'Evaporation: an introductory survey'. *Netherlands J. Agri. Sci.* **4**(1), 9–29.

Perkins, W. A. (1950). 'The rainfall of Ibadan'. *Farm and Forest* **10**, 26–32.

Phillips, R. W. (1950). *Improving Livestock under Tropical and Sub-tropical Conditions*. FAO.

Philpotts, H. (1967). 'The effect of soil temperature on the nodulation of cowpea (*Vigna sinensis*)'. *Aust. J. Exper. Agric. Animal Husbandry*, **7**, 372–6.

Picq, P. (1936a). 'Les chutes de grêle au Sénégal en 1934'. *Annals. du Phys. du Globe de la Fr. d'Outre-Mer.* **3**, 71–2.

Picq, P. (1936b). 'Notes sur l'origine des faibles visibilités et des brouillards d'hiver sur la côte du bas Sénégal'. *Annls. du Phys. du Globe de la Fr. d'Outre-Mer.* **3**, 156.

Picq, P. (1936c). 'Note sur divers phénomènes météorologiques au Sénégal'. *Bull. du Comité d'Etudes Historiques et Sci. de l'Afr. Occidentale Fr.* Gorée. **19**(1), 140–9.

Piery, M. (1934). *Traité de Climatologie Biologique et Médicale*. Masson et Cie, Paris.

Pinty, B., Szejwach, G. and Desbois, M. (1984). 'Investigation of surface albedo variations over West Africa from METEOSAT'. *Am. Met. Soc. Conf. on Satellite/Remote Sensing Applications. Clearwater Beach, Florida 1984*, 77–9.

Pople, W. and Mensah, M. A. (1971). 'Evaporation as the upwelling mechanism in Ghanaian coastal waters'. *Nature*, (London) **233**, 18–20.

Porges, L. (1964). *Eléments de Bibliographie Sénégalaise 1959–63*. Archives Nationales, Centre de Documentation, Dakar.

Portig, W. H. (1963). 'Thunderstorm frequency and amount of precipitation in the tropics, especially in the African and Indian monsoon regions'. *Archiv. für Met. Geophys. Bioklim.* **B13**, 21–35.

Priestly, C. H. B. (1966). 'The limitation of temperature by evaporation in hot climates'. *Agric. Met.* **3**, 214–46.

Puckridge, D. W. (1975). 'Biometeorological aspects of crop growth and development in arid and semi-arid zones'. *Progress in Biomet. Division C., Program Plant Biomet.* Vol. 1, 227–39. Amsterdam.

Pugh, J. C. (1952). 'Rainfall reliability in Nigeria'. *Proc. 7th Int. Geogr. Union Congress*, 36–41.

Pullan, R. A. (1962). 'The concept of the middle belt in Nigeria – an attempt at a climatic definition'. *Nig. Geogr. J.* **5**(1), 39–52.

Purseglove, J. W. (1968). *Tropical Crops. Dicotyledons* Vols. I and II. Longman.

Purseglove, J. W. (1972). *Tropical Crops. Monocotyledons* Vols. I and II. Longman.

Rainey, R. C. (1963). 'Meteorology and the migration of the desert locust'. *WMO* **138**, TP. 64. *Tech. Note* 54.

Rainey, R. C. (1969). 'Effects of atmospheric conditions on insect movement'. *Q. J. Roy. Met. Soc.* **95**, 424–33.

Ramusio, G. B. (1550). *Delle Navigationi et Viaggi nel qual se contiene la Descrittione dell' Africa*.

Ransom, W. H. (1960) 'Buildings for the storage of crops in warm climates'. Dept. Sci. Industrial Res.

Building Res. Station. Tropical Building Studies No. 2.

Ransom, W. H. (1962). 'Solar radiation: thermal effects on building materials'. BRS Tropical Building Studies No. 3.

Rapp, A. (1974). 'A review of desertization in Africa: water, vegetation and man'. Secretariat for Int. Ecology, Sweden (SIES), Rep. 1.

Rapp, G. M. (1952). 'Performance and properties of materials in hot climates'. BRAB. Res. Conf. Rep. 5, 109–20.

Rebert, J. P. (1978). Variabilité des Conditions de Surface dans l'Upwelling Ouest-Africain. Symp. sur le courant des Canaries, 11 April 1978, Las Palmas. Centre de Rech. Océanographique, Dakar-Thiaroye, Sénégal.

Reed, R. J. (1978). 'The structure and behaviour of easterly waves over West Africa and the Atlantic'. pp. 57–71 in Shaw, D. S. (ed.) Meteorology over the Tropical Oceans. Roy. Met. Soc.

Reiter, E. R. (1963). Jet-stream Meteorology. Univ. Chicago Press.

Reiter, E. R. (1979). 'Climatic fluctuations on the African continent'. Rivista di Meteorologia Aeronautica, 39, 5–14. Rome.

Rennick, M. A. (1976). 'The generation of African waves'. J. Atmos. Sci. 33, 1955–69.

Richards, P. (ed.) (1975). African environment: problems and perspectives. Int. Af. Inst. London.

Richards, P. (1985). Indigenous Agricultural Revolution. Hutchinson.

Riehl, H. (1965). Introduction to the Atmosphere. McGraw-Hill.

Riehl, H. (1979). Climate and Weather in the Tropics. Academic Press.

Riehl, H., Rossignol, D. and Lückefedt, W. (1974). 'Sur la structure et le maintien des lignes de grains d'Afrique de l'ouest'. ASECNA Publ. 35.

Rijks, D. A. (1965). 'The use of water by cotton in Abyan, southern Arabia'. J. Appl. Ecology, 2, 317–42.

Rijks, D. A. (1967). 'Water use by irrigated cotton in the Sudan. I. Reflection of shortwave radiation'. J. Appl. Ecology, 4, 561–8.

Rijks, D. A. (1973). 'Analysis of rainfall reliability in the Senegal river basin'. WMO 340, 32–41.

Rijks, D. A. and Walker, J. T. (1968). 'Evaluation and computation of potential evaporation in the tropics'. Experimental Agric. 4(4), 351–8.

Ripley, E. A. (1976). 'Drought in the Sahara : insufficient biogeophysical feedback?' Science 191, 100.

Ritchie, R. M. and Parker, B. F. (1965). 'Poultry in hot climates'. Agric. Engineering. St Joseph, Michigan. 46(2), 83–4.

Robertson, G. W. (1980). The Role of Agrometeorology in Agricultural Development. WMO 536.

Rodier, J. (1963). Bibliography of African Hydrology. UNESCO Natural Resources Res. II..

Rodier, J. (1964). Régimes Hydrologiques de l'Afrique

Noire à l'Ouest du Congo. ORSTOM, Paris.

Rossignol-Strick, M. (1983). 'African monsoons, an immediate climate response to orbital insolation'. Nature (London) 304, 5921, 46–9.

Russell, E. W. (1967). 'Climate and crop yields in the tropics. A review of progress in reducing some harmful effects of climate on crop production'. Cotton Growers Review, 44, 87–99.

Ruthenberg, H. (1976). Farming Systems in the Tropics. Oxford Univ. Press.

Rydings, H. A. (1961). The Bibliographies of West Africa. Ibadan Univ. Press.

Sadler, J. C. (1963). 'Utilisation of meteorological satellite cloud data in tropical meteorology' in Wexler, H. and Caskey, J. E. (eds.). Rocket and Satellite Meteorology. North Holland.

Sah, R. (1979). 'Priorities of developing countries in weather and climate'. World Development 7, 337–47.

Sahel Documentation Centre, (1977). Sahel : Bibliographic Bulletin. Michigan State Univ. Libraries.

Saini, B. S. (1973). Building Environment. An Illustrated Analysis of Problems in Hot Dry Lands. Angus and Robertson (Sydney).

Salou, O. A. (1986). 'An analysis of rainfall amount, duration and intensity in Port Harcourt, Nigeria'. J. Met 11 (110), 181–7.

Sampson, H. C. (1936). 'Cultivated crop plants of the British Empire and the Anglo-Egyptian Sudan'. Bull. Miscellaneous Information. Additional Series XII. Royal Botanic Gardens, Kew, London.

Samways, J. (1975). 'A synoptic account of an occurrence of dense harmattan dust in Kano in February 1974'. Savanna 4, 187–90.

Samways, J. (1976). 'Ill wind over Africa'. Geogr. Mag. 48(4), 218–20.

Sansom, H. W. (1966). 'The occurrence and distribution of hail in Africa'. Met. Mag. 95, 212–18.

Sargent, F. and Tromp, S. W. (1964). 'A survey of human biometeorology'. WMO 160, TP. 78. Tech. Note 65.

Sarle, C. F. (1946). 'Applications of climatology to building construction and agriculture'. Bull. Am. Met. Soc. 27, 210–15.

Saunders, J. H. (1961). The Wild Species of Gossypium and their Evolutionary History. Oxford Univ. Press.

Sayer, H. J. (1962). 'The desert locust and tropical convergence'. Nature (London) 194, 330–6.

Schmiedecken, W. (1979). ('Humidity and cultivated plants – an attempt at parallelizing zones of humidity and optimal locations of selected cultivated plants in the tropics'). Erdkunde 33, 266–74 (in German).

Schnell, R. C. (1975). 'Overgrazing and biogenic ice nuclei: a physical link in the Sahelian drought?' Am. Met. Soc. 12th Agric. Forest Met. Conf. Tucson, Arizona, April 1975, 29–30.

Schove, D. J. (1946). 'A further contribution to the

meteorology of Nigeria'. *Q. J. Roy. Met. Soc.* **72**, 105–10.

Schove, D. J. (1977). 'African droughts and the spectrum of time'. Chapter 4, in Dalby, D., et al. *Drought in Africa*.

Schroder, R. (1979). (Climatic geography of the Cameroons). *Deutschen Wetterdienstes, Seewetteramt, Einzelveröff-entlichungen*, **99** (in German).

Schumann, T. E. W. and Mostert, J. S. (1949). 'On the variability and reliability of precipitation'. *Bull. Am. Met. Soc.* **30**(3), 110–13.

Sealey, A. (1979). *Introduction to Building Climatology*. Commonwealth Association of Architects, London.

Seba, D. B. and Prospero, J. M. (1971). 'Pesticides in the lower atmosphere of the northern equatorial Atlantic Ocean'. *Atmos. Envir.* **5**, 1043–5.

Seck, A. (1962). 'Le "Heug" ou pluie de saison sèche au Sénégal'. *Annls. Géog.* **71**, 225–46.

Seck M. (1974). 'Artificial precipitations in Senegal'. Am. Met. Soc. *Int. Trop. Met. Meeting 1974, Nairobi*, Pt II, 111–13.

Seginer, I. (1969). 'The effect of albedo on evapo-transpiration'. *Agric. Met.* **6**, 5–31.

Sellers, W. D. (1965). *Physical Climatology*. Univ. Chicago Press.

Service Météorologique de l'Afrique Occidentale Française (1959). *Impracticabilité due aux Conditions Météorologiques d'un Choix d'Aérodromes d'Afrique Occidentale*. Dakar.

Servicio Meteorologico Nacional, Lisbon. *Anuarro Climatologico de Portugal, II, Territorios Ultramarinos* Vol. I (1947) onwards.

Shaw, L. H. and Durost, D. D. (1962). 'Measuring the effects of weather on agricultural output'. *U.S. Dept. Agric., ERS* – 72, Washington DC.

Sidikou, A. H. (1977). 'The Zamara people of Zarmaganda; their adaptive strategy and its limits in the face of the current drought (1965–76)' in Van Apeldoorn, J. (ed.). *The Aftermath of the 1972–74 Drought in Nigeria*. Ahmadu Bello Univ., Zaria.

Silverberg, R. (1969). *The Challenge of Climate*. Meredith Press.

Simmonds, N. W. (1962). *The Evolution of Bananas*. Longmans.

Simmons, A. J. (1977). 'A note on the instability of the African easterly jet'. *J. Atmos. Sci.* **34**, 1670–4.

Smith, A. C. (1967). 'The presence of primitive angiosperms in the Amazon basin and their significance in indicating migrational routes'. *Atlas do Simposia Sobre a Biota Amazonica*, **4**, 37–59.

Smith, D. E. (1937). 'Dust devils and desiccation in West Africa'. *Met. Mag.* **72**, 83.

Smith, D. E. (Compiler), (1961). *Weather of the Sierra Leone Peninsula*. Government Printer, Freetown.

Smith, G. (1984). 'Climate'. Ch. 2 in Cloudsley-Thompson, J. L. (ed.) *Key Environments: Sahara Desert*.

Smith, K. (1975). *Principles of Applied Climatology*. McGraw-Hill.

Smithsonian Institute, Miscellaneous Collections 79, 1927; 90, 1934; 105, 1947. *World Weather Records*. Washington.

Snijders, T. A. B. (1983). *A Study of the Variability in Space and Time of Upper Volta Rainfall*. Lisbon, Instituto Nacional de Meteorologia e Geofisica, II Int. Meeting on Statistical Climatology, Paper 3.5.

Soboyejo, A. B. O. (1971). 'Distribution of extreme winds in Nigeria'. *The Nig. Engineer*, **7**(3), 21–34.

Sonuga, J. O. (1977). 'Hydrological aspects of the drought event in Nigeria 1972/1973'. *Bull. Sci. Hydro.* 22, 487–502.

Souter, R. K. and Emerson, J. B. (1952). 'Summary of Available Hail Literature and the Effect of Hail on Aircraft in Flight'. National Advisory Committee for Aeronautics, Washington. *Tech. Note* 2734.

South African Council for Scientific and Industrial Research, (1957). *Symposium on Design for Tropical Living, 18 Oct. 1957*. SACSIR, Pretoria. Univ. Natal, Durban.

Sperling, R. (1967). 'Non-traditional building for warm climates'. *BRS Overseas Building Notes* 117.

Spottiswood, H. E. (1940). 'The general trend and distribution of rainfall in Nigeria'. *Nig. Field* **9**, 50–6.

Stamp, L. D. (ed.) (1956). *Natural Resources, Food and Population in Inter-tropical Africa*. Int. Geogr. Union Symp., London. Geogr. Publications.

Stanhill, G. (1962). 'The use of the Piché evaporimeter in the calculation of evaporation'. *Q. J. Roy. Met. Soc.* **88**, 80–2.

Stanhill, G. (1963). 'The accuracy of meteorological estimates of evapotranspiration in Nigeria'. *Inst. Water Engineers J.* **17**, 36–44, London and *Nig. Met. Serv. Tech. Note* 31.

Stanhill, G., Fuchs, M. and Oguntoyinbo, J. S. (1971). 'The accuracy of field measurements of solar reflectivity'. *Archiv. für Met. Geophys. Bioklim.* **19B**, 113–32.

Stanhill, G., Hofstede, G. H. and Kalma, J. D. (1966). 'Radiation balance of natural and agricultural vegetation'. *Q. J. Roy. Met. Soc.* **92**, 128–40.

State Hydrological Institute (1974). *Atlas of World Water Balance*. Gidrometeoizdat, Leningrad, and UNESCO Press, Paris.

Stebbing, E. P. (1938). 'The man-made desert in Africa'. *Indian Forester* **64**(6), 328–93.

Steele, W. M. (1972). *Cowpeas in Africa*. Unpublished Ph.D. thesis, Univ. Reading.

Steigner, J. M. and Ingham, M. C. (1971). 'Surface winds of the south eastern tropical Atlantic Ocean'. *NOAA Tech. Rep* NMFS, SSRF – 643.

Steinhorst, R. K. and Morris, J. W. (1977). 'World climate patterns in grassland and savanna and their relation to growing seasons'. *Bothalia* **12**, 261–5.

Stephenson, P. M. (1963). 'An index of comfort for Singapore'. *Met. Mag.* **92**, 338–45.

Stern, R. D., Dennett, M. D. and Gabbutt, D. J.

(1981). 'The start of the rains in West Africa'. *J. Clim.* **1**, 59–68.

Stranz, D. (1975). 'Über den Regen in Afrika und die Trockenheit der letzten Jahre im Sahel (1967–74)'. (On the rains in Africa and the drought in the Sahel in recent years 1967–74). *Deutschen Wetterdienstes, Seewetteramt, Einzelveröffentlichungen*, No. 88. Hamburg.

Stranz, D. (1977). 'Der Monsunregen in Westafrika'. *D. Wetterdienstes, Seewetteramt, der Wetterlotse*, **29**, 137–42.

Stranz, D. (1981). 'Die Regenzeit 1980 in der Sahelzone'. *D. Wetterdienstes, Seewetteramt, der Wetterlotse*, **33**, 37–41.

Stranz, D. (1986). 'Niederschalagsdefizit nur im Sahel?' (Rainfall deficit only in the Sahel?'). *D. Wetterdienstes, Seewetteramt, der Wetterlotse*, **38** (473/4), 72–81.

Suchel, J–B. (1978a). 'Les températures au Cameroun. Essai d'analyse sommaire au moyen de quelques cartes'. *Univ. Dijon Cent. de Rech. de Climatologie, Cahier 6, Pap.* 1.

Suchel, J–B. (1978b) 'Une méthode graphique de classification des climats Camerounais, en fonction de l'écologie humaine'. *Univ. Dijon Cent. de Rech. de Climatologie, Cahier 6, Pap.* 3.

Sulman, F. G. (1976). *Health, Weather and Climate.* S. Karger, Basel.

Sulman, F. G., Levy, D., Levy, A., Pfeifer, Y., Superstine, E. and Tal, E. (1974). 'Air-ionometry of hot, dry, desert winds (Sharav) and treatment with air ions of weather-sensitive subjects'. *Int. J. Biomet.* **18**(4), 313–18.

Summerfield, R. J., Minchin, F. R. and Roberts, E. H. (1978). *Report No. 17, Reading University – IITA Tropical Grain Legume Physiology Project* 197.

Sumner, E. J. and Tunnell, G. A. (1949). 'Determination of the true mean vapour pressure of the atmosphere from temperature and hygrometric data'. *Met. Mag.* **78**, 258–63.

Swami, K. (1970). 'Importance of daily and synoptic climatic analyses in ecological studies : an example from Nigeria'. McGill Univ., Dept. Geog. *Clim. Bull.* **8**, 40–57. Montreal.

Swami, K. (1973). 'Moisture conditions in the savanna region of West Africa'. McGill Univ., Dept. Geog., *Savanna Res. Series* 18. *Climatological Res. Series* 8.

Swan, A. D. (1958). 'The West African monsoon'. *Ghana Met. Dept., Dept Note* 8. Accra.

Swartman, R. K. and Ogunlade, O. (1967). 'Solar radiation estimates from common parameters'. *Solar Energy*, **11**, 170–2.

Sykes, S. (1972). *Lake Chad.* Methuen.

Tabor, H. (1962). 'Solar energy'. *The Problems of the Arid Zone. Arid Zone Res.* **18**, 259–70.

Tandoh, S. E. (1973). 'Probability distribution of annual rainfall in Ghana'. *Ghana Met. Serv. Dept. Note* 21.

Tannehill, I. R. (1947). *Drought, its Causes and Effects.*

Princeton Univ. Press.

Terjung, W. H. (1967). 'The geographical application of some selected physioclimatic indices to Africa'. *Int. J. Biomet.* **11**, 5–19.

Terjung, W. H. (1968). 'World patterns of the distribution of the monthly comfort index'. *Int. J. Biomet.* **12**, 119–51.

Tetzlaff, G., Peters, M. and Adams, L. J. (1985). ('Meteorological aspects of the Sahel drought'.) *Die Erde*, **116**, 109–20 (in German).

Thom, E. C. (1959). 'The discomfort index'. *Weatherwise* **12**, 57–60.

Thompson, B. W. (1965). *The Climate of Africa* (atlas). Oxford Univ. Press.

Thornthwaite, C. W. (1954). 'A re-examination of the concept and measurement of potential evapotranspiration'. *Publ. in Climatology* 7(1), 200–99. C. W. Thornthwaite Associates, Laboratory of Climatology, Centerton, New Jersey, USA.

Thornthwaite, C. W. and Mather, J. R. (1955). 'The water balance'. *Publ. in Climatology* 8(1), 104pp.

Thornthwaite, C. W. and Mather, J. R. (1957). 'Instructions and tables for computing potential evapotranspiration and the water balance'. *Publ. in Climatology* **X**(3), 181–311.

Thornthwaite, C. W. and Mather, J. R. (1962). 'Average climatic water balance data of the continents. Pt. I, Africa'. *Publ. in Climatology* 15(2).

Tout, D. G. (1979). 'A comparison between the 1949–1965 rainfall at Kabara and Timbouctou, Mali'. *Weather* 34, 175–84.

Townsend, D. (1977). 'Analyses of intense rainfall including the determination of maximum probability precipitation and return periods'. Dept. Hydrometeorological Serv., *Tech. Rep.* 2. Banjul, Gambia.

Trewartha, G. T. (1962). *The Earth's Problem Climates.* Methuen.

Tromp, S. W. (1963a). 'Human biometeorology'. *Int. J. Biomet.* 7(2), 145–58.

Tromp, S. W. (1963b). *Medical Biometeorology.* Elsevier.

Tromp, S. W. (1980). *Biometeorology.* Heyden.

Tromp, S. W. and Weihe, W. H. (1966). *Biometeorology.* Vols. I and II. Pergamon.

Tullot, I. F. (1951). *El Clima de la Posesiones Españolas del Golfo de Guinea.* Instituto de Estudios Africanos, Madrid.

Tuzet, A., Moser, W. and Raschke, E. (1984). 'Estimating global radiation at the surface from Meteosat-data in the Sahel region'. *J. Rech. Atmos.* **18**(1), 31–9.

United Nations, (1985). *UN Demographic Year Book 1983.* UN, New York.

UN Department of Economic and Social Affairs (1971). *Design of Low-cost Housing and Community Facilities.* Vol 1. *Climate and House Design.* UN, New York.

UNESCO. *Bibliography of African Hydrology.* UNESCO.

Urdahl, T. H. (1952). 'Dehumidification'. *BRAB Res. Conf. Rep.* 5, 149.

US Agency for International Development (1979). *Weather-Crop-Field Relationships in Drought-prone Countries of sub-Saharan Africa.* USAID, Office of Foreign Disaster Assistance, Washington DC, NOAA and Univ. Missouri.

US Air Force Environmental Technical Applications Center (1968 and 1969). *Worldwide Airfield Climatic Data, Vol. IX, (1 and 2). Africa (Northern half), Africa (Southern half).* Washington DC.

US Building Research Advisory Board (1953). *Preliminary Bibliography. Housing and Building in Hot-Humid and Hot-Dry Climates.* BRAB.

US Department of Commerce (1967). *World Weather Records, 1951–1960. Vol. 5. Africa.* ESSA, Washington, DC.

University College, Freetown, Sierra Leone (1963–4). *Thunderstorm Survey, 1963 Jan.–1963 Dec.*

University of Sierra Leone, Fourah Bay College, Department of Geography (1963). Meteorological station at Kortright, Freetown. *Monthly* and *Quarterly Reports.* Mimeo.

University of Sierra Leone, Fourah Bay College, Department of Geography (1963). Meteorological Station at Kortright, Freetown. *Summary tables* Jan. 1960–Sept. 1963. Mimeo.

Ussher, A. K. L. (1969). *Climatic Maps of Ghana for Agriculture.* Ghana Met. Serv., Legon.

Vasic, R. S. (1977). 'Use of climatology as a forecasting aid – 1'. In *Lectures on Forecasting of Tropical Weather.* WMO 492.

Viltard, A. and de Felice P. (1979). 'Statistical analysis of wind velocity in an easterly wave over West Africa'. *Mon. Weath. Rev.* **107**, 1320–7.

Vitkevich, V. I. (1963). *Agricultural Meteorology.* Office of Tech. Services, Washington.

Vittori, M. (1968). 'Note sur les invasions polaires en altitude sur le Sénégal et la Mauritanie'. *ASECNA Publ.* **10**, Dakar.

Voiron, H. (1964). 'Quelques aspects de la météorologie dynamique en Afrique occidentale'. *La Météorologie,* 215–31. (Also in *ASECNA Publ.* **3**).

Waddy, B. B. (1952). 'Climate and respiratory infections'. *Lancet* **263**, 674–7.

Waitt, A. W. (1961). 'Review of yam research in Nigeria, 1920–1961'. *Federal Dep. Agric. Res. Monograph* 31. Ibadan.

Walker, H. E. (1946). 'Housing schemes in West, East and South Africa. Report and plans'. *Public Works Dep., Lagos. Tech. Paper* 12.

Walker, H. O. (1957a). 'The weather and climate of Ghana'. *Ghana Met. Dep., Dep. Note* 5, Accra.

Walker, H. O. (1957b). 'Estimates of evaporation and potential evapotranspiration'. *Ghana Met. Dep. Dep. Note* 6, Accra.

Walker, H. O. (1959). 'Temperature variations in the coastal districts of Ghana'. *Ghana Met. Dep. Dep. Note* 13, Accra.

Walker, H. O. (1962). 'Weather and climate'. In Wills, J. B. (ed.). *Agriculture and Land Use in Ghana.* Oxford Univ. Press.

Walling, D. E., Forster, S. and Warzel, A. (1964). *Challenges in African Hydrology and Water Resources.* Proc. Harare Symposium. Int. Ass. Hydrological Sciences. No. 144.

Walter, M. W. (1958). 'Percentage of average seasonal values of rainfall'. *Ghana Met. Dep., Dep. Note* 10, Accra.

Walter, M. W. (1959a). 'Some new presentations of the seasonal rainfall of Ghana'. *Ghana Met. Dep., Dep. Note* 12, Accra.

Walter, M. W. (1959b). 'Dependability of rainfall in Ghana'. *Ghana Met. Dep., Dep. Note No.* 14, Accra.

Walter, M. W. (1967). 'The length of the rainy season in Nigeria'. *Nig. Geogr. J.* **10**(2), 123–8.

Wang, Jen-Yu. (1963). *Agriculture Meteorology.* Pacemaker Press, Milwaukee.

Ward, R. C. (1967). *Principles of Hydrology.* McGraw-Hill.

Ward, R. C. (1971). 'Measuring evapotranspiration: a review'. *J. Hydrology* **13**, 1–21.

Warner, J. (1968). 'A reduction in rainfall associated with smoke from sugar-cane fires – an inadvertent weather modification?' *J. Appl. Met.* **7**, 247–51.

Waterlow, J. C. (1982). 'Nutrient needs for man in different environments'. pp. 271–286 in Blaxter, K. and Fowden, L. (eds.). *Food, Nutrition and Climate.*

Webb, C. G. (1960). 'Thermal discomfort in an equatorial climate'. *J. Inst. Heating and Ventilating Engineers,* **7**, 1–8.

Webb, T. L. and Aardt, J. H. P. van (1957). 'Deterioration of materials under tropical conditions'. pp. B1–B26 and J6–J8 in *Symposium on Design for Tropical Living.* South African Council for Sci. and Industrial Res.

Weihe, W. H. (1979). 'Climate, health and disease'. In *World Climate Conf.* WMO 537, 313–68.

Weihe, W. H. (1982). 'The significance of African climate in human health'. pp. 438–56 in *Proceedings of the Technical Conference on the Climate of Africa.* Arusha. WMO 596.

Weisse, L. (1936). 'Les vents de sable au Soudan'. *Annls. du Phys. du Globe de la France d'Outre-Mer.* **3**, 149–50 and 154–6.

Weisse, L. (1937). 'Note sur la répartition mensuelle, journalière et horaire de la pluie à Dakar'. *Publ. du Comité d'Etudes Historiques Sci. de l'Afr. Occidentale Fr.,* Paris, Sér. B., **3**, 71–6.

Weisse, L. and Barberon, J. (1937). 'Note au sujet du grain du 28 au 30 juin, 1933'. *Publ. du Comité d'Etudes Historiques et Sci. de l'Afr. Occidentale Fr.* Paris Sér. B. **3**, 47–61.

Wellman, F. L. (1961). *Coffee.* Leonard Hill, London.

Welter, L. (1930). 'Bibliographie météorologique de l'Afrique occidentale française'. *Bull. du Comité d'Etudes Historiques et Sci. de l'Afr. Occidentale Fr.,* Dakar, **XIII** 475–82.

Wexler, R. (1946). 'Theory and observations of land and sea breezes'. *Bull. Am. Met. Soc.* **27**, 272–87.

Whalley, W. B. and Smith, B. J. (1981). 'Mineral content of harmattan dust from northern Nigeria examined by scanning electron microscopy'. *J. Arid Envir.* **4**, 21–9.

Whyte, R. O. (1963). 'The significance of climatic change for natural vegetation and agriculture'. In *Change of Climate with Special Reference to the Arid Zones. Proc. UNESCO/WMO, Rome Symp. Arid Zone Res.* XX. UNESCO.

Williams, A. (1970). 'The estimation of wind loading in Nigeria'. *J. Construction in Nigeria*, **4**(1), 37–41.

Williams, C. N. and Joseph, K. T. (1970). *Climate, Soil and Crop Production in the Humid Tropics.* Oxford Univ. Press.

Williams, G. J. (1964). 'Some observations on the rainfall of the Freetown Peninsula'. *J. W. Afr. Sci. Ass.*, **9**(2), 140–50.

Williams, G. J. (1970). *Kortright Meteorological Station, Summary of Records, 1955–67.* Dept. Geog., Fourah Bay College, Univ. Sierra Leone.

Williams, G. J. (1971). *A Bibliography of Sierra Leone, 1925–1967.* Africana Publishing Corp., New York.

Williams, M. E. J. and Faure, H. (eds.) (1979). *The Sahara and the Nile.* Balkema, Rotterdam.

Winstanley, D. (1973). 'Recent rainfall trends in Africa, the Middle East and India'. *Nature*, **243**, 464–5.

Winstanley, D. (1976). 'Climatic change and the future of the Sahel'. In Glantz, M. H. (ed.). *The Politics of Natural Disaster.* Praeger.

Wise, C. G. (1944). 'Climatic anomalies on the Accra plain' *Geography* **29**, 35–8.

Witwatersrand University, (1961). *Climatological Atlas of Africa.* Government Printer, Pretoria.

Woodacre, B. (1960). 'Regeneration of a line squall'. *Brit. W. Afr. Met. Serv. Tech. Note* 14. Lagos.

WMO (1953). 'World distribution of thunderstorm days'. WMO *Tech. Publ.* 21(6), I and II.

WMO (1956). *International Cloud Atlas.* WMO.

WMO (1963). *Guide to Meteorological Instrument and Observing Practices.* 2nd edn, Supp. 1. WMO.

WMO (1964a). *Weather and Man – the Role of Meteorology in Economic Development.* WMO 143, TP. 67.

WMO (1964b). *Proceedings of the Symposium on Tropical Meteorology, Rotorua, New Zealand, 1963.* New Zealand Met. Serv., Wellington.

WMO (1965a). 'Case studies of synoptic situations affecting locust movements in East and West Africa, India, Iran, Israel, Pakistan, Syria, Turkey, UAR and USSR'. *Proc. WMO/FAO Seminar on Meteorology and the Desert Locust, Teheran.*

WMO (1965b). 'Meteorology and the desert locust'. WMO 171, TP. 85, *Tech. Note* 69.

WMO (1965c). *Catalogue of Meteorological Data for Research.* WMO 174, TP. 86.

WMO (1966). 'Measurement and estimation of evaporation and evapotranspiration'. *WMO* 201. *Tech. Note* 83.

WMO (1973). 'Agroclimatology in the semi–arid areas south of the Sahara'. *Proc. of Regional Tech. Conf., Dakar, Feb. 1971.* WMO 340.

WMO (1977). *Lectures on Forecasting of Tropical Weather, including Tropical Cyclones, with Particular Relevance to Africa.* WMO 492.

WMO (1978). 'The West African monsoon experiment'. *GARP Pub.* Series 21.

WMO (1979). *Proceedings of World Climate Conference (of experts on climate and mankind).* WMO 537.

WMO (1981a). *Weather Reporting, Messages Météorologiques, Vol. A. Observing Stations.* WMO 9.

WMO (1981b). 'On the assessment of the role of carbon dioxide on climate variations and their impact'. *Rep. Joint WMO/ICSU/UNEP Meeting of Experts, Villach, Austria.* WMO.

WMO (1982) *Proceedings of Technical Conference on Climate – Africa. Arusha, Jan. 1982.* WMO 596.

WMO and Economic Commission for Africa (1969) *The Role of Meteorological Services in Economic Development in Africa.* WMO.

Yoshida, S. (1977). 'Rice'. pp. 57–67 in Alvim, P. de T. and Kozlowski, T. T. (eds.). *Ecophysiology of Tropical Crops.* Academic Press.

Zachariah, K. C. and Condé J. (1981). *Migration in Africa: Demographic Aspects.* Oxford Univ. Press, New York.

Zukovskij, P. M. (1962). *Cultivated Plants and their Wild Relations.* Royal Afr. Bureau, Farnham.

Author index

Page numbers in bold type refer to
names associated with tables or figures.

Subject index

Page numbers in bold type refer to material in tables or figures.